REGIOMONTANUS
ON TRIANGLES

Non est mortale quod opto

IOHANNE. S. de REGIO MONTE dictus
alias MVLLERVS.
Insignis Mathematicus et de
Re Typographica Norimbergensium
Optime meritus.
Nat. A. 1436. d. 6. Juny, Den. A. 1476. d. 6. July. Ac. XLI.
Ex collectione Friderici Roth-Scholtzii Norimberg. Leckner f.

"John of Regiomont, otherwise known as Müller," unknown artist, 1726.

REGIOMONTANUS
ON TRIANGLES

De triangulis omnimodis by Johann Müller,
otherwise known as Regiomontanus,
translated by Barnabas Hughes, O.F.M.,
with an Introduction and Notes

THE UNIVERSITY OF WISCONSIN PRESS
MADISON · MILWAUKEE · LONDON
1967

Published by the University of Wisconsin Press
Madison, Milwaukee, and London

U.S.A.: Box 1379, Madison, Wisconsin 53701
U.K.: 26–28 Hallam Street, London, W. 1

Printed in the United States of America
by North Central Publishing Company, St. Paul, Minn.
Library of Congress Catalog Card Number 66-22861

Samueli Barchas
Urbano in Lege Peritissimo
Astronomico Doctissimo
Atque Amico Carissimo
Gratissime D.D.D. Interpres

PREFACE

Five hundred years ago, in June of 1464, John Müller, Regiomontanus, completed *On Triangles*. In its one hundred and thirty-seven pages he gathered together the trigonometric knowledge of his predecessors and enriched it with his own improvements. As described by A. Wolf,

> Regiomontanus systematically summed up the work of both the Greek and Arab pioneers in plane and spherical trigonometry. His own special contribution was the application, to the solution of special triangles, of algebraic methods of reasoning derived from Diophantus, though without the use of abbreviations.[1]

In tribute to the man who laid the foundations[2] of modern trigonometry, this translation has been prepared.

There is a further reason for this translation. It is the belief of the translator that instructors in the mathematical sciences should be familiar with the sources upon which their subjects have been founded. Fifteenth-century Latin, however, is not an easy diet for those with but several years of a classical Latin background that has, perhaps, decayed with the passage of time. And so in translation the *Triangles* is offered here for those who wish a deeper appreciation of the source of modern trigonometry, of "the first work that may be said to have been devoted solely to trigonometry."[3]

Sufficient thanks cannot be expressed to the translator's good friend, Samuel I. Barchas, who spontaneously and at great expense purchased a first edition of *De Triangulis omnimodis* simply that it might be translated by this high school teacher. Mr. Barchas is indeed a mathematical "Good Samaritan." Particular appreciation is due also to Universal Microfilm Services, Inc., of Phoenix, Arizona, for xeroxographing the *De Triangulis* and the *vitae* of Adam and Gassendus, which made the translator's work easier. In addition, assistance was gladly given by the New York Public Library, the Stadtbibliothek of Nürnberg, and particularly by the Stanford University Library. Finally, gratitude is due Professor Ernst Zinner, Reverend Joseph T. Clark, S.J., and Sister Mary Claudia Zeller, O.S.F., for invaluable information, to my superiors for their encouragement and approval, and to the University of Wisconsin Press for preparing the publication of this work.

Fr. Barnabas Hughes, O.F.M.

Saint Mary's High School
Phoenix, Arizona
June 1965

[1] A. Wolf, *A History of Science, Technology, and Philosophy in the 16th and 17th Centuries* (1950), p. 189.
[2] George Sarton, *Six Wings: Men of Science in the Renaissance* (1957), p. 25.
[3] David Eugene Smith, *History of Mathematics* (1958), I, 260.

CONTENTS

INTRODUCTION

INTRODUCTION

The reputation of Regiomontanus and his influence upon his contemporaries and those who followed him in the next hundred years have been subjected to close attention.[1] But there is still room for further study. Five men in particular knew of him and his writings: Columbus, Beheim, Novara, Rhaeticus, and Copernicus. Columbus was born in the same year as Regiomontanus and used the astronomical tables of Regiomontanus even after he reached the West Indies.[2] Martin Beheim was in Lisbon in 1484, where "by passing himself off as a pupil of [Regiomontanus] he managed to enter the most learned and courtly circles."[3] Rhaeticus had mastered the *Triangles* before he allied himself with Copernicus.[4] Copernicus, previous to this, had studied in Bologna under Dominicus Maria Novara, himself a pupil of Regiomontanus.[5] Moreover, there is some indication that Regiomontanus contributed to the foundation of the heliocentric theory.[6]

The testimony of scholars about Regiomontanus and his *Triangles* is indeed interesting. The opinions of Wolf and Smith have already been cited in the Preface. In 1795, Charles Hutton wrote, ". . . he enriched trigonometry with many theorems and precepts. Indeed, excepting for the use of logarithms, the trigonometry of Regiomontanus is but little inferior to that of our own time."[7] A century earlier, Edward Sherburne wrote, "[The *Triangles*] is still a Book of good accompt, as containing in it divers extraordinary Cases about plain Triangles."[8]

In the sixteenth century, too, came significant testimony from England, France, and Denmark. William Borough, for instance, invites his readers to compare his work, *Discourse on the Variation of the Cumpas* (1581), with that of Regiomontanus, evidently his accepted authority.[9]

Tycho Brahe established his reputation with the publication in 1573 of *De nova stella*. In this short work of a little more than one hundred pages (and only twenty-seven of these are on the appearance of the new star in Cassiopea), he uses Regiomontanus' *Triangles* as his nearly exclusive authority for

[1] Lynn Thorndike in *A History of Magic and Experimental Science* (1932), V, 332–77, offers a broad analysis of the impact of Regiomontanus. A more detailed survey has been prepared by Sister Mary Claudia Zeller, O.S.F., in *The Development of Trigonometry from Regiomontanus to Pitiscus* (1944).

[2] Charles Singer, *A Short History of Scientific Ideas to 1900* (1959), p. 196.

[3] Samuel Eliot Morison, *Admiral of the Ocean Sea* (1942), I, 70.

[4] Zeller, *The Development of Trigonometry . . .*, p. 55.

[5] John Kepler, *Tabulae Rudolphinae* (1627), preface, p. 3.

[6] Leonardo Olschki in *The Genius of Italy* (1949), p. 374, states ". . . he went so far as to express his doubts about the validity of the Ptolemaic cosmology as a whole and of the geocentric system in particular." This author does not give any primary source for his statement. Cf. also: Arthur Koestler, *The Sleepwalkers* (1959), p. 209. Ernst Zinner in *Entstehung und Ausbreitung der Copernicanischen Lehre* (1943), pp. 135-36, indicates that this question is still open.

[7] Charles Hutton, *A Mathematical and Philosophical Dictionary* (1795), II, 132.

[8] Edward Sherburne, "Catalogue of Astronomers" (1675), p. 41, appendix to his *Sphere of Marcus Manilius*.

[9] Marie Boas, *The Scientific Renaissance, 1450–1630* (1962), p. 235.

establishing the location of the new star. Six times he cites the *Triangles* by book and theorem. Finally he writes,

> We have found the longitude and latitude of this new star with the help of the infallible method of the doctrine of triangles. Exactly how we went about doing this in finding the sides and angles of the triangles needed is indicated in the references. There is no place here for further explanation of them, for they are quite long and would only enlarge this work too much. A good part of the propositions are from the Fourth Book of [the *Triangles* by] Regiomontanus. This work was used because everything is closely tied together geometrically.[10]

It would certainly seem that Tycho Brahe held Regiomontanus in the highest esteem. Only a single side reference to Copernicus kept the full credit for the method of locating the position of the new star from going to the *Triangles* of Regiomontanus. Elsewhere[11] Brahe refers to him as "Clarissimus & praestantissimus Germanorum Mathematicus Johannes de Monteregio Francus . . . ."

Fifty years before Brahe, the renowned French mathematician Oronce Finé indicated his high regard for Regiomontanus by referring to him as "Matematico accuratissimo."[12]

Early laudatory testimony can be found in many places. Peter Apian (1495–1552) in his *Cosmographia* (1574 printing) refers to Regiomontanus' birthplace: "Mons Regius, vul. Kunigsperg, the birthplace of John Regiomontanus who restored the science of mathematics." None of the other famous men mentioned in this work is given such praise.[13] In a similar vein speak Philip Melanchton (1497–1560)[14] and Erasmus Rheinhold (1511–1553).[15]

The 1492 edition of the *Alfonsine Tables* was prompted by the correspondence between Santritter and Moravus which was printed along with the manual. In a letter to Moravus, Santritter wrote,

> Who can be patient with those unlearned persons who do not know the first things about mathematics. They carp at and condemn the astronomical works of John of Monte Regio, things that he found

only with the most diligent study, the greatest work and the most adroit efforts of the mind. . . . They ought to admire in John his proficiency in each of the languages [Latin and Greek] and his most rigid demonstrations of his statements.[16]

Despite the laudatory remarks from the past five centuries, it must be pointed out that some experts in the fields of mathematics and astronomy do not agree with the opinions presented. Among the moderns, it is noteworthy that Hooper[17] does not even mention Regiomontanus by name in his chapter on the development of trigonometry. In the last century, while Delambre[18] does call him "le plus savant astronome qu'eût encore produit l'Europe," he feels that the stature of Regiomontanus as a mathematician is perhaps exaggerated. In particular he questions whether or not Regiomontanus introduced the tangent concept, and offers good reason for supposing that he did not. At least, Delambre concludes, the mathematical reputation of Regiomontanus cannot be determined at the time of writing, in 1819.

Whether or not Regiomontanus understood the tangent concept at the time he wrote the *Triangles* deserves attention here. There is clear evidence that he knew, used, and appreciated the tangent when he wrote his *Tabulae directionum* in 1467, which he described as a "fruitful table" (*tabula fecunda*).[19] But did he have the tangent in 1462 when he began the *Triangles?* Zinner remarks,

[10] Tycho Brahe, *De nova stella* (1573), B4^v. Also in *Opera omnia* (1648), p. 357.

[11] Brahe, *Opera omnia*, p. 35.

[12] Oronce Finé, *Della geometria*, I, Ch. xiii, 16, in *Opere divise* (1587), which is the Italian translation of his collected works originally known as *Protomathesis* (1532).

[13] Peter Apian, *Cosmographia* (1574; ed. Gemma Frisius [1508–1555]), fol. 33.

[14] John of Sacrobosco, *De sphaera* (1563; ed. in 1531 by Philip Melanchton), A–4.

[15] George Peurbach, *Theoricae novae planetarum* (1542; ed. Erasmus Rheinhold), p. 103 verso.

[16] *Tabule astronomice Alfonsi regis* (1492), A–3.

[17] Alfred Hooper, *Makers of Mathematics* (1948).

[18] J. B. J. Delambre, *Histoire de l'Astronomie du Moyen Age* (1819), pp. 292–323, 347–65.

[19] Zeller, *The Development of Trigonometry* . . . , p. 34.

"There is no application of the tangent in the *Triangles*," but notes that Regiomontanus was using the tangent concept the next year (1465) in Rome and afterward in Hungary (1467). Finally, Zinner observes that Peurbach had outlined a tangent table in 1455.[20] It seems likely, then, that Regiomontanus knew of the tangent function when he wrote his *Triangles*. Why he did not use it is another question.

TRIGONOMETRY BEFORE 1464

It is quite difficult to describe with certainty the beginnings of trigonometry.[21] There is just not enough evidence. In general, one may say that the emphasis was placed first on astronomy, then shifted to spherical trigonometry, and finally moved on to plane trigonometry.

In particular, it seems certain[22] that the Babylonians of the old period (before 1600 B.C.) had some knowledge of chords for astronomical purposes. The trigonometry of chords employed the ratio of a chord of a circle to the diameter of that circle to determine the central angle. Hipparchus (*ca.* 180–125 B.C.) formulated a table of chords. But it was Menelaus of Alexandria (*ca.* 100 A.D.) who first formulated the theorem basic to all triangles: The product of the three ratios of the consecutive segments of the sides of a plane triangle made by any rectilinear transversal equals unity. His successor Ptolemy (*ca.* 150 A.D.) developed the heritage of the past and left his masterpiece, *Syntaxis*, or *Almagest*, for the future. In this work he brought a significant measure of perfection to determining his tables of chords. These tables were used throughout Europe without substantial improvement until Regiomontanus published his *Tabula fecunda*.

The next advance in trigonometry was in the East. The Hindus had the works of Menelaus and Ptolemy, but they improved upon their predecessors by considering the half-chord and the radius of the circle. Thus, they discovered the sine ratio upon which modern trigonometry is based. They took another step forward and performed calculations on a new ratio, based on measurements of shadows, from which the tangent function was de-rived.[23] Abū-l Wafā' (940–998) was the first to generalize the sine law to spherical triangles. He used the *umbra* (tangent) as a real trigonometric line and arrived at the relation $\tan a{:}1 = \sin a{:}\cos a$. For this he saw the utility of setting the radius equal to unity. Subsequently al-Bīrūnī (973–1048) wrote the sine law for plane triangles. All of these advances were brought to Europe by the transla-tors.

From the considerable work of the European translators, the following are the most significant to the development of mathematics in the West. Adelard of Bath (*ca.* 1116–1142) translated the astronomical tables of al-Khowarizmi (*ca.* 845) into Latin, thereby introducing the sine and tangent functions into part of Europe. John of Seville (*ca.* 1135–1153) translated al-Farghānī's *Elements of Astronomy* (*ca.* 861); Regiomontanus had a copy of this. He also had a copy of al-Battānī's *The Motion of the Stars* (*ca.* 920), which was translated by Plato of Tivoli (1134–1145). Regiomontanus must have had a Greek translation of the *Almagest*. There is also evidence[24] that he had some of the works of Jābir (or Geber, *ca.* 1225), who improved Menelaus' theorem concerning segments of the sides of a triangle by showing it applicable to four quantities as well as six, and that he had something[25] of al-Zarqālī (*ca.* 1075).

[20] Ernst Zinner, *Leben und Wirken des Johannes Müller von Königsberg genannt Regiomontanus* (1939), pp. 29–30, 64, 101, 107, 115.

[21] This resumé is based principally upon the following: John David Bond, "The Development of Trigonometric Methods down to the Close of the XVth Century," *Isis*, IV (October 1921), 295–323; George Sarton, *Introduction to the History of Science* (1927–1948), Vol. I, II (Pt. 1), III (Pt. 1); J. F. Scott, *A History of Mathematics* (1958), Ch. 3; David Eugene Smith, *History of Mathematics* (1958; 2 vols.).

[22] Otto Neugebauer, *Vorlesungen über Geschichte der Antiken Mathematischen Wissenschaften* (1934), I, 168. See also O. Neugebauer and A. Sachs, eds., *Mathematical Cuneiform Texts* (1945), pp. 38–41.

[23] Louis C. Karpinski, "The Unity of Hindu Contributions to Mathematical Sciences," *Scientia*, XLIII (June 1928), 382.

[24] Pedro Nunes, *Tratado da Sphera* (1537), fol. 39, in *Obras*, I, 62: ". . . Joannis de Monte Regio qui Gebrum imitatus est . . . ."

[25] This was probably Peurbach's copy of al-Zarqālī's *Canons or Rules on the Tables of*

Plate I. A sketch of Regiomontanus, possibly by Kepler.

While Regiomontanus had access to many of the translated works, he did not have them all; and this is significant. The tangent function was brought into Europe by Adelard of Bath, but it was slow to find its way into central Europe. Rather it stayed north in Paris and England. There is no doubt that in Paris the tangent functions (*umbra versa, umbra recta*) were known, at least to Dominicus de Clavasio (*fl.* 1346). And in England John Manduith (*fl.* 1310), whom Sarton calls "the real initiator of western trigonometry," knew and used these functions in his *Small Tract*. Moreover, Richard Wallingford (*ca.* 1292–1335) wrote *Quadripartitum de sinibus demonstratis*, using these functions. But apparently none of these works penetrated into central and south-central Europe. In this regard, one might say that a wall separated the Paris-English schools from the rest of Europe.

There is, however, one unexplained exception to the previous observation. Campanus of Novara (*ca.* 1260–1280) wrote a true table of tangents[26] for each degree, o to 45. In view of the meanderings of Regiomontanus about Italy and central Europe where he took every opportunity to hunt out scientific works, it is surprising that he apparently never came across this work.

Thus, when Regiomontanus organized his material for *On Triangles*, he was well familiar with the heritage of Ptolemy and the works of some of the Hindu-Arabic scholars. He knew of the tables of chords and their determination, the trigonometric ratios (his knowledge of the tangent function is questionable, at best), and the sine and cosine laws, all of this for both plane and spherical trigonometry.

THE CONTRIBUTION OF REGIOMONTANUS

What Nasir ad-Dīn had done two centuries previously for the East, Regiomontanus[27] did for the West: "He constructed a uniform foundation and a systematic ordering of trigonometric knowledge."[28]

The foundation is laid in geometry and the geometric method — definitions, postulates, and theorems — with Euclid his major authority. The first part of Book I (theorems 1 through 19) treats magnitudes and ratios, and the remaining theorems (20 through 57) of Book I propose geometric solutions for right, isosceles, and scalene triangles. There are seven exceptions to this method of solution. Theorems 20, 27, and 28 mention or use the sine function explicitly. The solutions for the four cases of oblique triangles are handled in theorems 49, 50, 52, and 53. While the sine function is not mentioned in any of these four theorems, reference is made to theorem 27 where it is used.

The systematic ordering of trigonometric knowledge may be said to begin with theorem 1 of Book II. Here Regiomontanus states the *Law of Sines*:

> In every rectilinear triangle the ratio of [one] side to [another] side is as that of the right sine of the angle opposite one of [the sides] to the right sine of the angle opposite the other side.[29]

He uses this law to solve two cases of the oblique triangle problem in theorems 4 and 5 of Book II: When two angles and any side or two sides and the angle opposite one of them are given, the remaining parts can be found.

Toledo. See M. Curtze, "Urkunde zur Geschichte der Trigonometrie . . . ," *Bibliotheca Mathematica* (1900), ser. 3, I, 338. If this is the case, it is difficult to understand why Regiomontanus did not utilize the *umbra* (cotangent, here) in his own work. For al-Zarqālī devoted several paragraphs to its determination and use (Curtze, pp. 342–43, 352). In fact, John de Lineriis (1300–1350), professor of mathematics at Paris and a follower of al-Zarqālī, defined clearly the *umbra recta* and *umbra versa* in his own *Canons on the Tables of the Primum Mobile* (Curtze, p. 399). Another question is, why was Regiomontanus not familiar with this work of John de Lineriis?

[26] Neither Braunmühl nor Zinner could find any evidence of the use of this table in Regiomontanus' writings. See Zinner, *Leben und Wirken . . .*, p. 107.

[27] Only the contributions of the *Triangles* are considered here.

[28] Zeller, *The Development of Trigonometry . . .*, p. 19.

[29] Zinner (*Leben und Wirken . . .*, pp. 65–66) shows that Regiomontanus was most probably not familiar with the trigonometric work of Levi ben Gerson, and that the statement of the *Sine Law* may be attributed to Regiomontanus. Zeller (*The Development of Trigonometry . . .*, p. 25) suggests that Regiomontanus was dependent upon al-Bīrūnī for the statement of this law.

Book II is particularly noteworthy for two things. First, in theorems 12 and 13 Regiomontanus offers algebraic solutions for finding the lengths of the sides of a triangle. Both of these solutions employ quadratic equations whose solutions Regiomontanus assumes are quite familiar to the reader. The algebra is literary rather than syncopated or symbolic. The second noteworthy aspect of Book II is theorem 26:

> If the area of a triangle is given together with the rectangular product of the two sides, then either the angle opposite the base becomes known or [that angle] together with its known [exterior] angle equals two right angles.

This is the first implicit statement of the trigonometric formula for the area of a triangle.[30]

Finally, scattered throughout Book II are a number of theorems which a modern trigonometry text would classify as exercises, such as theorem 8: "If the ratios of three sides are given and if the perpendicular is known, each side can be measured."

Book III again is an elementary foundation, for it is a spherical geometry developing much detail for what will come in Book IV. (One must remember that Regiomontanus was primarily an astronomer and that while the *Triangles* is a work on trigonometry, the author in his own preface considers it a necessary tool for astronomy.[31]) By theorem 16 he is ready for the *Law of Sines* for spherical triangles which he carries into theorem 17. Theorems 25, 26, and 27 treat right-angled spherical triangles. And theorems 28 through 34 give the six cases for solving oblique spherical triangles. Among these is theorem 29 which deals with the ambiguous case.

Book V continues the solution of problems of spherical triangles. Here in theorem 2 is contained the *Law of Cosines* for spherical triangles, disguised in the terminology of the versed sine. In modern notation[32] this theorem states

$$\frac{\text{vers sin } A}{\text{vers sin } a - \text{vers sin } (b-c)} = \frac{1}{\sin b \sin c},$$

which can be reduced to

$$\cos A = \frac{\cos a - \cos b \cos c}{\sin b \sin c}.$$

Apparently Regiomontanus first found the cosine law when, as a young man in Vienna, he was studying the *Astronomy* of al-Battānī.[33] He recognized its importance, and thus reworked the law into its first practical formulation.

In summary, Regiomontanus laid a solid foundation in plane and spherical geometry for a complete trigonometry. Besides offering a number of original theorems (including among them an implication of the trigonometric formula for the area of a triangle), he used algebra twice to solve geometric problems, and he presented the first practical theorem for the *Law of Cosines* in spherical trigonometry. From his work, his great successors, Copernicus and Rhaeticus, sought assistance and inspiration for their own trigonometries.

Just what influence the *Triangles* had on the mathematicians of the sixteenth century is explored in some detail by Braunmühl.[34] After the untimely death of Regiomontanus in 1474, Bernard Walter took control of his possessions, books, manuscripts, and instruments. When Walter died in 1504, these were scattered. Willibald Pirkheimer (1470–1530), a leading citizen of Nürnberg, managed to retrieve some of the works of Regiomontanus, notably the *Triangles*. Pirkheimer made his home a center of learned activity where whatever was available, including the Regiomontanus manuscripts, was common property. In his circle was John Werner (1468–1528), who wrote a book on spherical trigonometry. While Werner's book was never published, it did find its way into the hands of a fellow cleric of Werner, George Hartmann (1480–1545). Thence it went to George Rhaeticus (1514–1576).

From this, one may reasonably conclude that, after Pirkheimer made available to his friends the manuscript of the *Triangles*, Werner probably saw it. In writing his own book, he may have borrowed from Regio-

[30] Zeller, *The Development of Trigonometry* . . . , p. 25.
[31] See "Text and Translation," p. 27 below.
[32] Zeller, *The Development of Trigonometry* . . . , p. 30.
[33] Zinner, *Leben und Wirken* . . . , p. 66.
[34] A. von Braunmühl, *Vorlesungen über Geschichte der Trigonometrie* (1900), I, 133, 141.

montanus whatever was helpful. Eventually Rhaeticus had Werner's book. There is then a direct line from Regiomontanus to Rhaeticus and, consequently, to Copernicus, for Rhaeticus instructed Copernicus.

While this line of speculation is interesting, Braunmühl notes that Rhaeticus did personally inscribe a copy of the *Triangles* for Copernicus. And Copernicus did study the work thoroughly. This copy has been preserved and it shows numerous marginal notations in Copernicus' handwriting.

Other testimony to the influence of the *Triangles* in the sixteenth century is not wanting.[35] Francis Maurolyco (1494–1575) took some of the definitions and propositions from Regiomontanus' work for his translation and commentary on Euclid's *Elements*. John Blagrave (d. 1611) cited Regiomontanus as the source for sections of his *The Mathematical Jewel*. And Adrian Metrius (1571–1635) took from the *Triangles* the figure and proof for the sine law for spherical triangles, as Book V of his *Universal Astrolabe* shows.

THE TEXT AND PLATES

The text[36] used for the translation, as shown by the title page and colophon, is the edition published posthumously in octavo by John Petreus of Nürnberg for John Schöner, in 1533. Peter Gassendus testifies that this is the first edition.[37] Its title is, "Doctissimi viri et mathematicarum disciplinarum eximii professoris Joannis de Regio Monte de Triangulis Omnimodis libri quinque." A second, enlarged edition was published by Daniel Santbech at Basil in 1561, "Joannis Regiomontani, mathematici praestantissimi de Triangulis planis et sphaericis libri quinque, una cum tabulis sinuum."

Despite the fact that the second edition has eleven more pages plus the table of sines and chords which the first edition refers to but does not contain, a careful comparison of the two editions reveals no striking dissimilarities.[38]

There were no more editions of the *Triangles* published in either the sixteenth or the seventeenth century.[39] Zinner,[40] however, notes that it was translated into German by Matthew Beger of Reutlingen (d.

1661). That translation is in the city library of Reutlingen (Mss, No. 1873 and 1879).

In the translation, an effort was made to steer a path between the Scylla of literalness and the Charybdis of liberality. An effort was made to utilize the words and expressions that Regiomontanus used. Occasionally, however, this was not possible. For instance, on page 38, line 44, the text reads ". . . quod fit ex gb in bh." Literally, this translates into ". . . that which becomes from *GB* in *BH*." When the reference to Euclid is consulted, the expression becomes ". . . the product of *GB* and *BH*." Consonant with the tenor of the age, the author from time to time would omit the subject of the sentence or the verb or the object. Wherever an additional word or phrase not justified by translation seemed desirable, it was added in brackets. Finally, his various expressions to close the proofs of theorems (*quod intendebamus; qui est quod libuit absoluere; verum igitur est, quod theorema proposuit*; etc.) are frequently translated simply as "Q.E.D." The symbols $\angle$, $\triangle$, and ° have been used in the translation for angle, triangle, and degree; and the capitalization and punctuation have been modernized.

Paragraphing is nearly nonexistent in the text. To make the reading of the translation easier, the English text has been arranged

[35] Zeller, *The Development of Trigonometry* . . . , pp. 72, 90, 104. The works cited here were not available to the translator.

[36] Zinner notes (see *Leben und Wirken* . . . , p. 230) that the original manuscript consists of 106 pages in quarto, that Werner wrote the title "De Triangulis omnimodis quinque volumina," but that another hand wrote the dedicatory, "To the Most Reverend Father in Christ and Lord Bessarion, Bishop of Frascati, Cardinal of the Holy Roman Church and Patriarch of Constantinople, John the German of Regiomont offers himself, a most devoted servant." The Appendix described on the title page has not been translated here.

[37] Peter Gassendus, *Tychonis Brahei vita* (1655), pp. 368–69.

[38] Zeller, *The Development of Trigonometry* . . . , p. 19.

[39] Louis C. Karpinski, "Bibliographical Check List of All Works on Trigonometry Published up to 1700 A.D.," *Scripta Mathematica*, XII (1946), 268.

[40] Zinner, *Leben und Wirken* . . . , p. 231.

thus: the first paragraph following the statement of the theorem indicates what is given and what is to be proved. Each major part of the proof receives a separate paragraph. Illustrative examples, which often show the mechanics for using the theorem, receive one or more paragraphs, depending on their length.

The footnotes in this translation fall under four categories. The first group clarifies certain difficult or archaic terminology. The second offers a few definitions of Latin phrases. The third indicates corrections to the figures (except where Regiomontanus has specifically directed the student, "When [it] is drawn"). And finally, the fourth encompasses corrections to the Latin text where the correction is directly pertinent to the proof. Hence, there are no footnotes for inverted letters, misspellings, and trivia such as the misnumbering of theorems. The references to the figures in the Latin are not consistent in their order of notation (i.e., line *AB* alternately appears as line *BA* with no geometric distinction intended); these remain as they appear in the Latin text.

Finally mention must be made of the authorities Regiomontanus used.[41] His proofs of the theorems rested upon three sources. The first is that of axiom or definition, many of which are enumerated at the beginning of the text. The second source is Euclid's *Elements*. Gassendus tells us that Regiomontanus was quite familiar with the Campanus edition of Euclid,[42] so it may be assumed that that was the text he used. References to Euclid are expressed in one of two ways: "per 29. & 34. primi elementorum Euclidis," as in theorem 1, page 32; or "per primam sexti" from the same place. Since the first example given is the first reference Regiomontanus made to any author and the second example follows in the same paragraph, it was concluded that the second example together with all other similar unidentified references must be to Euclid. Occasionally Regiomontanus gave an incomplete reference to Euclid: either the theorem number or the book number was omitted. This is recognized in the translation by the phrase "[not given]." The third authority is Regiomontanus himself. When he referred to a previous theorem, it was always in this fashion: "per 6 aut 7 huius," as in

theorem 29, page 68. The key word is "huius" (translated throughout as "above"), for this is a clear reference to theorem 6 and theorem 7 of the present book. However, as John Schöner observes in his dedication, it is somewhat annoying that Regiomontanus did not complete his references in Books II through V. Very ungraciously he left a blank; for example, in Book II, theorem 8, page 114, he wrote "ex processu igitur primi huius . . . ." This reference, of course, is to some theorem of the First Book ("above"). Such omissions have been supplied in this translation within brackets, for example, "Th. I.[44] above."

It is unfortunate that there is no contemporary picture of Regiomontanus.[43] The three plates included here are representative of posthumous work. The Frontispiece is from the collection of Frederick Roth-Scholtz, engraved on copper by an unknown artist in 1726. The inscription reads "John of Regiomont, otherwise known as Müller. An outstanding mathematician and renowned publisher of Nürnberg. Born June 6, 1436. Died July 6, 1476. In his 41st year." Plate I is a sketch very possibly by Kepler for the title page of his *Rudolfine Tables*. Plate II is the earliest likeness. It appears as the frontispiece to Regiomontanus' *Epitome* which was published in Venice in 1496.

THE BIOGRAPHY

Something should be known about the man who wrote the *Triangles*. His life has been commented on, more or less, by many of the historians of mathematics and astronomy. Although biographical articles had been written before, the first definitive biography of

[41] While Regiomontanus referred only to Euclid and himself as the authorities in the *Triangles*, he did mention in a letter to Bianchini his dependence on Menelaus, Theodosius, and Jābir (or Geber). See Zinner, *Leben und Wirken* . . . , p. 65.

[42] Gassendus, *Tycho Brahei vita*, p. 364.

[43] See Zinner, *Leben und Wirken* . . . , pp. 192 ff., for the information in this paragraph. The Frontispiece is reprinted here through the courtesy of Prof. Ernst Zinner. Plate I is from the Stadtbibliothek Nürnberg, and Plate II from Samuel I. Barchas.

Introduction

Regiomontanus was written by Peter Gassendus in 1651, as an appendix to his much larger biography of Tycho Brahe.[44] He used, besides the books and letters of Regiomontanus with their dates and contents, the incidental remarks of others whom he quoted by name: Starovolsius (p. 345), Cardanus (p. 361), Ramus (p. 361), Kepler (p. 367), Schöner (p. 368), and others. Briefly, Gassendus constructed his biography from the works of Regiomontanus and occasional comments by other astronomers and literary luminaries.[45] What follows is an abridged translation of the Gassendus biography.[46] It was taken from the final pages (345–373) of *Tychonis Brahei, Equitis Dani, Astronomorum Coryphaei, Vita. Accessit Nicolai Copernici, Georgii Peurbachii, & Joannis Regiomontani, Astronomorum celebrium, Vita.* Authore Petro Gassendo, Regio Matheseos Professore. Editio Secunda auctior & correctior. Hagae-Comitum. Ex Typographaei Adriani Vlacq, M.DC.LV.

With the exception of the more specific localization of Regiomontanus' place of birth, the few additional contributions on his life from more modern writers have been assigned to footnotes.

John Müller, Regiomontanus

John Müller was born at Unfinden, near Königsberg in Lower Franconia,[47] on June 6, 1436. Earlier writers sometimes refer to him as Johannes Germanus or Johannes Francus, the one because he was a German, the other because Franconia was sometimes called Eastern France. He learned his grammar at home until the age of twelve when his parents sent him to Leipzig for his formal education. Here both dialectics and the theory of spheres were favorites. These led him to the study of astronomy and whatever arithmetic and geometry were necessary for a better understanding of this science. These he mastered quickly; in particular, whatever was wrapped in theory drew his avid study. From Leipzig he graduated to the Academy at Vienna where he came under the influence of George Peurbach. This was in 1451 or 1452.

Peurbach had the greatest influence of anyone over Regiomontanus. And this was acknowledged with gratitude on many occasions. Peurbach recognized the remarkable genius of this young man and realized that this was the one student destined for great things. In particular he respected the enthusiasm of Regiomontanus for astronomy and saw in this ingenious youth an opportunity to rejuvenate that science. Since the respect was reciprocated, Peurbach promised to omit nothing that would bring the desires of Regiomontanus to fruition. From this time onward, the one was likened to a father, the other to a son.

His studies began with the more developed theory of planets, since he was already trained in the theory of spheres. Using the

[44] Melchior Adam included a seven column biography of "Joannes Mullerus Regiomontanus" in his *Vitae Germanorum philosophorum* (1615), using material which he obtained from Philip Melanchton. Melanchton, in turn, borrowed from Erasmus Rheinhold's *Oratio de Joanne Regiomontano mathematico* (1549) for his own *Selectarum declamationum* (1551). Another biographical article was included by Paul Jovius in his *Elogia doctorum virorum* (1556). And there are others.

[45] The Reverend Joseph T. Clark, S.J., has courteously supplied the translator with the information on the Gassendus biography.

[46] Several weeks after digesting the Gassendus biography, the translator happened upon a copy of Johann Friedrich Weidler's *Historia astronomiae* (1741). And in this on pp. 304–13 is a digest in Latin of the Gassendus together with Regiomontanus' *Index* and a catalogue of Regiomontanus' printed works. It was altogether satisfying to find that the biography as condensed here is practically parallel to the abstract by Weidler.

[47] This location is given by Zinner, *Entstehung . . .*, p. 101. D. E. Smith (*History of Mathematics*, I, 259) identifies the birthplace as Unfied. Florian Cajori in *A History of Mathematical Notations* (1928), I, 97, has an illustration from Regiomontanus' *Almanac* with the title, "Calender des Magister Johann von Kunsperk (Johannes Regiomontanus)." Charles Hutton in *A Mathematical and Philosophical Dictionary*, II, 130, names the birthplace as Koningsberg. Gassendus merely states, "Natus est Joannes . . . in oppido . . . cui Regius Mons nomen. . . . appellitatus potius fuerit Joannes de Monte Regio, vel de Regio Monte, ac Regiomontanus" (*Tycho Brahei vita*, p. 345).

Plate II. The earliest likeness of Regiomontanus.

former as the central idea for all of astronomy, he found it easier to begin the study of the Ptolemaic doctrine. While studying Ptolemy, he did not neglect geometry or practice in calculations. In these fields he built an enviable reputation for speed and nicety of demonstration. To reach an even greater perfection he studied all the mathematical works written in Latin that he could find. For this he had James of Cremona's translation of Archimedes, as well as the works in translation of Apollonius and Diophantus. Peurbach never had to encourage him. Indeed he was like a straining horse, eager for the finish line. Perseverance was a bit of a problem at the start, and Peurbach had to remind him that what was begun with great ardor must be persevered in.

Regiomontanus made every attempt to master what was available regarding astronomy; for there was much previous knowledge to be harvested in the science of astronomy; and it all had to be gathered in were the art to be entirely rebuilt. In particular he had to make himself familiar with the points of the Zodiac, the hinges of the Ecliptic, just under the Aplanes or Firmament. While he did not learn all of the fixed stars, he knew those which he could compare with the planets. He became familiar with the instruments of Hipparchus and Ptolemy, together with other instruments that would assist him in observing the celestial bodies. One of his earliest observations was that in comparing the position of Mars with the nearby fixed stars, he discovered that the tables of the time were two degrees off. Regiomontanus left written records of three lunar eclipses which he observed in his early years. On all three occasions he was able to correct the time predicted for the eclipse according to the *Alfonsine Tables*, some of his corrections being as small as one minute and others as great as seventy minutes.

About this time Peurbach received a copy of Cardinal Nicolas of Cusa's work, *On the Quadrature of the Circle*. While there is no written record of Peurbach's criticism, Regiomontanus gave his mentor credit for the criticism he himself developed. The essence of the Cardinal's idea was that, when two diameters in a circle were drawn at right angles to each other and one of these was extended at both ends, the end of the other diameter could be used as the center of another circle of radius equal to a third of the circumference of the original circle, and that the line segment included between the points of intersection of the new circle and the extended diameter of the original circle would be equal to half the circumference of the original circle. Regiomontanus showed that this line segment would in fact be less than what the Cardinal claimed.

After Peurbach, perhaps the greatest influence in the life of Regiomontanus was Cardinal Bessarion. Apart from being a successful diplomat and trouble-shooter for the Pope he was a scholar in his own right, particularly in astronomy. A Greek by birth, he mastered the Latin language so that he could produce a definitive translation of Ptolemy. Unfortunately his ecclesiastical duties kept him too occupied for this sort of work; hence, it was fortunate that he met Peurbach. Peurbach had spent much time in the study of Ptolemy, attempting to correct a Latin translation simply by analyzing the translation as such, for he knew no Greek. Nor, with Regiomontanus in mind, did he feel it necessary to master the language. For he had introduced his protégé to the Cardinal, and from this time onward Regiomontanus began his study of the Greek language. In a comparatively short time he became proficient in this new tongue. Not only as a reward but also as an incentive for further study, the Cardinal made available to Regiomontanus other scientific works written in Greek.

The preparation of an *Epitome* of Ptolemy's *Almagest* was but half completed when Peurbach died, April 8, 1461, at the age of 37. This was, perhaps, the greatest loss in the life of Regiomontanus. Of his teacher he wrote, "He was a man of the first caliber in habit and integrity of life, a scholar in every subject and superior to all in mathematics." On his death bed, Peurbach committed the Ptolemaic translation and its completion to Regiomontanus. This became a sacred trust for the fatherless student.

Leaving the remains of Peurbach in Vienna, he accompanied Cardinal Bessarion to Italy, where in Rome he brought the preparation of the *Epitome* closer to completion. It was at this time that he first met George Trebizond, an authority on Ptolemy and his commentator Theon. Moreover, he cultivated the friendship of every available learned man, particularly those versed in Greek. At the same time he kept busy with his astronomical observations, spending much time on these between December of 1461 and the following March. Many of these observations were made at Rome, and for the others he went to Viterbio particularly in the summer and fall of 1462. Among his other occupations was the collecting and copying of rare books, both in Greek and in Latin. And among these was a *New Testament*, which was his constant companion.

About this time Cardinal Bessarion was sent to Greece on Church affairs and Regiomontanus left for Ferrari. This place appealed to him, for John Blanchinus, of whom Peurbach had spoken so highly, was teaching here. He sought out the company of Theodore Gaza, among others, for further instruction in Greek. Here he read the orators and historians, the philosophers and poets. These gave to his mastery of Greek the polish that enabled Regiomontanus to go through and finish the entire text of Ptolemy and the Theonic commentary. Here he discovered that Trebizond had erred frequently and seriously. And he would not let the poor translator forget the mistakes.[48]

The following year saw Regiomontanus in Padua where the Academy invited him to lecture. In his initial lecture on al-Farghānī he sang the praises of Peurbach, and then he told his audience that he himself had read all of the ancient writings both in Latin and in Greek. The Arabic writings were still wrapped in a language unknown to him, but he indicated that he was about to learn it. After the lunar eclipse at Padua in April 2, 1464, he left for Venice to await Cardinal Bessarion. It was here during May and June that he finished his work, *On Triangles*. The remainder of June and most of July were spent in preparing the refutation of the *Quadrature* of Nicolas of Cusa. He also devoted some time attempting to rectify the current calendar that had Christians and

Jews celebrating the Pasch at different times.

Returning to Rome by December, he sought to improve his library of rare works, either by purchasing the books outright or by copying them. For he was anticipating a return to Germany from where he would have little opportunity to obtain the books he would want. It was at this time he encountered Trebizond and took him to task for his poor interpretation of the *Commentaries* of Theon. He made much about pointing out the most serious errors. Perhaps he pushed his criticisms too far. At any rate Regiomontanus left Rome for Venice again where he taught mathematics for a while.

From Venice he left for Budapest to accept the invitation of the king, Mathias Hunniades Corvinus. The king had returned victorious from war with the Turks, and part of the booty consisted of many scientific works in Greek which he had seized in Constantinople, Athens, and elsewhere in the Turkish domains. Regiomontanus accepted the opportunity and handsome salary offered him to be the librarian of the new Royal Library of Buda. Shortly after arriving in the capital city, he offered further services to the king who had become quite ill. While others had despaired of the king's life, Regiomontanus showed from astrology (in which he was quite adept) that the king was merely suffering from weakness of the heart brought on during a recent eclipse and that recovery was imminent. So it happened, and the king showered him with further gifts.

Another of his Hungarian friends was the Archbishop of Strigonium, to whom Peurbach many years before had sent a geometric gnomon. Regiomontanus spent some time with this ecclesiastic, instructing him in as-

[48] In a written criticism of Trebizond's work, Regiomontanus in direct address calls him, "impudentissime atque perversissime blatorator." (You are the most impudently perverse blabbermouth!) Cantor notes that such remarks were not uncommon among men of this caliber and were not considered serious insults, ordinarily. Considering what may have been the cause of Regiomontanus' death, however, one wonders how seriously Trebizond's sons took his remark. (See Moritz Cantor, *Vorlesungen über Geschichte der Mathematik* (1900), II, 257.)

trology[49] and the use of a set of tables which he had written.

The King of Hungary unfortunately embroiled himself in a war over Bohemia, and Regiomontanus found it to his advantage to leave for Nürnberg. Before he departed, he took time to observe that Jupiter was in the constellation Virgo on March 15, 1471. He reached Nürnberg by June 2nd for a lunar eclipse.

Regiomontanus had a particular affinity for Nürnberg. It was close to his home town. It had become a center of learning, and his library was appreciated. But two things in particular captured his enthusiasm. First was the printing press which had been set up. He saw its possibilities and was eager to spread the printed word of science. The second was that the city had become a center of the practical arts, and their practitioners could manufacture the astronomical instruments he desired. The financial means to take advantage of these two boons was soon provided from the close friendship that sprung up between Regiomontanus and Bernard Walter. This latter, besides being one of the most influential citizens of Nürnberg and quite wealthy, was a scholar in his own right, a patron of the arts, and an amateur astronomer.

Among the instruments constructed for Regiomontanus were astronomical staffs for measuring the altitudes of the sun, moon, and stars, then an astronomical radius to determine the distances of the stars. Next was an armilla such as Ptolemy and Hipparchus had used to note the location and movements of the stars. Finally he constructed other instruments, such as a torquet and a Ptolemaic meteoroscope. These instruments made it possible for Regiomontanus to make further corrections in the *Alfonsine Tables*. He wrote, in fact, "What discrepancy there is between Alphonse and the Heavens. His tables are frivolous." It was about this time that Regiomontanus divided the whole sine into one hundred thousand parts to facilitate his computations.[50]

The first books he printed were the *New Theories* of Peurbach and the *Astronomy* of Manilius. Then he published the first of his own printed works, the *New Calendar*. As described in his *Index*, it "contains the true conjunctions and oppositions of the stars together with their eclipses, the place of the stars from day to day, descriptions of the equinoctial and seasonal hours, and other useful pieces of information." His next book, the *Almanac*, enjoyed wide popularity, being distributed throughout Hungary, Germany, France, and England. With these four books in print, Regiomontanus then published his *Index of Books*.

Nowhere, perhaps, is the enthusiasm of Regiomontanus for the sciences better seen than in the *Index*. For he announced the titles of the books he intended to publish and, in some instances, the reasons for publishing the books. These books fall within two groups.

The first group consisted of the works of the ancients, particularly the Greeks. In preparing these for publication he would rely not only on his own abilities in the languages and sciences but also on the learning of his associates, as he wrote in a letter to M. Christian of Erfort. The following list is taken from the *Index* itself. "(1) A new translation of the *Cosmography* of Ptolemy, since that old translation of the Florentine James Angel is vicious. Although he meant well, he was quite weak in his knowledge of both Greek and mathematics. This opinion is concurred with by the pre-eminent Latin and Greek scholar Theodore Gaza and the scholar in mathematics and Greek, Paul Florentine. (2) Ptolemy's *Great Composition* (or *Almagest*) in a new translation. (3) The Campanus edition of Euclid's *Elements* with Hypsicles' *Ascensions*, which must be corrected in many places and to which a commentary of sorts will be added. (4) The *Commentary on the Almagest* of the eminent mathematician of Alex-

---

[49] Lynn Thorndike claims to have seen two editions of Regiomontanus' *Almanac* which show the extent to which astrology could enter into a person's private life. Both of these editions give the favorable hours for having one's hair cut and for taking a bath! (*A History of Magic . . .*, III–IV, 442). Astrology was in widespread use among the people of the times.

[50] Hutton says that Regiomontanus made the radius or whole sine into one million parts (*A Mathematical and Philosophical Dictionary*, II, 132).

andria, Theon. (5) Proclus' *Astronomical Hypotheses.* (6) A new translation of Ptolemy's *Tetrabiblos* and the one hundred fruits thereof. (7) The works of Julius Firmicus that are available, together with the writings of Leopold of Austria, the fragments—which are quite useful—of Anthony of Montulmo, and any other worthy writers on astrology. (8) The geometric works of Archimedes, namely, *The Sphere and Cylinder, The Measurement of the Circle, Conoids and Spheroids, The Spirals, Equilibrium, The Square of the Parabola, The Sand Reckoner,* together with the commentaries of Eutocius Ascalon on *The Sphere and Cylinder, The Measurement of the Circle,* and *Equilibrium.* The translation will be that of James of Cremona with some corrections. (9) *The Perspective* (Optics) of Vitello, an immense and noble work. (10) Ptolemy's *Perspective.* (11) *The Music* of Ptolemy together with Porphyry's exposition. (12) A new edition of *The Sphere* of Menelaus. (13) Theodosius' *Spherics, Habitations,* and a new translation of his *Days and Nights.* (14) The *Conics* of Apollonius of Perga, together with his *Serene Cylindrics.* (15) The *Spiritalia* of Heron, a truly pleasureful work on mechanics. (16) The *Elements of Arithmetic* of Jordan and his *Data Arithmetica.* (17) The *Quadripartitum of Numbers,* a work that abounds with various insights. (18) Aristotle's *Mechanics.* (19) The *Astronomy* of Hyginus with his chart of the heavenly bodies. (20) The *Rhetoric* of Tulliana. (21) Maps of the known world, of Italy, Spain, all of France, Greece, and Germany, together with histories collected from various sources that will discuss the geographies of the mountains and seas, the lakes and rivers, and of other particular places."

The second group of works would be his own books. Besides the *Calendar* and *Almanac* already mentioned, these would include the following. "(1) His *Great Commentary* upon Ptolemy's *Cosmography,* in which will be explained the construction and use of the meteoroscope with which Ptolemy himself estimated almost all the numbers in his works (despite the beliefs of some that the numbers of the longitude and latitude were reached by observation of the stars), and also will be included a description of the armilla together with the entire heavens, described on such a level that anyone can learn what never before was available to those understanding only Latin, because of the extremely poor translations. (2) A small *Commentary* against the translation of the Florentine James Angel, which will be sent to arbitrators (Gaza and Paul, mentioned above). (3) A *Defense* of Theon of Alexandria in six books, against George Trebizond, in which anyone will be able to see how superficial was his commentary on the *Almagest* and how poor his translation of that work of Ptolemy. (4) A small *Commentary* on the Campanus edition of Euclid's *Elements,* in which certain gratuitous statements will be disproved. (5) *The Five Equilateral Solids* — which have a place in nature and which do not — and this will be against the commentator on Aristotle, Averroes. (6) A *Commentary* on those Archimedian books which lack Eutocius' comments. (7) *Squaring the Circle,* against Nicolas of Cusa. (8) *Directions,* against the Archdeacon of Parma. (9) *On the Distinction of the Places in the Heavens,* against Campanus and John Gazulus Ragusinus, and in this work some statements from *The Hours* will be exposed. (10) *The Motion of the Eighth Sphere,* against Thebit and his followers. (11) *A Reformed Ecclesiastical Calendar.* (12) *A Short Almanac.* (13) *Triangles of Every Kind,* in five books. (14) *Astronomical Problems* with reference to everything in the *Almagest.* (15) *Comets,* their size, distance from the earth, and true position. (16) *Geometric Problems of Every Kind,* a particularly fruitful work. (17) *The Pannonian Game* or *Tables of Directions,* written at Strigonium and dedicated to the Archbishop. (18) *The Great Table of the Primum Mobile,* with many uses and grounded on solid reasons. (19) *Balances and Aqueducts,* with illustrations of the instruments necessary for these. (20) *Burning Mirrors* and many other things together with their wonderful uses. (21) *The Astronomer's Workshop,* in which many instruments for observing the heavens as well as instruments for a more earthly use will be described." He concludes the *Index* with the comment that if he can print all of this before he dies, death will not have any sting to it; for he wishes to leave all this to posterity that it not be without the necessary books.

The publication of the new *Calendar* brought him a request from the reigning Pope, Sixtus IV, to come to Rome and revise the Julian Calendar. As an inducement, the Pope appointed Regiomontanus the Bishop of Ratisbon. The esteem in which he was held at this time was recorded by Ramus.

Nürnberg gloried in Regiomontanus, his mathematics, his studies, his works. Tarento had its Archyta, Syracuse its Archimedes, Byzantium Proclus, Alexandria Ctesibius, and Nürnberg Regiomontanus. Archytas and Archimedes, Proclus and Ctesibius are dead. The mathematicians of Tarento and Syracuse, of Byzantium and Alexandria are gone. But among the Masters of Nürnberg, the joy of the scholars is the mathematician Regiomontanus.

It would be hard for him to leave Nürnberg, for he had established himself as a scientist of no mean ability. Ramus records that, besides constructing a perpetual motion machine, he made a mechanical fly that could leave a person's hand and, after buzzing about the room, return to one's hand. Larger than this was an eagle that could leave the city, fly to meet an approaching dignitary (the Emperor is mentioned), and accompany him to the city of Nürnberg.[51] The weight and responsibility of the pastoral office was not particularly attractive to Regiomontanus, for he much preferred the joys of study and disliked leaving unfinished the works he had begun. But at the insistence of Cardinal Bessarion, he was willing to leave the unfinished projects in the capable hands of his close friend and collaborator, Bernard Walter. And so toward the end of July, 1475, he left for Rome.

And in the Eternal City everything came to an abrupt end. He died within a year. For he had incurred the vehement hatred of Trebizond's sons and they took the earliest opportunity to poison him. Paul Jovius, however, says that he died during a plague. Kepler agrees with him and at the same time takes issue with Joachim Camerarius who lays his death to a comet which appeared the follow-

ing year. Whatever the cause, Regiomontanus died July 6, 1476, at the age of forty years and one month. His premature death was greatly mourned, even more so since he had accomplished so much in such a short time. It is reported that he was buried in the Pantheon. Typical of the eulogies sung in his praise was that of Jovius: "A man of wonderful skill, divine ingeniousness, he was certainly the most outstanding of all Astronomers, even of those who preceded him." Latomus composed the following in his memory.

When Jove beheld the spinning Spheres increas'd,
And a bright Tenth just added to the rest,
Began to dread his Palace wou'd be seen,
Naked, and open to the View of Men.
In vain, said he, we mov'd that Race from hence,
Since these high Battlements are no Defence.
Can none of us these bold Attempts oppose?
Or shall we sit content, invaded thus?
In vain, said Hermes, you the Fates defy;
'Tis their Decree, this Man shall never die.
If you'd restrain him, I advise you thus —
This Instant canonize him one of us.[52]

With the news of his death, Nürnberg went into public mourning. No one mourned more faithfully than Bernard Walter who continued the work of Regiomontanus until his own death, sometime after 1491 when he made his last observation in the skies.

Mention has already been made of what Regiomontanus himself published while he had the use of the printing press in Nürnberg: the *New Calendar*, the *Almanac* (both of these in 1464), and the *Index of Books* (in the following year or so). But most of his completed works were published by John Schöner of Nürnberg. The list of these books follows. (1) *Triangles of Every Kind.* (2) *On*

[51] Zinner discounts these stories as legends passed on from overzealous Nürnbergers to an unsuspecting Peter Ramus, who hastened to record them. See *Leben und Wirken* . . . , pp. 161–62.

[52] Translated by Benjamin Martin in his *Biographia philosophica* (1765), p. 156.

Sines and Chords, a joint work of Regiomontanus and Peurbach. (3) *The Rejection of the Manner of Erecting Themas* as proposed by Campanus. (4) The *Genethliacum,* which was enlarged by Schöner. (5) *Descriptions and Explanations* of various astronomical instruments including the Torquet, the Armilla, the Great Rule of Ptolemy, the Staff, and the Astronomical Radius. (6) The *Thesaurus,* a joint work of Regiomontanus and Walter. (7) *Dialogue* against Cusa and his *Quadrature.* (8) The *Solutions of all Geo-metric Problems.* (9) The *Great Table of the Primum Mobile.* (10) The *Disputations* against the *Theories* of Cremona. (11) The *Tables of Directions.* (12) The letter to John Driandrus on the *Composition of the Meteoroscope.* (13) The letter to Cardinal Bessarion in which Regiomontanus severely criticizes the work of Trebizond. His work on the *Commentaries* of Theon on Ptolemy was published by Joachim Camerarius.

Thus was Regiomontanus: his life, his work, his reputation.

TEXT
AND
TRANSLATION

DOCTISSIMI VIRI ET MATHE-
maticarum diſciplinarum eximij proſeſſoris
IOANNIS DE RE-
GIO MONTE DE TRIANGVLIS OMNI-
MODIS LIBRI QVINQVE:
Quibus explicantur res neceſſariæ cognitu, uolentibus ad
ſcientiarum Aſtronomicarum perfectionem deueni-
re:quæ cum nuſquã alibi hoc tempore expoſitæ
habeantur,fruſtra ſine harum inſtructione
ad illam quiſquam aſpirarit.

Acceſſerunt huc in calce pleracp D.Nicolai Cuſani de Qua
dratura circuli, Decp recti ac curui commenſuratione:
itemcp Io.de monte Regio eadem de re ἐλιγμα-
κα, hactenus à nemine publicata.

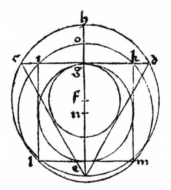

Omnia recens in lucem edita, fide & diligentia
ſingulari. Norimbergæ in ædibus Io. Petrei,
ANNO CHRISTI
M. D. XXXIII.

ON
TRIANGLES
OF EVERY KIND

JOHN, REGIOMONTANUS

In five books the author explains all those things necessary for one who wishes to perfect his knowledge of astronomy. Since these matters have never been developed before anywhere, one cannot aspire to learn this science without these ideas.

An Appendix contains the works of Nicolas of Cusa on the *Squaring of the Circle* and the measuring of straight and curved lines, together with their heretofore unpublished refutation by John, Regiomontanus.

All of these have been thoroughly edited with singular diligence and fidelity at the Publishing House of John Petreus, in Nürnberg.

A.D. 1533.

IOANNES SCHONE-

RVS CAROLOSTADIVS AMPLISS. SENA
torum Ordini ciuitatis Noricæ Dominis prudentiss. S. P. D.

TINAM prudentissimi Domini, ita Deo uisum fuisset, ut qua occasione nunc ego sum usus ad celebrandam rempubl. uestrã, dum uobis optimis scriptis doctissimi hominis Iohannis de Regio monte dedicandũ putaui, ea autor ipse in uestra florentiss. urbe excudendo cum hoc tum alia plurima ad illius immortalem gloriam uti potuisset; Nimirum ut absolutius hoc opus in lucem exiret, ita maiorem famã excitaret ciuitatis uestræ. Sed quia hanc non solum in ista parte nobis felicitatem, sed innumerabilibus alijs, sua illa morte abstulit, faciendum scilicet quod conceditur, quando id quod uolumus non licet: Hoc est conandum ut quantumcunꝗ quidem & qualecunꝗ illius uiri in manus nostras peruenerit, communicetur studiosorum utilitati. Nihil est illoꝝ enim quamuis imperfectum atꝗ inornatum aut etiam dissectum, quod non maximi pendi debere uideatur. Optandum certe, ut quia Regiomontanum ab officina, unde tot egregia opera emitterentur, quot indice præmisso indicauerat, in Italiam retraxerat uocatio, honestiss. ea quidem, sed cui obsecutus reuerteretur nunquam, ab eo relicta saltem monumenta cum ipsius tum aliorum ueterum potissimum laborum, conseruarentur; Sed hæc ipsa quædam ita uastauit calamitas, ut ex tanta tamꝗ splendida copia, qualem indices ostendunt, perpauculæ reliquiæ ad nos peruenerint. Quarum ipsarum non absꝗ præcipua utilitate sua & habent à nobis nonnihil studiosi, &, si Deus successum aspirarit conatibus meis, habituri sunt.   Hunc autem librum, cui de Triangulis omnimodis ipse autor titulũ indidit, clarissimus ordinis uestri uir Bilibaldus Pircamerus, illo tempore, quo tam speciosa suppellex Regiomontani parum diligenter conseruabatur, cum, ut sæpe dicere audiuimus, magna pecunia comparasset, non tam sibi quàm studiosis disci plinarum Mathematicarum; Hunc igitur ipsum librum, uisum Deo fuisset, ut ab eo, quem dixi clariss. uiro Bilibaldo Pircamero in lucem æderetur: quem ut uirtute & sapientia, ita literis quoꝗ & doctrina fuisse instructissimum scimus, de quo in præsentia neꝗ res fert, ut multum uerborum faciamus, neꝗ ipsius uirtutis ac eruditionis magnitudo paucis potest esse contenta, quem in uita, ut dilexerant, ue neratiꝗ fuerant docti õmnes, sic nunc ea defunctum deplorant atꝗ lugent. Magnum hoc illi tribuentes ac certiss. testimonium iudicij sui, cui in epistolæ breuita te quis etiam præstans eloquentia nedum infacundus ego satisfacturum se speret? Redeo igitur ad propositum. Ergo etiam hanc comoditatem præcidit siue Deus siue fortuna, satius esse iudicauimus, etsi minus pulcram, optimã tamẽ per se mercem contingere studiosis expositione nostra, ꝗ ut retenta omnino ea carerent. Et est primus sane liber ad eum modũ ab autore perultus, ut neꝗ ipso edente melius habiturus fuerit. Reliquis extrema manus & limæ labor non accessit, nã numeros præcedentium propositionum, quibus sequentia probabat, leui sane uel nulla potius legentium iactura festinabundus passim ascribere neglexit, in quibus neꝗ nos uoluimus ingenium industriamꝗ nostram ostentare, quamuis id facile fuerit, & hoc quisꝗ sibijpse ꝓstare possit, sed fide maxima curauimus de archetypo in aliquot exempla trãscribi, quæ patrocinio uestro Domini Prudentiss. uisum est tuta defensa'ꝗ publicare, Non parua in spe gauisuros uos tam honestã

A 2   uobis

JOHN SCHÖNER OF CAROLSTADT
Dedicates this Work to the Most Esteemed
and Learned Lord Senators of Noricum

It is to the honor of your city, Learned Lords, that these great works of that brilliant man John of Regiomontanus are dedicated to you. If God had pleased he would still be at work in your wonderful city, and he would be increasing its glory by many other accomplishments. But this pleasure and many others were taken from us with his death. What we would like to do cannot be done, but only that which is permitted us.

At long last this work was discovered, to the increasing glory of your city. As it has come into our hands, so must it be made available for the profitable study of all. None of his works — however inornate, imperfect, or even discarded they may seem — should be considered other than great. He was called to Italy, never to return. But as his *Index* shows, the work he had planned here was extensive. So it is certainly desirable that the monuments of his labors and of others be preserved.

But calamity has destroyed some of these. Of what the *Index* had promised, only a handful are left us. The scholars among us know the value of these works. And, if God grants success to my efforts, the scholars shall have them.

This work, titled *Triangles of Every Kind* by the author, was obtained by your illustrious colleague Willibald Pirkheimer at the time when the infamous thief of Regiomontanus kept it poorly. As has been told before, he got it at a great price, and not for himself but for learned mathematicians. This book, as it has pleased God, was brought to light by the well-known Willibald Pirkheimer. Adorned with virtue and wisdom, he was learned in letters and knowledge. Neither few nor many words are capable of acknowledging the heights of his virtue and erudition. Beloved by the venerable and learned alike in life, in death he is mourned and grieved by all. The great testimony of those who honor him is hardly equaled by this eloquence of mine. Yet I hope to fill some measure of it. To return to the point.

Either God or fortune has provided us with the opportunity to bless scholars with the rewards of this work. (We have already said too much, too poorly.) Not to publish it would be their loss. The author planned the first book. No editor could have done better. As far as the remaining books are concerned, there are not enough hours in the day nor revisors available for what should be done. The author neglected to make proper reference to preceding theorems by which he established subsequent theorems. This he could have done easily, for the references should have been obvious to him. Our task has been to transcribe with the greatest fidelity everything from the original text. It has been published with the safe blessings of your patronage.

uobis clientelam obtigiſſe,qui omnes bonas res cupiditate ac promptitudine in rep.ueſtra,quæ quidem eatenus abfuerint,inſtituere,conſtitutas autem diligentia & cura ſingulari conſeruare ſoleatis.Neq̃ parum hac uos confido credituros ornatum iri editione Regiomontani ſub nomine ueſtro ſcriptorum,quandoquidẽ illi uiuo V.ciuitati famam,quam deſtinarat,negatum fuit parere,Cum enim hãc mors ipſius excluſerit,una hæc ratio fuerit & inſtaurandæ illius & conciliandæ alterius generalis cuiuſdã benignitatis nimirũ & fauoris erga bonas artes ueſtræ. Qua his turbulentiſſ. temporibus inter has ſtulticias hominum complecti nos ui deamus penè deſertas illas ab omnib.mortalib. Qua de cauſſa quantas laudum quamq̃ præclaras materias deducere poſſem quis non intelligit? Quid enim præ conio literarum q̃ harum protectio dignius? Aut quo in argumento libentius illæ uires ſuas q̃ celebratione cultorum ac defenſorum ſuorum exercituræ uideantur,ſed mihi in ſuſcipiendo hoc onere conſiderandum ſcilicet,non ſolum quid cupiam,ſed etiam ac multo magis quid ualeam.Ne ſi parũ dextre quod uereor hanc rem adminiſtrauerim,& illæ iure de malo interprete,& V.Pr.detritore ſplẽdoris ſui cõqueri poſſint.Relinquam igitur hanc alijs prouinciam,qui gerant gnaui ter,cuiuſmodi ut ſcio fuiſſe,ita futuros multos arbitror.Hoc modo uos orabo,ne qua re patiamini opt.& ſanctiſſ.hoc uobis propoſitum extorqueri,quo retento immortalem profecto gloriam conſequemini,qua & ipſi frui & quam poſteris ue ſtris relinquere poſſitis.Qualia hæc ſint tempora uidetis,planè enim ut renaſcentes artes nemo magnopere reſpexit,illæq̃ ſuum tacitæ caput protulere,ita neglectas ſtertentib.hominib.interituras eſſe metuendum.Deus,ſolus enim Deus,con ſeruare poterit,hanc immittendo Principib. & Ciuitatib.mentẽ,ut rigare arentes & caducas fulcire cupiant.Quod ut hactenus à uobis ſtrenue factũ cernimus, ita uos õro obteſtorq̃ per hoc nomine quæſitã celebritatem,ne intermittere uell tis. Ad me & hunc laborem meum quod attinet,ſatis fuerit mihi comprehendi me à ueſtra benignitate in uulgari numero literatorum,quos etiam cũ uos omnes tueri fouereq̃ conſtet,nõ timeo ne uobis ego excidam,nẽue poſt hanc etiam qua ſi indicationẽ noſtri à uobis negligar. Valete Domini prudentiſſ. ex urbe ueſtra Norica pridie Iduum Sextil,anno ſalutiferi partus M. D. XXXIII.

Schöner's Dedication

With great expectation you have protected your honest clientele. With prompt willingness you have made it your business to establish in this city every good thing that had been abandoned, to restore and care for them with singular diligence and care. It is no small thing for me to hope that you believe that the writings of Regiomontanus have been adorned and edited under your name. For when he was alive, he brought no little fame to your city. But since death has brought his work to an end, your benevolent concern for the arts is sufficient reason for their restoration.

During these turbulent times when we are hemmed in by the stupidity of men, we see these arts nearly abandoned by all mortals. For this reason no one understands the praise I may evoke by the famous works I publish. For what is more worthy of the praise of the letters than their protection? And in what battle do the arts seem more freely to spend their strength than in the praise of culture and its defense?

When I took this task upon myself I considered not so much what I wanted to do, but what I could do. I had some fear of my dexterity — that it would give cause for complaint from my readers or from you. Now I feel that complaints will come only from the omnipresent knaves. I ask that you see that this work dedicated to you be not twisted about. Under your protection it shall bring you immortal glory, for yourselves and your posterity.

You know the times. No one really looks for a rebirth of the arts. They are so silent and neglected, it may well be feared that dolts will stamp them out. God alone can preserve them by inspiring the leaders of the city to water the seeds and nurture the plants. This we have seen you do with vigor. So I beseech you, in the name of fame itself, that you never stop your work. As for me and my labor, it suffices that I am numbered among the crowd of literates whom you protect and encourage. After this, Most Learned Lords, I fear neither your neglect nor your banishment.

Noricum, August 12, 1533.

Etſi uidebamur quibuſdam de indicijs conieƈturam certam feciſſe, cui autor epiſtolam dedicationis conſcri
pſiſſet, auƈtoritate cʒ doƈtrina præſtantiſſimo, ut ipſe ait, uiro, tamen quia in archetypo, quod manu ipſius de
ſcriptum eſſet, nominatim erat nullius præfixa mentio, maluimus relinquere huius etiam ɤ̃i uobis arbitriū
quàm noſtrum iudicium interponere. Gratiorem rati uobis diligentiam fidelitatemɋ́ noſtram, quàm in ali=
eno libro ingenium cʒ ſapientiam futuram. Valete.

V A M V I S hoſce triangulorum libellos poſt epitoma cõ
ſcripſerim, præpoſtero fretus ordine, poſterius quidē
opus texendo introduƈtorium, ɋ̃ artem ipſam tradi=
derim: nemini tamen triangulos noſtros prætereunti
aſtrorm diſciplina ſatis agnoſcetur. Quod ſi quiſpiam
inique faƈtū inſimulet, is niſi me animus fallit, iure qui
eſcet, ubi maiorum parere monitis & æquum & bonum arbitrabitur.
Sanè moribundo præceptori morem geſtum oportuit, qui abſolutis
nuperrime ſex luminariū libris, ſuperſtites ſeptē Ioanni ſuo reliquit,
imò mandauit ɋ̃ citiſſimum expediendos. Tantum nempe apud eū
ualuit Beſſarionis imperium, ut quod incolumis adhuc principi ſpo=
ponderat digniſſimo, iuxta iam moriturus explere curaret. Igitur iuſ=
ſa præceptoris capeſſenti mihi, plurimus triangulorū & planorum &
ſphæralium incidit uſus: quæ res iam pridem Georgio quoɋ̃ in pri=
mis ſex libris crebro occurrens, animum induxit triangulorum artem
conſcribere. Verum ut epitomati finem, ita triangulis dare initi=
um Deus ipſe uetuit, quo nunc aſpirante, orbitam uiri doƈtiſſimi
quoad potero ſeƈtabor: eo quidem libentius, quo doƈtrinã hanc ple=
riſɋ̃ placituram amicis arbitror, quorum quidem inſeruire cõmodis
bonam felicitatis meæ partem exiſtimo: eos autem, ut uirtus ipſa mo=
net, gratis amplexibus munus illud ſuſceptum ire non dubito, ſiquidē
ad alia demū altiora calcar addere pergūt. Si præterea magnis & ſci
tu iucūdiſſimis rebus ſtudere uelis, quiſquis ſiderū motus admiraris,
hæc triangulorum theoremata in primis legenda ſunt: quippe quorū
diſciplina omnibus Aſtronomicis, nonnulliſɋ̃ Geometricis quæſitis
ianuam pandit. Quemadmodum enim cæteras figuras inuicem tranſ=
mutandas ad triangulum uſɋ̃ reſolui oportet, ita reliquæ Aſtrono=
morū quæſtiones hiſce noſtris egebunt libellis. Reuera planetarum
æquatiões numerare, ipſasɋ̃ in tabulis collocare, ſed & eclipſibus lumi
nariū ſatisfacere, quãtæɋ̃ reliqs quinɋ̃ erraticis latitudines debeātur,
noſſe uolenti prior cõſulendus uenit liber. Qui demū in qualibet regi
one & aſcenſiones & arcus diurnos, deinde angulos ſphærales eclipſi
ſolari neceſſarios, mediationem cœli ac ortū obliquū ſtellis fixis eue=
nire ſolitum: poſtremo omnia quæ per figuras ſeƈtoris non ſine gra=

A 3 ui ſudo

TO READERS

The writer of the dedicatory epistle has written that the author is almost a paragon, a man of outstanding knowledge and authority. He almost makes it seem that we, to whom the dedication was made, would force this opinion upon you, the readers. On the contrary, we prefer to leave judgment about this book to you. We are sure, however, that you will be more grateful for our present diligence and fidelity than for some future choice of another book.

Although I wrote this work on triangles after the *Epitome*, the art which I have wished to pass on must be studied in the reverse order, as this introduction attempts to show. For no one can bypass the science of triangles and reach a satisfying knowledge of the stars. And if anyone has imagined that he could do such a thing, unless I am greatly mistaken, that person has nurtured only his own pride in preferring to ignore the advice of his betters.

When my teacher was dying, he left his student with six books on the stars just finished and seven for him to complete. He commanded me to do these as quickly as possible. Moreover, as a protégé of Cardinal Bessarion whose commands he took seriously, he was anxious, even on his deathbed, to fulfill them. And so not only did I fall heir to the command, but I obtained the use of many instruments: triangles, planes, and spheres. It was these things that George Peurbach used in those first six books which led to the idea of writing on triangles. As God would not let him finish the *Epitome*, He did not wish him to begin work on the *Triangles*. But now under His inspiration, I will follow as best I can the path of that learned man. Indeed I do this more than willingly for I am sure to please his many friends whose good will I cherish and whom I wish to assist. For their friendship spurs me to higher things. With open arms I welcome this duty.

You, who wish to study great and wonderful things, who wonder about the movement of the stars, must read these theorems about triangles. Knowing these ideas will open the door to all of astronomy and to certain geometric problems. For although certain figures must be transformed into triangles to be solved, the remaining questions of astronomy require these books. Indeed, to number the equations of the planets, to tabulate them, and to give a satisfactory accounting of the eclipses of the stars, to know how much latitude is due the other five wonders — to desire to know all this, one must study the former book. From this latter book, one can get help (briefly and easily though with some effort) to know in any region both the ascension and diurnal arcs,

ui sudore paßim exquiruntur, breuiter & ꝙ facillimũ aſſequi cupiet,
ex poſteriore libello comparabit auxilium. Quid memorem ſtellarũ
à terra uarias menſuratuꝗ incredibiles remotiones atꝗ corpulentias,
orbiumꝗ ſuorũ ſpiſſitudines? quos limites corporibus denſis reſolu=
ti uapores tranſilire non auſint: groſſicies inſuper Cometæ quandoli
bet apparentis, eiusꝗ à terra elongatio, nunquid nõ ſubtile uolet ſcru
tinium? His & mille alijs eiuſcemodi rebus hæc triangulorũ præcepta
iter monſtrabunt accuratiſſimũ, ſi prius obſeruationibus motuũ atꝗ
alijs primordijs parumper exercearis. Quod ſi in tanta rerũ ſciendarũ
copia, pleraꝗ dictu ambigua, aut factu forſitã ardua, lectoris noui de
terreant animũ, haud extemplo deſperandum, digna etenim talibus
medela obiectabit, ubi theoremate ꝙlibet tranſcurſo, ad numeros de=
ſcenderis exemplares. Ad hæc demũ accedit tabulæ ſinuũ non minus
utilis ꝙ noua compilatio, quæ abſꝗ faſtidio numerorum frangendo
rũ aut fractorum ad integros prolixa reductione per ſinũ arcus ſuos
ex arcuꝗ ſinũ offeret, præter cæteras eius generis tabulas id quidẽ faci
litatis habens, cꝗ per ſingula minuta expanditur: quantũꝗ uni ſecũdo
in quolibet tabulæ reſpondeat loco, diſcernitur: id autem certitudinis
cꝗ ſinus totus in ea ſex milia miliũ particularũ conſtituit. Interdũ ue=
ro primos duos quorumlibet numerorum characteres negligere nõ
dabitur uicio, ſi exactiſſimam operis præciſionẽ parui faciemus, quẽ
admodum canonibus ſuis cautum eſt. Quo tandem fieri oportet, ut
quæ cæteri longis ſcrutantur ambagibus, breui admodum nobis & iu
cunda inueſtigatione conſequi liceat. Hoc igitur ô pater optime, clien
tuli tui munus aſpernari nolis, paucula ꝗuis membrana contextũ, plu
rimis tamen atꝗ excelſis rebus ſolenne fundamentũ: Radicẽ ſcalæ ad
ſidera ducentis haud iniuria dixerim: ubi quidem immodeſti aliquid
ſi forte offenderis, iure tuo reſecabitur licet: ſi uero quicꝗ egregij au=
toritas tua ſummaꝗ huiuſcemodi ſtudiorũ peritia confirmandũ duxe
rit, tuo nomini conſecratũ eſto, qui quemadmodum durã hac tempe=
ſtate Chriſtianæ ſalutis accepiſti prouinciã, ita murmura ſua Philoſo
phi moderni te imperatore miſſa facient. iamdudũ enim quaſi cœlis
errantibus, ſideribusꝗ orbitas ſuas oblitis perculſi, ſpectatiſſimũ Phi
loſophiæ genus ſocordia præteriere. Perge igitur ut cœpiſti felici=
ter, ô mundi decus, terrenam prius compeſcere turbam, dehinc ſuo cœ
leſtia lumina reducas itineri: ne ut antehac cultores deludãt ſuos, quo
tandẽ immortali poſteris gloria nimirũ celebraberis. Vale.

De tri

to become accustomed to necessary spherical angles of the solar eclipse, to know the zenith of the heavens and the obliquity of the fixed stars. All of this he will learn through proper diagrams. To measure the distance of the stars from the earth together with their incredible movements and weights, to understand the extent of their orbits, to know the limits that the atmospheres of these dense bodies dare not exceed, to learn the size and occurrences of any comets and their distances from the earth: all of these things are ready for deep scrutiny. These and a thousand other things the *Triangles* will show most accurately, if first you give a little time to observing movements and other fundamentals.

Among such an abundance of things to learn, some may seem ambiguous or hard to understand. A new student should neither be frightened nor despair. Good things are worthy of their difficulties. And where a theorem may present some problem, he may always look down to the numerical examples for help. A useful table of sines has been added, just as good as a new compilation. By a long method of reduction, it has simplified integers and fractions to whole numbers, proceeding from arcs to their sines and vice versa. There are other useful tables carried out to minutes. "One second" in any table means that the whole sine has been broken down into six million parts. One should not overlook the first two digits of any number if he wishes exact work in the smallest detail, just as his rules caution. Now, it may well be that what others will take longer to do we can handle more easily and pleasantly. And so, most worthy patron, do not spurn your client. Nor should anyone avoid this small volume. It is a foundation for so many excellent things. I do it no injury by calling it the foot of the ladder to the stars. And if perchance I may offend someone somehow in a moment of excess, you have the right to correct it.

And if your excellence and skill in these studies leads [you] to confirm anything, this work is dedicated to you. For during these troubled times of the Christian era you have undertaken a difficult duty with the result that the mouthings of the modern philosopher — at your command — have been silenced. For some time now, like the wandering heavens or like stars that seem to have forgotten their orbits, a group of wide-eyed students have bypassed the proven path of philosophy for foolishness.

Go on happily, then, as you have begun, O wonder of the world. First to encompass the earthly mass, then to stride through paths of the stars. As the stars have not before disappointed their admirers, so now may you enjoy glory that is both great and immortal.

DE TRIANGVLIS

LIBER PRIMVS.

DIFFINITIONES.

Ognita uocabitur quantitas, quam mensura famosa, aut pro libito sumpta secundum numerum metitur notum. Quantitas mensu rare dicetur aliã quantitatem, quæ in alia continetur secundũ nume rum notum, aut quæ in alia quantitate quoties unitas in numero no to reperitur. Numerus autem notus habebitur, dũ inter eius unita tes discretionem ponit intellectus. Proportionem datam appellabimus, quan do aut denominatio sua data est, aut ipsa uel sibi æqualis proportio terminos ha bet cognitos. Proportiones æquales sunt, quibus una communis est denomina tio. Quantitatum altera ad alterã data dicitur, dum mensura per quam altera eaꝗ nota est, & reliquam notam efficit. Quantitates quotlibet inter se datas ap pellabo, quas una communis mensura notas reddit. Differentia quantitatum inæqualium uocatur portio maioris, qua minorem superat. Costa quadrati est linea recta, ex cuius in se ductione quadratum ipsum nascitur. Secundum quan titatem lineæ circulus quilibet describitur, dum semidiameter eius ipsi lineæ æqua lis statuitur. Arcus est pars circumferentiæ circuli. Linea uero recta sibi con terminalis corda sua uocari solet. Arcu & corda sua dimidiatis medietatem cor dæ dimidij arcus sinum rectum nuncupabimus. Complementum arcus cuiu libet dicitur, quæ sibi & quadranti interest differentia. Complementum autem anguli differentia ipsius ad angulum rectum diffinitur. Si quamlibet termina lem trianguli lineam basim intellexeris, duas reliquas usitato nomine latera uoca ri licebit. In triangulo tamen æquicrurio latera dicimus duas æquales lineas, & tertiam reliquam basim. Sed & in omni triangulo linea quæ perpendicularẽ sustentat, basis uulgo geometrarum nuncupari solet. Triangulus æquilaterus dicitur, quem tres æquales claudunt lineæ. Aequicrurius, cuius duntaxat duæ æquales sunt lineæ terminales. Varius aũt triangulus, qui tres inæquales habet lineas. Casus perpendicularis uocatur portio basis, perpendiculari & alterutro laterum intercepta. Multiplicatio numeri per numerum est cuiusdam numeri, in quo multiplicatus continetur, quoties unitas in multiplicante procreatio. Diuisio autem numeri per numerum sit, quando numerus elicitur, in quo unitas quoties diuisor in ipso diuiso reperitur.

Commu

CONCERNING TRIANGLES

THE FIRST BOOK

DEFINITIONS

A *quantity* is considered *known* if it is measured by a known or arbitrarily assigned measure a known number of times.* One quantity is said to *measure* another when the former is contained in the latter a known number of times or when the former is found in the latter quantity as often as the unit value is found in that known number. A *number* is *known* as long as one can recognize how many times it contains a particular unit value. A *ratio* is described as *given* when its designation† is given or when the ratio itself, or a ratio equal to it, has known terms. *Ratios* are *equal* when they have the same designation. One quantity is said to be given in [terms of] another‡ as long as the measure by which one is known permits the other to be known. Any number of quantities are given in [terms of] each other as long as one common measure makes them known. *Quantities* are *unequal* as long as one is greater than the other. The *side of a square* is the straight line from whose self-replication§ the square was produced. A *circle* can be drawn according to any length of line when its radius is set equal to that line. An *arc* is a part of the circumference of a circle. The straight line conterminous with the arc is usually called its *chord*. When the arc and its chord are bisected, we call that half-chord the *right sine*** of the half-arc. Furthermore, the *complement* of any *arc* is the difference between [the arc] itself and a quadrant. The *complement* of an *angle* is the difference between [the angle] itself and a right angle. Regardless of which boundary line of a triangle is called the *base*, the other two lines are called by the common name, the *sides*. In an *isosceles triangle*, we call the two equal lines the sides and the third one remaining the base. But in all triangles, the line to which the perpendicular is drawn is, in the common parlance of geometry, called the base. An *equilateral triangle* has three equal sides. The *isosceles triangle* has only two

equal sides. And the *scalene* has three unequal sides. The *segment*†† determined by the perpendicular denotes that portion of the base intercepted by the perpendicular and one of the two sides. *Multiplication* of a number produces another number which contains the multiplicand as often as the multiplier contains unity. *Division* of one number by another gives a third number which contains unity as often as the dividend contains the divisor.

Secundum is a difficult word to translate gracefully. "According to" conveys the sense in an intuitive manner but is not a direct mathematical statement. The word is used in describing the quotient that results when a quantity is divided up ("measured") evenly by a divisor ("measure"); for example, 9 is the number "according to" which a line or quantity of 27 feet is "measured" by 1 yard. Where possible, then, a *secundum* phrase will be rendered as "the number of times" when rectilinear measurements are involved. When a circle in a plane is under discussion, a *secundum* phrase is usually translated as "with . . . as radius," or when a circle is to be inscribed in a sphere from a pole, *secundum* is used as "with . . . as polar line." However, in some instances, where the excess verbiage which would be required is not warranted by the Latin, *secundum* must be translated strictly as "according to."

†A "designation" (*denominatio*), as it is used in this translation, indicates that common fraction or ratio which is expressible as a single word. Hence, *duplum* (double), *subduplum* (half), *triplum* (triple or threefold), *subtriplum* (a third), and so on, are designations. This usage of *denominatio* has as its only common modern counterpart the monetary denominations (nickel, dime, quarter). Therefore, it has seemed more appropriate, to avoid confusion particularly with the two forms of "denominator" (i.e., the bottom number in a fraction versus the name or designation of the entire ratio), to employ this slightly less accurate, but also less ambiguous, term. The term will be noted in its other forms as "designator" and "to designate."

‡*Dati inter se*, "given between themselves" or "given relative to each other," implies "expressed in the same units," and for modern clarity will henceforth be translated by that or by an equivalent phrase, such as "given in terms of each other."

§*In se ducere*, here used as "to replicate itself"

Communes animi conceptiones.

AEquales quantitates æqualiter mensurare . In duabus quoqʒ autquotli-
bet æquis quantitatibus mensuram eandem æqualiter contineri.   Vnita-
tis ad quemlibet numerum,& econtra,datam esse proportionem. Omnis nume
ri partem esse unitatem ab ipso denominatam.   Si à duabus quantitatibus inæ
qualibus æqualia aut idem commune abstuleris,relictis inter se atqʒ totis eandem
haberi differentiam.   Et si ex æqualibus quantitatibus inæquales portiones
secueris,relictas & defectas alternatim æquales sortiri excessus.   Omnem proportionem datam in numeris reperiri.

THEOREMA I.
Omnis datæ lineæ quadratum erit cognitum.

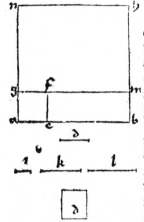

Ex data linea a b quadratũ describatur a h, quod
dico cognitũ iri.Mensura eм,pquã ipsa a b nota habeᵗ
sit linea d,cui ex duabus costis quadrati a h, q̃ sint a b
& a n,abscindantur duæ æquales lineæ a e & a g,pro
ducaturcʒ g m quidem æquedistans lineæ a b & e f
æquedistans ipsi a n,erit itacʒ superficies a f quadrata
per 29.& 34.primi elementoᴚ Euclidis. cũcʒ a b sit no
ta per mensurã lineaſē d aut a e sibi æqualē,sit k numerus secundũ quē d mēsura uel a e mēsurat lineã a b.
& l numerus alius ad quem se habeat k,sicut unitas ad
ipsum numerum,eriticʒ numerus l quadratus per octauam noni, cuius radix quadrata per uigesimam septimi numerus k declarabitur. Quoniam uero a b nota
ponitur per mensuram d,aut a e sibi æqualcm secundũ
numerum k,erit a e in a b quoties unitas in k nume-
ro per diffinitionem,quare proportio a e ad a b est,sicut unitatis ad k numerũ
Vt autem a e ad a b, ita per primam sexti quadratum a f ad paralellogramũ a
m.quadrati ergo a f ad paralellogramum a m, & unitatis ad k numerum eadē
est proportio.item per 34.primi b m æqualis est a g, quæ æquatur utricʒ linearum a e & d.Proportio igitur b m ad b h costam quadrati a h est ut a e ad a
b,uel a f ad a m, per primam autem sexti b m ad.b h est ut paralellogramum
a m ad quadratũ a h.quare paralellogramũ a m ad quadratũ a h pportionē ha
bebit eam quam quadratum a f ad ipsum paralellogramum a m. Erat autem
a f ad a m sicut unitatis ad k numerum : quare & a m ad a h erit ut unitatis
ad k numerum, & ideo ut k numeri ad l numerum. Per æquam igitur propor-
tionalitatem a f ad a h ut unitatis ad l numerum.Ex diffinitione itacʒ a f qua
dratum æquale quadrato mensuræ famosæ d mensurabit quadratum a h lineæ
a b secundum numerum l.& ideo notum habebitur quadratum a h,quod erat
demonstrandum.   ▽Opus breuissimum.   Numerus secundum quem nota
est linea,in se multiplicetur,&productus erit numerus secundum quem quadratum eius notum habebitur.   Vt si a e siue d mensura fuerit in a b secundum
numerum 5.multiplicatis 5 in se,producuntur 25.quadratellum igitur a f in
quadrato a h secundum numerum 25.reperitur.& similiter in reliquis.

Quadra

AXIOMS

Equal quantities have equal measures. And in two or more equal quantities, the same measure is contained an equal number of times. The ratio of unity to any number (and vice versa) is a given ratio. The whole part of every number is designated by [the number] itself.* If equals (having the same common property) are subtracted from two unequals, their respective differences are unequal in the same order. If unequals are subtracted from equals, the remainders are unequal in the opposite order. Every ratio can be expressed in numbers.

Theorem 1

The square of any given line is known.

If square AH is drawn from the given line AB, then the square is known.

The measure of AB is length D, and the sides of AH are AB and AN. On these two sides, measure equal lengths AE and AG. Draw GM parallel to line AB, and EF parallel to AN. Then this, according to I.29 and I.34 of Euclid's *Elements*, is the square AF. Since AB is known by the linear measure D or its equivalent AE, let K be the number of times that D, or AE, measures line AB. And let L be another number to which K is as unity is to K itself. The number L, then, is a square, according to [Euclid] IX.8, and the square root of L, by [Euclid] VII.20, is the number K. But since AB is known to be measured by D, or its equivalent AE, K times, then AE will be in AB as often as unity is in the number K by definition. Therefore the ratio of AE to AB is as that of unity to the number K. Furthermore, as AE is to AB, so also is the square AF to the parallelogram AM, by [Euclid] VI.1. Therefore the ratio of square AF to parallelogram AM is the same as that of unity to K. Likewise, by [Euclid] I.34, BM is equal to AG, which is equal to each of the lines AE and D. The ratio of BM to BH (the side of square AH) is as that of AE to AB, or AF to AM. Moreover, by [Euclid] VI.1, BM is to BH as parallelogram

AM is to square AH. Hence, parallelogram AM is in proportion to square AH as square AF is to parallelogram AM. Now AF was to AM as unity was to the number K; hence, AM to AH is as unity to the number K and therefore as the number K to the number L. Because of equal proportions, AF is to AH as unity is to the number L. Thus, by definition the square AF, which is equal to the square of the known measure D, measures the square AH, drawn from AB, L times. And therefore the square AH is known. Q.E.D.

To put it briefly: Take the number according to which the line is known [by the chosen measure] and multiply it by itself. The product will be the number according to which the square of the line will be known [by the square of the chosen measure]. For instance, if measure AE (or D) were contained in AB 5 times, then multiplying 5 by itself gives 25, which is the number of times the little square AF is contained in AH. And so on in other cases.

---

because it applies to the definition of a geometrical form, will hereafter be translated as "to be squared." The association between the definition and its subsequent use is apparent.

**A right sine should not be confused with the sine of a right angle. *Rectus* is used only to emphasize that the sine referred to is not a versed sine, *sinus versus* (one minus the cosine or, geometrically, that portion of the radius between the sine of an angle and its arc). *Sinus totus*, the whole sine, is, of course, the sine of 90°.

††*Casus, -us*, in its closest meaning ("coincidence"), is here and subsequently translated as "segment" (of the base), indicating either one of the two portions of the base intercepted by the perpendicular and the two sides or the one long portion — the base extended — between the more distant side and the perpendicular if the perpendicular falls outside the triangle.

*Because the proofs which follow often refer to this axiom, the abstruse phraseology here should be clarified. In Latin, one-half is rendered as *dimidia pars* (the halved part or ½); hence, *unitas pars* indicates the "whole part" or ¼. Then *unitas pars omnis numeri* is the whole part of every number (N) or, in fractions, $N/1$. Therefore, the axiom means simply that $N/1 = N$.

II.

Quadrati noti costa non ignorabitur.

In figuratione superiori quadratũ a h statuatur notum per mensurã quadratam d, dico ꝗ costa eius a b nota ueniet. Inter unitatẽ enim & numerũ 1, secundum quem mensura quadrata d est in quadrato a h, medius proportionalis sit k numerus, quem constat esse radicem quadratã numeri 1, processus autẽ præmissæ docuit parallelogrammũ a m esse mediũ proportionale inter quadratũ inter duo quadrata: a f quidem mensurans, & a h mensuratum. Cũꝗ sit proportio a f ad a h, sicut unitatis ad 1 numerum, id enim ex hypothesi pendet: erit proportio a f quadrati ad a m parallelogrammũ, tanꝗ unitatis ad k numerũ, quoniam utraꝗ harum ꝓportionum medietas est suæ totius, sed quadrati a f ad parallelogrammum a m proportio est, ut lineæ a e ad lineam a b per primã sexti, quare proportio a e, & ideo lineæ d ad a b, sicut unitatis ad k numerum. Vnitas igitur in k numero quoties d linealis mensura in costa a b reperitur, unitas autẽ in k est secundũ ipsummet k numerũ. Omnis eĩ numeri pars est unitas ab ipso denominata, quare & mensura linealis d continebitur in costa a b secundum numerum k. ex diffinitione igitur costam a b notam effecimus, quod expectabatur ostendendum. Tenent autẽ hæc omnia, dum 1 numerus secundũ quem mẽsuramus quadratum ꝓpositum, quadratus occurret. tunc enim reperibilis est medius proportionalis inter eum & unitatem: ꝗ si numerus 1 non quadratus fuerit, nullus erit medius proportionalis inter eum & unitatem, neꝗ unꝗ costa quadrati nota habebitur, stando in terminis quemadmodũ diffiniti sunt. Cum autẽ sæpenumero accidat numeros secundũ quos quadrata nostra metimur esse non quadratos ne prorsus ignoremus propinquũ ueritati (ut sunt scibilia humana) laxius posthac utemur uocabulo quantitatis notæ, ꝗ initio diffinierimus. Quantitatem igiꝷ oẽm *Quanti*
quæ aut nota præcise fuerit, aut notæ quãtitati fermè æqualis, uniuoce notam ap *nota lax*
pellabimus, pulchrius eqdẽ arbitror scire propinquũ ueritati, ꝗ ueritatem ipsam *definitur*
penitus negligere: non modo enim contingere metam, uerumetiã propinque ac *& mag*
cedere uirtuti dabitur. Non libuit autẽ hoc pacto superius diffinire quantitatẽ no *usitate.*
tam per præcisum & ꝓpinquum, ne suspecta lectori diffinitio nostra redderetur, fluctuante uocabulo propinqui id agente: nam etsi præcisum pro uero ponere soleamus, ꝓpinquum tamen ueritati uix diffinitionem lectori satis facturam accipi et. Ad rem ipsam demum redeundo, quoties numerus occurret non quadratus, siue integer, siue fractus fuerit, accipiemus loco eius numerum quadratum ipsi ut libet ꝗ ꝓpinquissimum, siue integer, siue fractus fuerit, inter quem & unitatẽ medium ꝓportionalem eliciemus, & procedendo ut supra notam concludemus costam quadrati, quod mensuratur secundum numerum quadratum pro libito acceptum Quemadmodũ autem numerus non quadratus noster, numero quadrato assumpto propinquus est, ita & costa quadrati nostri costæ alterius quadrati præcise cognitæ propinqua, & ideo nota habebitur. ▽Operatio facillima. Extrahe radice numeri secundum quem mensuratur quadratum ꝓpositum: si quadratus fuerit numerus huiusmodi, aut radicem numeri quadrati sibi propinqui, si non quadratus occurret, ipsa enim radix elicita quadrati tui costam notificabit.

III.

Si quotlibet quantitates inter se datæ fuerint, aggregatũ ex eis notum habebitur.

B Tres

Theorem 2

The side of a known square can be found.

If square *AH* in the preceding figure is known by the square of measure *D*, then the side *AB* of the square may be found.

Between unity and the number *L*, which is the number of times the square of measure *D* is contained in square *AH*, there is a mean proportional, the number *K*, which, it is agreed, is the square root of *L*. Moreover, the preceding proof showed that parallelogram *AM* is the mean proportional between the one square and the two squares: *AF* is the measurer and *AH* is the measured. Now since the ratio of *AF* to *AH* is as that of unity to the number *L*, it certainly follows from the hypothesis that the ratio of square *AF* to parallelogram *AM* will be as that of unity to the number *K*. Since each of these proportions is a mean of its [respective] whole, and furthermore since the ratio of square *AF* to parallelogram *AM* is as that of line *AE* to line *AB* by [Euclid] VI.1, then the ratio of *AE*, and therefore line *D*, to *AB* is as that of unity to the number *K*. Unity, therefore, is found in the number *K* as many times as the linear measure *D* is found in side *AB*. Moreover, unity is in *K* exactly *K* times, for the whole part of every number is designated by [the number] itself. In side *AB*, therefore, the linear measure *D* is contained *K* times. Thus, from the definition, we have determined side *AB*, as was expected.

All of this holds true provided that the number *L*, according to which the square was measured, is a square, for then *L* is found to be the mean proportional between the given square and unity. But if the number *L* were not a square, then there would be no mean proportional between the given square and unity, and not one side of the square will be found, if one stays within the terms as they were defined. However, since it often happens that the numbers we use

to measure our squares are not squares yet we certainly know (as human affairs are knowable) a close approximation, then hereafter we shall use the term "of known quantity" more loosely than we defined it at the start. Therefore, we will consider any quantity to be known whether it is precisely known or is almost equal to a known quantity. I personally think it nicer to know the near-truth than to neglect it completely, for it is worthwhile not only to reach the goal but also to approach close to it. Now to define a known quantity by both "precise" and "approximate" is not desirable, lest our above definition appear to the reader as ambiguous and vague. But, indeed, even if we are accustomed to take preciseness as the truth, yet enough things are accepted as definitions by the reader which are barely approximations of the truth.

At last, let us return to the point. Whenever an integer or fraction occurs which is not a square, we take in its place the closest square number, whether it be an integer or a fraction. Then between this [approximation] and unity we determine a mean proportional, and by proceeding as above, we ascertain the side of the square which is measured according to the arbitrarily chosen square number. Just as our nonsquare number is close to the substituted square number, so also is the side of our given square close to the precisely known side of the substituted square; therefore, the side of our square will be known.

The mechanics: Take the square root of the number according to which the given square is measured if that number is a square. Or if that number is not a square, take the square root of a number close to it. Then this root will be the side of your square.

Theorem 3

If several quantities are given in terms of each other, their sum may be found.

Tres quantitates a b c, aut quotlibet inter se datæ sint, quarum aggregatum sit h, dico φ ipsum h aggregatum fiet notum. Quoniam enim inter se datæ sunt

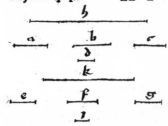

quantitates illæ, mesurabit eas communiter una quantitas quæ sit d, mensuret igitur quantitatem a quidem secundum numerum e, & b secundum numerum f : quantitatem autem c secundum numerum g, ex his tribus numeris coaceruatis resultet numerus k. Cum igitur sit d in a, quoties unitas in e numero per diffinitionem quantitatis notæ, erit pportio a ad d mensuram, sicut e numerum ad unitatem, similiter erit pportio b ad d tanqz numeri f ad unitatem, & c ad d ut g ad unitatem: quare per. 24. quinti bis assumptam proportio aggregati ex tribus quantitatibus a b c, quod est h ad d mensuram, erit ut aggregati ex tribus numeris e f g, qui est k ad unitatem, unitas ergo in k numero quoties d mensura in aggregato h continebunt, diffinitio itaqz quantitatis notæ concludet propositum.

IIII.

Duarum inæqualium inter se datarum quantitatum, differentiam cognitum iri.

Sint duæ quantitates inter se datæ, a b quidem maior, & c minor, quarū differentia sit a e, quam prædico futuram notam. Communis enim mesura ambas metiens datas quantitates sit d, quæ mesuret quantitatem a b quidem secundum numerum f g, & quantitatem c secundum numerum h, erit itaqz proportio a b ad d sicut numeri f g ad unitatē, d autem ad b c sicut unitatis ad h, per æquā igitur a b ad c sicut f g ad h. Quemadmodum ergo a b maior est c quantitate, ita f g maior est h numero, separatoqz numero k g æquali ipsi h, ex f g differentia eorum habebitur f k, & erit diuisim proportio a e ad e b, siue ad c sicut f k ad k g siue ad h. Quantitas autem c ad d mensuram, ut h ad unitatem. per æquam igitur a e ad d sicut f k ad unitatem, & ideo d mensura in a e differentia, quoties unitas in f k numero continebitur : quare per diffinitionem a e differentia nota redditur, quod erat deducendum. ▽ Operaberis autem hoc pacto. Duorum numerorum secundum quos mensurantur quantitates datæ, minorem ex maiore demas, relictus enim numerus cum mesura comuni no tam efficiet differentiam inter se datarum quantitatū. Qd' si huiusmodi duo numeri in unitate sola differant, differentia ipsarum quantitatum mensuræ communi reperietur æqualis, modus autem id demonstrandi à pristino non discrepabit.

V.

Omnes duæ inter se datæ quantitates proportionem habent eam, quam duo numeri secundum quos ipsæ mensurantur, unde manifestum proclamabimus, omnem proportionem datam in numeris reperiri.

Sint duæ quantitates a & b inter se datæ, quas ex diffinitione comunis quantitas

If three quantities *A*, *B*, and *C* (or any number of quantities), whose sum is *H*, are given in terms of each other, this sum *H* may be found.

Since these quantities are given in terms of each other, there is a common measure *D* which measures quantity *A* according to number *E*, *B* according to *F*, and *C* according to *G*. When these three numbers, *E*, *F*, and *G*, are added together, the number *K* results. Now since *D* is in *A* as often as unity is in *E*, according to our definition of a known quantity, the ratio of *A* to the measure *D* will be as that of *E* to unity. Similarly, there will be a ratio of *B* to *D* as that of *F* to unity and *C* to *D* as that of *G* to unity. Wherefore, by [Euclid] V.24 applied twice, the ratio of the sum *H* of the three quantities *A*, *B*, and *C* to the measure *D* will equal the ratio of the sum *K* of the three numbers *E*, *F*, and *G* to unity. Therefore unity is in *K* as much as *D* is contained in the sum *H*. Hence, the definition of a known quantity concludes the theorem.

Theorem 4

When two unequal quantities are given in terms of each other, their difference may be found.

If two quantities are given in terms of each other, with *AB* larger than *C* by *AE*, then this difference *AE* may be found.

Let *D* be the common measure of both quantities: *AB* is measured according to the number *FG*, and *C* according to the number *H*. Hence, the ratio of *AB* to *D* is as that of the number *FG* to unity. Furthermore, *D* is to *BC** as unity is to *H*. By simultaneous solution,† *AB* is to *C* as *FG* is to *H*. Hence, just as *AB* is greater than the quantity *C*, so *FG* is greater than the number *H*. When a number *KG*, equal to *H*, is taken from *FG*, their difference *FK*‡ will be obtained [and when *EB*, equal to *C*, is taken from *AB*, their difference *AE* remains]. By division, the ratio of *AE* to *EB* (or *C*) will be as that of *FK* to

KG (or *H*). Now quantity *C* is to the measure *D* as *H* is to unity. Thus by simultaneous solution, *AE* is to *D* as *FK* is to unity, and therefore measure *D* is in the difference *AE* as often as unity is in the number *FK*. Then by definition, the difference *AE* becomes known. Q.E.D.

The mechanics: Of the two numbers according to which the given quantities are being measured, subtract the smaller from the larger, for the number left [multiplied by] the common measure will be the difference between the given quantities themselves. But if the two numbers differ only by unity, the difference between the quantities themselves is found equal to the common measure. The proof of this statement will not differ from the foregoing demonstration.

Theorem 5

Any two quantities, given in terms of each other, have the same ratio as that of the two numbers according to which the quantities are measured.§ Hence, we may state that every given ratio may be found in numbers.

If two quantities *A* and *B* are given in terms of each other, so that by definition a common quantity *D*

*For *BC* read *C*.

†*Per aequam* — "through the equal [quantity or equation]" — is most succinctly, if not literally, translated as "through simultaneous solution" of the equations.

‡In the figure, for line *fhg* read *fkg*.

§See note, p. 31. The reader should bear in mind throughout this proof and some of the subsequent theorems that Regiomontanus was primarily an astronomer, versed in physical measurement; hence, he has taken considerable care to relate physical measurement to numerical expression. The reader will often find, therefore, this curious distinction between "number" and "quantity," the latter referring to an actual length of line, size of angle or arc, or an area, while the former is the value remaining after the quantity is "measured" (divided) by a "measure" common to all the quantities — thus a kind of reducing to the lowest terms.

titas d menſuret a quidem ſecundum e numerū, b uero ſecundum f, dico ꝙ pro
portio a ad b eſt ſicut proportio numeri e ad numerū
f. quoniam enim d menſurat a ſecundum e numerū,
erit per diffinitionem d in a quoties unitas in e numero, & ideo a ad d ꝓportio tanꝗ e ad unitatem. Item d
in b quoties unitas in f numero reperitur, d menſuran
te b ſecundum f numerum, quare proportio d ad b ſi
cut unitatis ad f numerum. per æquam igitur a quantitatis ad b quantitatem,
tanꝗ numerū e ad numerum f erit ꝓportio, quod libuit concludere.

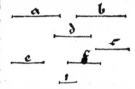

V I.

Proportio duarum quantitatum data, ex altera earum præſcita,
reliquā ſuſcitabit cognitam.

Altera duarum quantitatum a & b notam ad inuicem proportionem habentium, ſit cognita: dico, ꝙ & reliqua nota dabitur. Aut enim ꝓportio illa data
eſt per denominationem, aut per ſibi æqualem proportionem. Sit primo data per
denominationem, ponaturꝗ numerus c denominator huiuſmodi proportionis,
cunꝗ altera duarum quantitatū nota ſubijciatur, ſit antecedens a nota per menſuram d ſecundum numerū e, quo diuiſo per numerum c denominatorē propor
tionis, exeat numerus f, erit itaꝗ per diffinitionē diuiſionis c in e, quoties unitas in f numero, & ideo ꝓportio c ad e ſicut unitatis ad
f, permutatimꝗ c ad unitatem ſicut e ad f. Proportio
nem aūt c ad unitatem denominat ipſemet numerus c.
denominabit igitur & ꝓportionem e ad f. cunꝗ deno-
minet etiam proportionē a ad b, erit per diffinitionem
æqualium proportionū a ad b ſicut e ad f, & conuer-
ſim b ad a ſicut f ad e, ſed a ad d menſuram ſicut e numerus ad unitatem. per
æquam igitur b quantitas ad d menſuram ſicut f ad unitatem: quare d menſura in b quantitate, quoties unitas in f numero continebitur, ex diffinitione ergo
b quantitas reliqua nota redditur. Qꝺ ſi b conſequentē attuleris datam, ſit hoc ꝑ
d menſuram ſecundum numerū f, qui multiplicatus per numerum c denomina
torem proportionis, producat numerū e, ſecundū quem oportebit eſſe notā quan
titatem antecedentē a, erit enim per diffinitionē multiplicationis f in e, quoties
unitas in c: quare proportio e ad f ſicut c ad unitatem. Proportionis autē c ad
unitatem denominator eſt ipſemet numerus c, omnis enim numeri pars eſt unitas ab ipſo denominata, quare per conuerſionē diffinitionis c numerus denominabit proportionē e ad f, denominabat aūt & ꝓportionem a ad b, æquales igit
ex diffinitione ſunt proportiones a ad b & e ad f, ſed b conſequentis ad d menſuram ſicut numeri f ad unitatē, ꝙ b nota ſupponatur per d menſuram ſecundum numerū f, per æquam ergo fiet proportio a ad d menſuram ſicut numeri e
ad unitatem. a itaꝗ continebit d menſuram, quoties e numerus unitatē, & ideo
per diffinitionem notæ quantitatis concludemus propoſitum. ▽ Si autem data
ſit ꝓportio quātitatū per ſibi æqualē, illa ſcilicet ſibi æqualis terminos habebit
cognitos, qui aut erunt numeri, aut ex præmiſſa proportionē habebunt adinuicem, ſicut numeri, qui ſint k & l, ita ꝙ a ad b ꝓportio ſit tanꝗ k numeri antece
dentis ad l numerum conſequentem. Poſitaꝗ primum quantitate antecedente a
nota per menſuram d ſecundū numerum e, multiplicetur e per l, & productus

B   2   ſcilicet

measures A according to number E and B according to F, then the ratio of A to B is the same as the ratio of number E to number F.

Since D measures A according to the number E, by definition D is in A as often as unity is in the number E; hence, the ratio of A to D is as that of E to unity. Similarly D is in B as unity is in the number F, for D measures B according to F. Therefore the ratio of D to B is as that of unity to the number F. Then by simultaneous solution, quantity A is to quantity B as number E is to number F. Q.E.D.

Theorem 6

If one of two quantities in a given ratio is known, the other can be found.

If one of the two quantities A and B, in a known ratio to each other, is given, then the other quantity can be found. The ratio is given either by a designation or by a second ratio set equal to the first.

In the first case, where the ratio is given by a designation, let the number C be the designator of this ratio. Now since one of the two quantities is known, let the antecedent A be known, through the measure D, according to the number E. E, divided by the designator C, leaves the number F, and therefore by the definition of division, C is in E as many times as unity is in the number F. Then the ratio of C to E is as that of unity to F; by rearrangement, C is to unity as E is to F. The ratio of C to unity is designated by the number C itself, and hence C will designate the ratio of E to F. And since C also designates the ratio of A to B, then, by the definition of equal ratios, A is to B as E is to F; inversely, B is to A as F is to E. But A is to the measure D as the number E is to unity. By simultaneous solution, B is to D as F is to

unity. Whence measure D is in quantity B as many times as unity is contained in the number F. And therefore by definition the remaining quantity B has been found.

Now if the consequent [term] B had been given, it would be known, through the measure D, according to the number F. F, multiplied by the designator, C, of the ratio, gives the number E, according to which the antecedent quantity A is known. Since, by the definition of multiplication, F is in E as many times as unity is in C, then the ratio of E to F is as C to unity. The designator of the ratio of C to unity is the number C itself, since the whole part of every number is the number itself. Thus, by the converse of the definition, number C will designate the ratio of E to F, and it already designated the ratio of A to B. Therefore by definition the ratios, A to B and E to F, are equal. But the consequent [term] B is to the measure D as the number F is to unity, because B was given such that the measure D was contained in it F number of times. Thus, by simultaneous solution, the ratio of A to the measure D is as that of E to unity. Therefore A contains D as many times as E contains unity, and thus, by the definition of a known quantity, we conclude the proposition.

If, however, the ratio of the quantities were given by another [ratio] equal to itself — specifically, if the latter ratio, of equal value to the former, has known terms K and L, which either are numbers or, from the preceding [theorem], are in a ratio to each other — then the ratio of A to B is as that of the antecedent number K to the consequent number L. First, assume that the antecedent quantity A is known, through the measure D according to the number E. E is multiplied by L, and the product,

$$\frac{a}{b} = c$$

scilicet m numerus diuidatur per k numerum antecedentem, ut exeat numerus f, erit itaq; per secundam partem uigesimæ septimi proportio e ad f sicut k ad

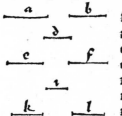

1, & conuersim f ad e sicut l ad k, & ideo sicut b ad a, sed e ad unitatem sicut a ad d mensuram ex diffinitione notæ quantitatis, per æquam igitur f ad unitatem sicut b quantitas ad d mensuram: quare b quantitas continebit d mensuram quoties f numerus unitatem. per diffinitionem ergo quantitatis notæ inferemus propositum. Si demum b quantitas consequens ponatur nota, sit hoc per mensuram d secundū numerum f, ducto q; f in k numerum, producto q; diuiso per l, exeat numerus e, erit q; per secundam partem uigesimæ quinti ut supra: proportio l ad k sicut f ad e, & con uersim k ad l sicut e ad f. sed erat k ad l sicut a ad b: quare e ad f sicut a ad b, f autem ad unitatem sicut b ad d mensuram, quoniā d mensurat b secundū numerū f, per æquam igitur a ad d mensuram sicut numerus e notus ad unita tem, erit itaq; d in a, quoties unitas in e numero noto, diffinitio ergo quantitatis notæ quod reliquū est cōcludet. ▼ Opus habeto bimembre. Si proportio da ta offeratur per denominationē, & antecedens quantitas fuerit nota, diuide nume rum quantitatis antecedentis per numerū denominatorē pportionis, & exibit nu merus quantitatis consequentis. Si aūt consequentem habeas quantitatem notā, multiplica numerum eius per numerū denominatorē proportionis, & producetur numerus quantitatis antecedentis. In exemplo: Si fuerit a .24. proportio autem eius ad b quantitatē tripla, ecce denominatorē proportionis 3. per quem diuido 24. & exibunt 8. pro quantitate consequente b. Si autem b sit 8. pportio uero a ad b quit; tupla, multiplicabo 8. per 5. denominatorē proportionis, & producent 40. pro a quantitate. Qd' si proportio data fuerit per sibi æqualem proportionē, & antecedens quantitas fuerit data, multiplicabis numerū quantitatis anteceden tis per numerum consequentē, & productum inde diuides per numerū anteceden tem, exibit enim numerus secundum quē quantitas consequens nota habebitur. Si uero' quantitas consequens data fuerit, numerū eius per numerū antecedentem multiplica, & productū per numerum consequentē partiaris, qui enim exibit nu merus, quantitatem antecedentē notificabit. Vt si a fuerit 8. & proportio eius ad b sicut 4. ad 5. multiplicabo 8. per 5. producuntur 40. quæ diuidam per 4. exibunt 10. pro quantitate b. Sed pportio a ad b sit, ut 3 ad 7, sit q; quantitas b 28 . mul tiplicabo 28 per 3, producendo 84. quæ diuido per 7. exeunt 12. erit igitur a quā titas 12. Ita in cæteris.

VII.

Si fuerint duæ quantitates inter se datæ, quarum altera per mensu ram nouam sit cognita, & reliqua per eandem nouam mensuram no ta ueniet.

Vt sermo tam breuior q; lucidior appareat, ueterem diffiniuimus mensuram eam, per quā ambæ quantitates cōmuniter notæ sunt, nouam uero, per quam alte ra earum tantum. Hæc autē à præmissa in hoc discrepat, q; illa alteram duntaxat quantitatū supponit datam, hæc uero ambas quantitates subijcit notas per unam mensuram cōmunem, & insuper alteram earum per mensuram aliam. Sint igitur duæ quantitates a & b datæ per mensuram cōmunem d, quā dicemus ueterem, altera the number M, is divided by the antecedent number K so that number F is found. Thus, by the second part of [Euclid] VII.20, E is to F as K is to L, and inversely F is to E as L is to K, and therefore as B is to A. But E is to unity as A is to D, from the definition of a known quantity. By simultaneous solution, F is to unity as quantity B is to the measure D; whence quantity B contains the measure D as many times as F contains unity. Therefore, by the definition of a known quantity, we conclude the proposition.

Finally, if the consequent quantity B is assumed known, let it be measured, through D, F number of times. Multiplying F by K and dividing the product by L yields number E. Now, by the second part of [Euclid] V.20, it will be as above: The ratio of L to K is as F to E, and inversely, K to L is as E to F. But K was to L as A was to B. Hence, E is to F as A is to B. Furthermore, F is to unity as B is to the measure D, because D measured B according to the number F. By simultaneous solution, therefore, A is to the measure D as the known number E is to unity. And thus D is in A as many times as unity is in the known number E. Consequently, the definition of a known quantity concludes what is left.

The mechanics are of two types: If the ratio is given by a designation and the antecedent quantity is known, then divide the number of the antecedent quantity by the designating number of the ratio, and the number of the consequent quantity will appear. If, however, the consequent quantity is known, then multiply its number by the designator ⌐f the ratio, and the number of the antecedent quantity will be found. For example, if the ratio of A to B, where A is 24, is threefold, the designator of the ratio is 3, by which one divides 24 to get 8 for the consequent quantity B. If, however, B is 8 and the ratio of A to B is fivefold, then multiply 8 by 5 (the designator of the ratio) and 40 will result for the quantity A. If the ratio is given by another [ratio] equal to itself, and the antecedent quantity [in the first ratio] is known, then multiply the number of the antecedent quantity [in the first ratio] by the consequent number [in the second ratio] and divide this product by the antecedent number [in the second ratio]. This will yield the number according to which the consequent quantity [in the first ratio] is known. However, if [in the first ratio] the consequent quantity is given, then multiply its number by the antecedent number [in the second ratio] and divide the product by the consequent number [in the second ratio]. The result is the number which will identify the antecedent quantity [in the first ratio]. For example, if A is 8 and its ratio to B is as 4 to 5, then multiply 8 by 5 to obtain 40, which is then divided by 4 to produce 10 as quantity B. On the other hand, let A to B be as 3 to 7, where quantity B is 28; then multiply 28 by 3 to obtain 84, and divide 84 by 7 to get 12, the value of quantity A. And so on.

Theorem 7

If two quantities are given in terms of each other and one of them is known by a new measure, then the other may be found by this same new measure.

To be as clear and concise as possible, let the old measure be defined as that by which both quantities are conjointly known, while the new is defined as that by which only one of them is known. This theorem differs from the preceding [theorem] in that the [preceding] assumed only one of the quantities to be given, while this one permits both quantities to be known by one common measure and, in addition, one of them by a different measure. So, let the quantities A and B be given, both measured by the old measure D,

$$\frac{a}{x} = \frac{c}{d}$$

$$x = \frac{ad}{c}$$

altera infuper earum(uerbi gratia) a data fit per menfuram nouam c :dico, φ &
b quantitas per eandem c menfurā nota profiliet.Metiatur enim d menfura du=
as quantitates fubiectas a & b fecundū numeros e & f ,erit itacʒ per quintā hu=
ius proportio duarum quantitatum a & b ,tancʒ duo-
rum numerorū notorum e & f,quæ cum fit nota , erit
etiam proportio quantitatum a & b nota.Eſt autem
a quantitas data per mēfurā c,quare per præmiſſam &
b quantitas per eanden notificabitur menfuram, quod
uolebamus inferre. ⁊ Operatio huius habita propor
tione duarum quantitatum propofitarum in terminis notis,ab opere præceden=
tis non diſſonabit.

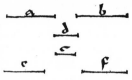

VIII.

Si utracʒ duarum quantitatum ad tertiam data fuerit,ipfæ inter fe
datæ habebuntur.

Duarum quantitatū a & b utracʒ ad tertiam quantitatē c data intelligaᵗ:
Dico, φ ipfæ inter fe reddentur notæ.Quoniā enim utracʒ quantitatū a & b ad
quantitatē c data eſt,menfurabit eas cōmuniter una menfura,quæ fit d ,fimiliter
duæ quantitates b & c ,menfuram habebunt unam cōmunem,quæ fit e . Si itacʒ
d & e menfuræ æquales fuerint,eas tancʒ unam nō iniuria reputabimus, ficʒ du
as quantitates a & b ,quantitas una cōmunis metietur:unamquācʒ fecundū nu=
merum fuum,per diffinitionē ergo inter fe datæ comprobantur. Si uero duæ mē
furæ d & e diuerfas fe obtulerint,duæ quantitates dictæ,etfi datæ habeantur feor
fum,inter fe tamen nondū datæ funt.Menfuret igitur d quantitates a & c fecun
dum numeros f & g ,menfura autē e duas quantitates b & c fecundū numeros
h & k .Placeat itacʒ duas quantitates a & b inter fe
notas efficere per menfuram d ,quam pro libito primā
e uero fecundam nuncupabimus:fimiliter ʒcʒ a quan
titatem primam uocare licebit, b autem fecundam, φ
hanc cum c quantitate prima menfura,illam uero fe=
cunda metiatur,hoc enim pacto fermonis uitabitur cō
fufio.Reperiatur itacʒ numerus l, ad quem fe habeat g
numerus ficut k ad h,ʒd facile fiet,fi ad uigefimam fe
ptimi recurres.Cuncʒ fit proportio h ad κ,ficut quan
titatis b fecundæ ad quantitatem e tertiam,ex quin=
ta huius, erit & l ad g tancʒ b ad c, fed g numerus ad unitatē, ficut quantitas c

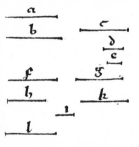

ad menfuram d primam,cum c quantitas nota fit per menfuram d fecundū nu=
merum g .Per æquam igitur l numerus ad unitatem,ficut b quantitas ad d mē
furam:quare b continebit d menfuram quoties l numerus unitatem,& ideo per
diffinitionem quantitatis notæ, b quantitas nota habetur per menfuram d,quā
continet fecundum numerum l .erat autem & quantitas a per eandem menfurā
nota.ex diffinitione igitur inter fe datarum quātitatum conſtabit a & b inter fe
datas eſſe.Qď fi idem per menfurā e confequi libeat,ficut ipfam e menfurā primam,ita & b quātitatem primā dicemus,ad numerum autem l fe habeat k nu=
merus,tancʒ g ad f,reliqua ut antehac procedent. Vniuerfaliter enim menfura,
per quā duas quantitates ad tertiā datas inter fe notas efficere conamur,prima uo
cabitur,reliqua uero fecūda.Similem denicʒ ordinem duabus quātitatibus dictis

B 3   deputa=

and furthermore, let *A* be given by the new measure *C*; then quantity *B* can also be measured by *C*.

Let *D* measure the two given quantities *A* and *B* according to the numbers *E* and *F*. Now by Th. 5 above, the ratio of the two quantities *A* and *B* will be as that of the two known numbers *E* and *F*. Since this latter ratio is known, so also the ratio of the quantities *A* and *B* will be known. Moreover, quantity *A* is given by measure *C*; therefore *B* will be identified by both the old and the new measures. Q.E.D. The examples of the previous theorem can be used here.

Theorem 8

If each of two quantities is given in terms of a third, they can be expressed in terms of each other.

If each of two quantities *A* and *B* is known in terms of a third quantity *C*, then they can be known in terms of each other.

Since each of the quantities *A* and *B* is expressed in terms of *C*, one measure *D* measures [*A* and *C*] commonly; similarly the two quantities *B* and *C* have one common measure *E*. Now if *D* and *E* are equal measures, they may rightly be considered as one, and thus one common quantity measures the two quantities *A* and *B*, each according to its own number. Therefore by definition it is proved that they are expressed in terms of each other.

But if the two measures *D* and *E* are not the same, then the two given quantities, even though they are known individually, are not yet known in terms of each other. Therefore, let *D* measure quantities *A* and *C* according to the numbers *F* and *G*, and let *E* measure the two quantities *B* and *C* according to the numbers *H* and *K*. Therefore, we wish to make *A* and *B* known in terms of each other through the measure *D*, which we will arbi-

trarily call the first [measure] while we will call *E* the second [measure]. Similarly, one may also call *A* the first quantity and *B* the second, because the first measure [*D*] measures the former [*A*] together with quantity *C*, while the second measure [*E*] measures the latter [*B*] together with quantity *C*. In this way, confusion between the expressions will be avoided. Number *L* is found such that number *G* is to number *L* as *K* is to *H*; this is easily done if you refer to [Euclid] VII.20. Since the ratio of *H* to *K* is as that of the second quantity *B* to the third quantity *E*,* from Th. 5 above, then *L* to *G* will be as *B* to *C*. But number *G* is to unity as quantity *C* is to the first measure *D* because quantity *C* is known by the measure *D* according to number *G*. By simultaneous solution, number *L* is to unity as quantity *B* is to the measure *D*; whence *B* contains the measure *D* as many times as number *L* contains unity. And therefore, by the definition of a known quantity, quantity *B* is known by the measure *D*, which it contains *L* times. Furthermore, quantity *A* was known by the same measure. Thus, by the definition of quantities given in terms of each other, it is established that *A* and *B* are [now] expressed in terms of each other.

But if one wished to pursue the same goal through the measure *E*, then, just as *E* will be called the first measure, so *B* will be called the first quantity. Furthermore, let *K* be to *L* as *G* is to *F*, and the rest proceeds as above. For in general that measure by which we attempt to interrelate two quantities that have been expressed in terms of a third will be called the first [measure], while the other [measure] will be called the second [measure]. We will assign a similar rank to the two given quantities,

*For *E* read *C*.

deputabimus eam notando primā, cui & quantitati tertiæ prima respondet mensura, reliquā uero secundā. Ad numerum autem reperiendum, qui uidelicet notificaturus est q<del>u</del>ātitatem secundā se habeat numerus quātitatis tertiæ primus, sicut numerus eiusdem quātitatis tertiæ secundus ad numerum quātitatis secundæ, primum autem numerum quātitatis tertiæ duos habentis numeros, uoco eum secundum quem prima mensura in ipso continetur numero, reliquū autem secundum. ¶ Operationem sic habebis. Numerum quātitatis secundæ duc in numerum primum quātitatis tertiæ, productum uero per numerū secundum eiusdem quātitatis tertiæ partiaris, exibit enim numerus quātitatis secundæ quæsitus, secundum quē uidelicet mensura prima in ipsa secunda quātitate continetur, ut in exemplo. Sit quantitas a 5. cubitorum, dum c est 9. cubitorum, item b 7. pedum, dū c est 26. erit itacp d cubitus unus, & e pes unus. Volo scire quot cubitos habeat quantitas b, multiplico 7 per 9, producuntur 63, quæ diuido in 26, exeunt $2\frac{11}{26}$. Quantitas igitur b duos complectet cubitos, & undecim uigesimassextas unius cubiti, sicp duas quātitates a & b inter se notas reddidimus, utendo mensura d cubitali. Nō aliter operabimur, si per mensurā e pedalem ipsas nouisse libeat quantitates. Constat igit ex hoc, cp 7 pedes $2\frac{11}{26}$ cubitis æquipollent, diuisocp numero 7 pedum in numerum cubitorum sibi æquipollentiū, exibit numerus pedum correspondentiū uni cubito, qui numerus erit $2\frac{8}{9}$. hoc quoniā multis in locis utile est, prætereundi non erat consilium. Possumus autem & breuius theorema præsens stabilire, si ad præmissam confugerimus. Libeat enim duas quantitates a & b inter se notas efficere per mensurā d, quoniā itacp duæ quantitates b & c datæ sunt per mensurā e ueterem, quarū altera uidelicet c data est per mensurā nouā d, erit & b quantitas ex præmissa per eādē nouā mensurā d cognita, sed hypothesis subiecit a quātitatem notā per d mensurā. Duæ igitur quātitates a & b, quas una communis mensura d metitur secundū numeros notos, cognitæ per diffinitionem declarantur, quod pollicebamur ostendendum.

I X.

Si duarum quantitatum utracp ad tertiam data fuerit, summā earū atcp differentiam, si inæquales fuerint, cognoscemus.

Duarū quantitatū a & b utracp sit data ad c quantitatem: Dico, cp summa ex eis conflata cognoscetur, cum differentia earū si inæquales fuerint. Si em inter se datæ fuerint, tertiā & quartam huius consulemus. si uero non inter se, sed ad tertiam duntaxat, quemadmodū supponitur, datæ sint, per præmissam reddemus eas inter se datas, quo facto, per tertiā & quartā præallegatas, quod reliquū est absoluemus.
¶ Operationem autem huius ab operationibus dictarum propositionum pender nemo dubitabit.

X.

Quotlibet quantitates ad aliam datæ, inter se non erunt ignotæ.

Tres quantitates a b c, aut quotlibet datæ ponantur singulatim ad quantitatem e: Dico, cp ipsæ inter se notæ uenient. Quoniā enim a & e inter se datæ sunt mensurabit eas communiter mensura una, quæ sit d, similiter b & e communem habebunt mensurā, quæ sit d, sed & duæ quantitates c & e per mensurā h notæ intelli

by indicating as the first [quantity] that to which, along with the third quantity, the first measure corresponds, while the other quantity will be the second [quantity]. Moreover, the first number [G] of the third quantity [C] is to the unknown number [L], which will identify the second quantity [B], as the second number [K] of that third quantity [C] is to the number [H] of the second quantity [B]; now, that [number G] according to which the first measure [D] is contained in the number [C] itself is called the first number of the third quantity (which has two numbers); the other [K] is the second [number of the third quantity].

The mechanics: Multiply the number of the second quantity by the first number of the third quantity, and divide the product by the second number of the third quantity. This will give the desired number of the second quantity — the number of times the first measure is contained in that second quantity. For example, let quantity A be 5 cubits, while C is 9 cubits; likewise, B is 7 feet, while C is 26 feet. D equals one cubit and E one foot. The question is, how many cubits are there in quantity B? Multiplying 7 by 9 gives 63, which is divided by 26 to give $2\,\frac{11}{26}$. Therefore B consists of 2 cubits and $\frac{11}{26}$ths of one cubit. Thus, by using the cubit measure D, we have expressed quantities A and B in terms of each other. We will not work any differently if we wish to obtain these same quantities by the foot measure E. It can be established from the fact that 7 feet equal $2\,\frac{11}{26}$ cubits. Dividing the number of 7 feet by the number of cubits equal to it, one obtains the number of feet corresponding to one cubit, namely $2\,\frac{8}{9}$. Since this procedure is useful for many things, this instruction was not to be neglected.

One can prove the present theorem more briefly if he has recourse to the previous one. Let it be desired to express the two quantities A and B in terms of each other through measure D. Then, since the two quantities B and C are given through the old measure E, and since one of them, C, is given by the new measure D, then, from the previous theorem, B will be known through the same new measure D. But the hypothesis gives quantity A in terms of D. Therefore the two quantities A and B, which are measured by the common measure D according to known numbers, are declared found by definition, which was promised.

Theorem 9

If each of two quantities is given in terms of a third, both their sum and [their] difference, should they be unequal, may be found.

If each of two quantities A and B is expressed in terms of a third quantity C, then their sum and their difference (if they are unequal) may be found.

If these two are given in terms of each other, we consult Ths. 3 and 4 above. If, however, they are given not in terms of each other but only in terms of a third quantity, as was proposed, then, by the previous theorem, we can find them in terms of each other. When this is done, then, by Ths. 3 and 4 above, we can determine what remains. It should be obvious that the mechanics of this theorem follow from those of the previous one.

Theorem 10

Any number of quantities expressed in terms of one other quantity can be known in terms of each other.

Let A, B, and C (or any number of quantities) be individually given in terms of some quantity E; then these can all become known in terms of each other.

Since A and E are expressed in the same terms, they have the common measure D; likewise B and E have the same measure D.* And furthermore, C and E are known through measure H.

*For D read G.

intelligantur. Si igitur tres mensuræ d g h æquales fuerint tres quantitates pro‑
positas inter se datas, ex diffinitione conuincemus: si ue‑
ro inæquales occurrant, non erunt dictæ quantitates in‑
ter se datæ. Volenti ergo tres quātitates propositas inter
se notas efficere, eligenda est una trium mensurarum, per
quā id facere placet, sitǭ d talis mensura, quam, ut circa
octauā diffiniuimus, primā dicemus, quātitatem quoǭ a
primā trium quantitatū statuemus. Cum itaǭ duæ quātitates a & b ad tertiā
e quantitatem datæ sint, erunt ipsæ inter se notæ, quemadmodū octaua huius do‑
cuit per mensurā d communem. Item duabus ǭtitatibus a & c datis, existenti‑
bus ad quantitatem e tertiā, eas inter se notas, efficiet octaua huius per mensurā
d cōmunem. Tres itaǭ quantitates a b c, mensura una communis notas reddi‑
dit, quare per diffinitionem inter se notarum quantitatū ueritas cōnstabit propo‑
sitionis. Non aliter procedendū erit, si plures ǭ tres huiusmodi occurrent quanti‑
tates, neǭ refert quācunǭ cōmunium elegeris. Ad summā igitur huius theorema
tis processus totus non est, nisi octauæ huius tenor repetitus. ▽ Opus ǭǭ huius
ab operatione illius octauæ, quoties oportuerit resumpta non differet.

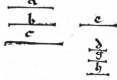

XI.

Aggregatum ex quotlibet quantitatibus ad aliam datis, cum diffe
rentia duarum quarumlibet si qua fuerit, nō ignorabit Geometra.

Si enim quantitates illæ inter se notæ fuerint, tertiam & quartā huius repete‑
mus, ǫ si non inter se, sed ad aliam tantū, quemadmodū supponitur, datæ extite‑
rint, posteaǭ ex præcedenti theoremate inter se notas reddiderimus eas, ad tertiā
huius & quartā confugiemus. ▽ Opus autem, cū & facile sit, & ab opere dicta‑
rum propositionum pendeat, prætereundum censeo.

XII.

Si utriusǭ duarum quantitatum ad tertiam data fuerit proportio,
earum inter se proportionem patefieri.

Sint duæ quantitates a & b, quarū utraǭ ad quantitatem c proportionem
habeat cognitā: Dico, ǫ pproportio a ad b nota ueniet. Oportet enim notas esse p
portiones a & b ǭtitatum ad c quantitatem, aut per denominationes, aut sibi
æquales pportiones. Sit hoc primo per denominationes. cum itaǭ proportio a
ad c nota sit, erit denominatio eius nota. sit ergo denominator huius proportiōis
numerus d, similiter denominator proportionis b ad c notæ sit numerus e, erit
autem proportio a ad c, sicut d ad unitatem, nam
utriusǫ harū proportionū denominator est ipse nu‑
merus d. denominat enim pproportionem a ad c quē
admodū posuimus, sed & proportionē sui ad unitatē
per cōmune animi conceptionē, omnis enim nu‑
meri pars est unitas ab ipso denominata. per eadem
ǭǭ media erit proportio b ad c tanǭ e ad unitatē,
& conuersim c ad b sicut unitas ad e, erat autem c

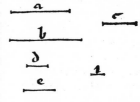

a ad c, ut d ad unitatem. per æquam igitur pproportio a ad b sicut d ad e, sed pro
portio d ad e nota est per diffinitionem: quare & pproportio a ad b nota redditur.
Liquet itaǭ ex dictis proportionem primæ quantitatis ad secundam æqualē esse
proportioni

Now if the three measures D, G, and H are equal, we conclude from the definition that the three given quantities are known in terms of each other. But if they are unequal, then the given quantities are not known in terms of each other. Therefore since you want to make the three given quantities known in terms of each other, choose one of the three measures that will do it. D is such a measure, which, as we defined in Th. 8 above, we call the first measure. Also, we consider A to be the first quantity of the three quantities. Now since the two quantities A and B are expressed in terms of a third quantity E, they are known in terms of each other by the common measure D, as Th. 8 above states. Likewise the two given quantities A and C, being expressed in terms of a third quantity E, become known in terms of each other through the common measure D, by Th. 8 above. Thus, since one common measure makes the three quantities A, B, and C known, then, by the definition of quantities known in terms of each other, the truth of the proposition is established. One proceeds no differently if more than three quantities of this kind are given, nor does it matter which common measure is chosen. In summary, therefore, the procedure of this theorem is not complete unless the course of Th. 8 above is recalled.

The mechanics of this theorem do not differ from those of Th. 8 above, repeated as often as necessary.

Theorem 11

Geometry can find the sum of any number of quantities expressed in terms of another quantity, together with the difference of any two of these if a difference exists.

If the quantities are expressed in terms of each other, then we return to Ths. 3 and 4 above. If they are expressed not in terms of each other but in terms of some other quantity as was proposed, then, after we have found them in terms of each other by the preceding theorem, we have recourse to Ths. 3 and 4 above. The work here is so obvious from what has been said that further comment is unnecessary.

Theorem 12

If the ratio of two quantities to a third is given, their ratio to each other may be found.

If two quantities A and B are given such that each of them has a known ratio with quantity C, then the ratio of A to B can be found. One must know the ratios of quantities A and B to quantity C either by designations or by ratios equal to the ratios themselves.

First, let the ratios be known by designations. Since the ratio of A to C is known, its designation will be known. Thus, let D be the designator of that ratio. Similarly, let the designator of the ratio of B to C be the number E. Moreover, A is to C as D is to unity. Now the designator of both of these ratios [i.e., of A to C and of D to unity] is the number D itself, for D designates the ratio of A to C, as we set it, and furthermore, [D] designates the ratio of itself to unity by the axiom that the whole part of every number is designated by the number itself. In the same manner, the ratio of B to C is as that of E to unity, and inversely C is to B as unity is to E. Now, since* A was to C as D was to unity, then, by simultaneous solution, the ratio of A to B is as that of D to E. But the ratio of D to E is known by definition. And thus the ratio of A to B becomes known.

It is evident from what has been said that the ratio of the first quantity to the second quantity is equal

*In Latin, for c read *cum*.

proportioni denominatoris primi ad denominatorem secundum, quod pro corollario non inutili reputabimus. Quòd si a & b quantitatū ad c proportiones datæ offerantur per sibi æquales ‚pportiones, oportebit eas in numeris reperiri p diffinitionem proportionis datæ & quintā huius. Sit itaq‚ ‚pportio a ad c sicut numeri e ad numerū f, quātitatis uero b ad c sicut numeri g ad numerū h, multi-
plicatoq‚ f per g, & producto diuiso per numerū h, exeat numerus k, eritq‚ per
secundā partem uigesimæ septimi, ‚pportio h ad g, sicut f ad k. Cum igitur e nu-

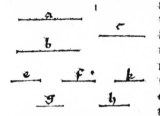

merus ad f se habeat, sicut a quātitas ad c quantitatem, f autem ad k sicut h ad g, & ideo conuersim argumentando, sicut c quantitas ad b quātitatem, erit per æquam proportio e numeri ad k numerum, sicut a quantitatis ad b quātitatem, ‚pportio autem e numeri noti ad k numerum notū data
est. quare & proportio a quantitatis ad b quātitatem cognita elicis‚, q‚d libuit explanare.   ▼ Opus
primū. Denominatorem primæ proportionis per denominatorem proportionis
secundæ partiaris, exibit enim denominator ‚pportionis quā habēt dicti denominatores, primus uidelicet ad secundū, quæ etiam proportioni primæ quātitatis ad
secundā communis existit. Aut denominatores ipsos serua pro noticia proportionis, satis enim nouisti primæ quantitatis ad secundā, si eam in numeris notis repe-
risti. Vt si ‚pportio a ad b fuerit quintupla, b uero ad eandē c quātitatem septu-
pla: cum primæ proportionis denominator sit 5, secundæ uero 7, erit proportio a
ad b sicut 5 ad 7.   ▼ Opus aliud. Numerū consequentem primæ proportionis
duc in numerū antecedentem secundæ proportionis, & productū diuide per numerum consequentem secundæ proportionis, exibit enim numerus ad quē se habet numerus antecedens primæ proportionis, sicut prima quantitas ad secundā.
Vt si a ad c fuerit proportio sicut 5 ad 7. b autem ad c sicut 3 ad 8. multiplicabo 7 per 3. producentur 21. quæ diuido per 8, exeunt 2 ⅝, a igitur habebit se ad b
sicut 5 ad 2 ⅝.

XIII.

Si quotlibet quantitatū ad aliam datæ fuerint proportiones, omnium duarum ex eis proportio manifestabitur.

Tres quantitates a b c, aut quotlibet plures ad quantitatem aliā d proportiones habeant cognitas. Dico, q‚ quælibet duæ ipsarum proportiones sortientur
datas. Placeat enim primum proportionem a ad b reddere notā. Cum itaq‚ utri-
usq‚ quātitatum a & b ad quātitatem d proportio sit no-
ta, earum inter se proportio non ignorabitur ex præmissa
Similiter de omnibus duabus reliquis prædicabimus. Nihil enim alieni præsens addit præmissæ, nisi q‚ processum
eius ingeminat.   ▼ Opus q‚q‚ huius opus illius est, quoties oportuerit repetitū.

XIIII.

Si utriusq‚ duarum quantitatum ad tertiam data fuerit proportio,
fueritq‚ altera earum nota, reliqua quoq‚ notam se offeret.

Vtriusq‚ duarum quātitatū a & b ad quātitatem c data sit proportio, sitq‚
a quantitas nota. Dico, q‚ & b quātitatem notam fieri oportet. Cū enim utraq‚
earum

to the ratio of the first designator to the second designator. This is a useful corollary.

Now if the ratios of quantities *A* and *B* to *C* are given by ratios equivalent to the given ratios, then they must be found in numbers by the definition of given ratios and by Th. 5 above. Hence, let the ratio of *A* to *C* be as that of number *E* to number *F*, while quantity *B* to *C* is as number *G* to number *H*. Then multiply *F* by *G* and divide the product by the number *H* to obtain the number *K*. Thus, by the second part of [Euclid] VII.20, the ratio of *H* to *G* will be as that of *F* to *K*. Since number *E* is to *F* as quantity *A* is to *C*, and since *F* is to *K* as *H* is to *G* — and by inversion as quantity *C* is to quantity *B* — then, by simultaneous solution, the ratio of number *E* to number *K* is as that of quantity *A* to quantity *B*. Since the ratio of the known number *E* to the known number *K* is given, the ratio of quantity *A* to quantity *B* has thus been found. Q.E.D.

The mechanics [of the] first [type]: Divide the designator of the first ratio by the designator of the second ratio to obtain the designator of the ratio of the given designators — namely, of the first to the second; this ratio is also equal to the ratio of the first quantity to the second. Or keep the designators [as they are] for identification of the ratio, for that is enough to give the ratio of the first quantity to the second, if that ratio is to be found in known numbers. For instance, if the ratio of *A* to *B** were fivefold, while that of *B* to the same quantity were sevenfold, then the designator of the first ratio is 5 while that of the second is 7, and the ratio of *A* to *B* is as 5 to 7.

The mechanics of the other type: Multiply the consequent number of the first ratio by the antecedent number of the second ratio. Then divide the product by the consequent number of the second ratio. This yields a number [such that] the antecedent number of the first ratio is to this [new] number as the first quantity is to the second. For instance, if the ratio of *A* to *C* were as that of 5 to 7, and if *B* to *C* were 3 to 8, then multiplying 7 by 3 and dividing their product by 8 gives 2⅝. *A* is to *B*, then, as 5 is to 2⅝.

Theorem 13

If the ratios of any number of quantities to one other [quantity] are given, then the ratio of any two of these becomes known.

If three quantities *A*, *B*, and *C*, or any greater [number of quantities], have known ratios to some other quantity *D*, then the ratios between any two of these can be considered known.

First, let us agree to determine the ratio of *A* to *B*. Now since the ratio of each of these quantities *A* and *B* to the quantity *D* is known, their ratio to one another is known from the previous theorem. The same can be said of all the remaining pairs of quantities. This really adds nothing new to what has gone before except that the procedure is enlarged.

The mechanics here are the same as the mechanics above, repeated as often as is necessary.

Theorem 14

If the ratio of each of two quantities to a third is given and one of the quantities is known, then the other also can be found.

If the ratio of each of the two quantities *A* and *B* to a quantity *C* is given and quantity *A* is known, then quantity *B* can be found.

Since each

*For *B* read *C*.

earum ad c datam habeat proportionem,proportio ipſarū inter ſe per 12.huius nota ueniet.quare per ſextam huius e quantitate nota exiſtente,& b quātitas nota emerget,quod expectabatur declarandum. ☞ Operatio aūt huius ex operibus duodecimæ & ſextæ huius commiſcetur.Nam poſtʠ ex duodecima huius proportionem huiuſmodi quantitatum reperies,per ſextam tandem, quæ prius ignota erat,notam efficies quantitatem.

X V.

Si quotlibet quantitates ad aliam quandam proportiones habuerint datas,quarū una quælibet ſit noʦa,reliquæ omnes notæ proſiliēt.

Sint tres aut quotlibet quantitates a b c, quarum unaquæʠ ad c quantitatem proportionem habeat dataʒ:ſitʠ una earum quæcunʠ(uerbi gratia) a data.Dico,ʠ reliquæ omnes notæ occurrent.Erit enim per 13.huius proportio a quantitatis ad ſingulas alias data,quare per ſextam huius a quantitate nota exiſtente,ſingulæ reliquæ innoteſcent,quod erat concludendū. ☞ Operationem ex tredecima & ſexta huius facile comparabis.

X V I.

Quod ſub duabus inter ſe datis rectis lineis continetur, parallelogramum rectangulum latere non poterit.

Sit parallelogrammū rectangulum a b c d,duabus inter ſe datis contentum lineis a b & a d.Dico,ʠ ipſum prodibit cognitum.Quoniā enim duæ lineæ a b & a d inter ſe notæ ſunt,menſurabit eas communiter una quantitas quæ ſit h. menſuret itaʠ lineam a b ſecundum numerum k,& lineam a d ſecundum numerū l,& abſcindanʒ ex duobus lateribus pallelogrami ˏppoſiti:duæ lineæ a g & a e,quarum utraʠ menſuræ communi h æqualis exiſtat,productis lineis e n g dem æquediſtante ipſi a b & g f,æquediſtante lateri a d,eritʠ per 29.34. primi & diffinitionem quadrati ſuperficies a f quadrata.cunʠ h ſiue a g ſibi æqualis menſuret latus a b ſecundum numerum k,erit a g in a b quoties unitas in k,& ideo proportio a g ad a b,ſicut unitatis ad k numerum,quare ˏp primā ſexti proportio quadratelli a f ad parallelgrammum a n,ſicut unitatis ad k numerum.Vnitatis autem ad k numerum proportio data eſt per animi conceptionem,quare & ˏpportio quadratelli a f ad parallelogrammum a n nota perhibebitur. Item quoniā a e æqualis ipſi h menſurat latus a d ſecundum numerū l,erit a e in a d,quoties unitas in l numero.quare proportio a d ad a e eſt ut numeri l ad unitatem.proportio autem a d ad a e per primā ſexti eſt tanʠ parallelogrāmi a c ad parallelogrammum a n.parallelogrāmum ergo a c ad parallelogrāmum a n ſicut numerus ad unitatem.ſed numerus l ad unitatem proportionem habet datam ex communi animi conceptione,unde & proportio a c ad a n ſcita ueniet.Iam igitur duarum quantitatum ſuperficialium a f & a c,utraʠ ad parallelogramū a n proportionem habet datam,quare per 12.huius earum inter ſe conſtabit proportio.

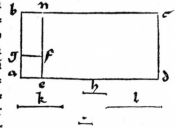

C &quo

of these forms a given ratio with *C*, then their ratio to each other is known by Th. 12 above. Hence, because by Th. 6 above quantity *E* is known, quantity *B* becomes known, as was expected.

The mechanics for this are taken from Ths. 12 and 6 above. For after the ratio of the two quantities is found by Th. 12 above, Th. 6 gives the value of the unknown quantity.

Theorem 15

If any number of quantities form given ratios with another particular quantity, and if any one of these quantities is known, then all the others may be found.

If any three quantities *A*, *B*, and *C* are given [such that] each one forms a given ratio with quantity *C*,* and if some one of them — *A*, for example — is known, then all the other quantities may be found.

By Th. 13 above, the ratio of quantity *A* to each of the others individually is known. And because quantity *A*† is known by Th. 6 above, the remaining quantities are individually found.

One may easily provide the mechanics here from those of Ths. 13 and 6 above.

Theorem 16

The product‡ of two straight lines given in the same units reveals [the area of] a rectangular parallelogram.

If rectangular parallelogram *ABCD* is given, bounded by two lines *AB* and *AD* expressed in the same units, then [their product] will determine the [area of the rectangular parallelogram] itself.

Since the two lines *AB* and *AD* are expressed in terms of each other, one common quantity *H* can measure them. Line *AB* is thus measured by *H* according to the number *K* and line *AD* according to the number *L*.

Now on these two sides of the given parallelogram, measure off two lines *AG* and *AE*, each of which is equal to the common measure *H*. When line *EN*, paralleling *AB*, and line *GF*, paralleling side *AD*, are drawn, then by [Euclid] I.29 and I.34 and the definition of a square, the area *AF* is a square. Since *H*, or its equal *AG*, measures *AB* according to the number *K*, then *AG* will be in *AB* as many times as unity is in *K*. Hence, the ratio of *AG* to *AB* is as that of unity to *K*. Therefore by [Euclid] VI.1, the ratio of small square *AF* to parallelogram *AN* is as that of unity to the number *K*. The ratio of unity to the number *K* is given by axiom, and therefore the ratio of small square *AF* to parallelogram *AN* is made known. Similarly, since *AE*, or its equal *H*, measures side *AD* according to the number *L*, then *AE* will be in *AD* as often as the unit measure is in *L*. Whence the ratio of *AD* to *AE* is as that of number *L* to unity. Moreover, the ratio of *AD* to *AE*, by [Euclid] VI.1, is the same as the ratio of parallelogram *AC* to parallelogram *AN*. Therefore parallelogram *AC* is to parallelogram *AN* as number *L*§ is to unity. But the ratio of *L* to unity is given by axiom. Hence, the ratio of *AC* to *AN* becomes known. Now then, of the two areal quantities *AF* and *AC*, each has a given ratio to parallelogram *AN*; hence, by Th. 12 above, their ratio to each other is established.

*For *C* read *D*.
†For *A* read *E*.
‡*Quod sub duabus lineis continetur*, sometimes with *rectangulum* added (e.g., on pages 92 and 130), refers to the rectangular area contained under two sides; hence, the product. Again it should be recalled that this is a text primarily for applied trigonometry rather than for the more contemporary theoretical trigonometry, and physical qualities and measurements are apparent throughout.
§In Latin, for *numerus* read *numerus l*.

& quoniam quantitas a f nota eſt:in omnibus nᷓ menſurationibus notam ſup
poni oportet menſuram,per ſextam huius parallelogramū a c notum e nuncia=
bitur,quod erat peragendum. Conſtat autem hoc in ͵ppoſito quadratellum a f eſ
ſe menſura ſuperficialem,ᵱ coſtam eius a e menſuræ lineali h æquale initio ſta=
tuerimus.   ▼ Idem alio tramite conſequemur.Prolongetur utraᷓ linearum c b
& d a uerſus ſiniſtrā,donec duæ lineæ b p & a q ſibi,& lineæ a b æquales ueni

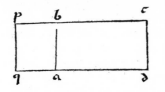

ent.continuatiſᷓ terminis earum p & q ,per
lineam p q claudetur quadratum q b per 29.
33.primi,& diffinitionem quadrati,cuius cum
hypotheſis notā dederit coſtam a b ,ipſum ᷟᷓ
ᵱ primā huius notum habebitur.eſt autē ex pri
ma ſexti ͵pportio quadrati q p ad parallelogra
mum a c,tanᷓ q a ſiue a b ad a d .propor=
tio autem a b lineæ notæ per hypotheſim ad a d notam per diffinitionem data,
unde & proportio quadrati q b ad parallelogramū a c data erit. per ſextā igit
huius(quadrato q b noto exiſtente)ueritatem theorematis inferemus.   ▼ Ad=
huc aliter & ad operationem aptius.Reſumpta figuratione prima,numerus k in
numerū l ductus,efficiat numerū l m .Cū itaᷓ,ut ſupra memoratū eſt , propor

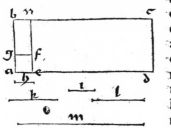

tio quadratelli a f ad parallelogramū a n eſt,ſi=
cut unitatis ad k numerum ,a n autē ad a c ſi=
cut unitatis ad l numerum .ſuperius enim erat
a c ad a n tanᷓ numeri l ad unitatem,unitatis
demū ad l numerum ſicut numeri k ad numerū
m .eſt enim k in m ,quoties unitas in l ex diffini
tione multiplicationis. per æquā igiſ ͵pportiona
litatem erit a f quadratelli ad a c parallelogra
mum,ſicut unitatis ad m numerum.quare a f in
a c,quoties unitas in m numero reperitur. ᵱ dif
finitionem itaᷓ notæ quantitatis parallelogramum a c notum effecimus , in eo
enim menſura ſuperficialis a f continetur ſecundum numerum notum,qui eſt m
quod libuit abſoluere.   ▼ Opus autem docebimus unicum, tametſi demonſtra=
tione freti ſimus uaria.Duos numeros duorum laterum parallelogrami notorum
in ſe multiplicabis,alterum uidelicet in alterum, ͵pducetur enim numerus paralle
logrami ſecundū quem menſura ſuperficialis,quadratū ſcilicet menſuræ linealis
in ipſo continebitur parallelogramo.Vt ſi latus a b 5,& latus a d 7,pedes cō=
plectatur lineales,ductis 5 in 7,creabuntur 35.tot igitur pedes quadrati parallelo
gramum a c conſtituent.Ita in cæteris operabere.

XVII.

Ex dato latere quolibet parallelogrami rectanguli cogniti,reliquū
latus emerget notum.

Sit parallelogramum rectangulū a b c d cognitum,cuius etiam latus unū
quodcunᷓ fuerit notum habeatur,ſitᷓ(uerbi gratia) a b .Dico, ᷟ reliquū latus

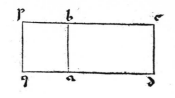

eius a d ſcitum erit. Eductis namᷓ lineis d a
& c b ad puncta q & p,donec utraᷓ linearū
a q & b p æquabitur lineæ a b datæ,comple
atur quadratum q b protracta linea p q, erit
itaᷓ per primā ſexti ͵pportio q b quadrati ad
a c parallelogramum ſicut lineæ q a ad lineā
a d.eſt

Since, by Th. 6 above, the areal quantity AF is known — and a known measure need not be given in all measurements — parallelogram AC is declared* known. Q.E.D. Moreover, it is established by the proposition that the small square AF is the areal measure because its side AE was initially set equal to the linear measure H.

This can be proven in another way. Extend each of the lines CB and DA toward the left until the two lines BP and AQ are equal to each other and to line AB. When their ends P and Q are joined by line PQ, a square QB is formed, by [Euclid] I.29, I.33, and the definition of a square. Since the hypothesis gave the side of square AB, then the [area of the square] itself will also be known by Th. 1 above. Now by [Euclid] VI.1, the ratio of square QP† to parallelogram AC is the same as that of AQ, or AB, to AD. Moreover, the ratio of line AB, known by hypothesis, to AD, known by definition, is given. Hence the ratio of square QB to parallelogram AC will be given. Therefore, by Th. 6 above (because square QB is now known), we may infer the truth of the theorem.

Another proof is offered that is more pertinent to the practical mechanics. Referring to the first figure, multiply number K by number L to obtain the number LM‡. Therefore, since, as was mentioned above, the ratio of the small square AF to parallelogram AN is as that of unity to number K, and furthermore since AN is to AC as unity is to L, or, as before, AC is to AN as number L is to unity, and finally since unity is to number L as number K is to number M, for K is in M as many times as unity is in L from the defi-

nition of multiplication, then, by simultaneous solution of these proportions, small square AF is to parallelogram AC as unity is to number M. By definition of a known quantity, therefore, we have made parallelogram AC known, for the areal measure is contained in it M number of times. This ends the solution.

Despite the varying proofs, only one method will be presented for the mechanics. Multiply the two numbers together that correspond to the two known sides of the parallelogram to find the number of times the areal measure — namely, the square of the linear measure — is contained in the parallelogram. For example, if side AB is 5 linear feet and AD is 7 linear feet, then multiply 5 by 7 to get 35; thus that many square feet make up the area of the parallelogram. And so on in similar problems.

Theorem 17

From any given side of a rectangular parallelogram [of] known [area], one can determine the other side.

If [the area of] rectangular parallelogram $ABCD$ is known [and] any one side is also known — for example, AB — then its other side AD can be found.

Extend line AD to Q and line CB to P, until each of the lines AQ and BP is equal to the given line AB. Complete the square QB by constructing line PQ. Now by [Euclid] VI.1 the ratio of square QB to parallelogram AC will be as that of line QA to line AD.

---

*For *e nunciabitur* read *enunciabitur*.
†For QP read QB.
‡For LM read M.

a d ꞏeſt autem proportio quadrati q b ad ipſum parallelogramum a c data per
diffinitionem,ꝙ utraꝗ ſuperficierum q b & a c data ſit. q b quidem ꝑ primam
huius ꞏeſt enim quadratum lineæ a b datæ,parallelogramum autem a c notum
ſubiecit hypotheſis.Proportio igitur & lineæ q a ad lineam a d nota redditur,
ſed q a & a b ſunt coſtæ quadrati q b æquales ,unde & proportionem a b ad
a d notam eſſe oportet,cunꝗ altera illarum,ſcilicet a b nota ſupponatur,erit per
ſextam huius reliqua linea ſcilicet a d nota,ſicꝗ reliquum parallelogrami latus
a d notum exegimus,quod libuit attingere.　▼ Idem aliter & ad operationem
accomodatius.Quoniã latus a b notũ ſupponiſ,menſuret ipſum h famoſa quan
titas ſecundum numerum k,ſitꝗ utraꝗ linearum a g & a e ex parallelogrami
noſtri lateribus abſumptarum æqualis menſuræ h ,& ducatur e n quidẽ æquedi=
ſtans lateri a b, g f uero lateri a d æquediſtans,eritꝗ ſuperficies a f quadrata
per 29.& 34.primi & diffinitionem quadrati,quæ quidem ſuperficies menſurabit

parallelogramũ a c ſecundum numerũ notum,
qui ſit m ꞏquoniã parallelogramũ ſupponitur co
gnitum,hunc numerum in poſtremo per nume=
rum k partiamur,ut exeat l numerus.Quia itaꝗ
ꝓportio a b ad h,ſiue ad a g eſt, ut numeri k
ad unitatem,h menſurante lineam a b ſecundũ
k numerum:erit per primã ſexti parallelogrami
a n ad quadratellum a f,ſicut numeri k ad uni=
tatem,quadratelli autem a f ad parallelogramũ

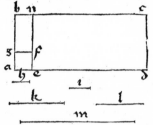

a c,ſicut unitatis ad numerum m,ꝙ parallelogramũ ipſum quadratello menſu=
retur ſecundum numerũ m,quare per æquam proportio parallelogrami a n ad
parallelogramum a c eſt,ut numeri k ad numerum m ꞏeſt autem a n ad a c,
tanꝗ a e ſiue h ſibi æqualis ad lineã a d per primã ſexti ꞏ Vnde & h ad a d eſt
ut k numeri ad numerum m,ſed k ad m ſicut unitatis ad l numerum per diffi=
nitionem diuiſionis,quare proportio h ad lineã a d eſt,ſicut unitatis ad l nume
rum,menſura igitur h in linea a d,quoties unitas in numero l continetur, qua=
re linea a d nota concluditur,quoniã menſura h famoſa continetur in ea ſecun=
dum numerum l notum,reliquum ergo parallelogrami latus effecimus cognitũ,
quod intendebamus.　▼ Opus breue.Numerum parallelogrami noti in nume=
rum lateris noti partiaris,& exibit numerus lateris reliqui quæſitus.Vt ſi paralle=
logramum a c offeratur 36.pedum ſuperficialium,habens latus a b 4 pedum li=
nealium,diuidã numerum 36 in numerum 4.& exibunt 9.Latus igitur reliquum
a d ,nouem pedes complectetur lineales.

XVIII.
Ex data proportione laterum parallelogrami rectanguli cogniti, utriuſꝗ lateris pendebit noticia.

Sit parallelogramum rectangulum a b c d cognitum,cuius latera a b & a
d proportionem habeant adinuicem notã.Dico,ꝙ utrunꝗ ipſorum notum habe
bitur.Reſumpta enim prima figuratione præcedentis,erit per primã ſexti propor
tio a c parallelogrami ad a p quadratum,ſicut lineæ d a ad lineã a q ꞏ eſt autẽ
proportio lineæ d a ad lineã a q data,ꝙ a q æqualis habeatur lineæ a b ꞏqua=
re & proportio parallelogrami a c ad quadratum a p nota redditur,cunꝗ no=
tum ſubiecerimus parallelogramum a c,erit per ſextã huius & quadratum a p
　　　　　　　　　C 2　　　cognitum

Therefore the ratio of square *QB* to parallelogram *AC* is given by definition, since each of these areas *QB* and *AC* is given; for indeed, by Th. 1 above, *QB* is the square of the given line *AB*, and the hypothesis has given parallelogram *AC*. Therefore, the ratio of line *QA* to line *AD* becomes known. But *QA* and *AB* are the equal sides of square *QB*, and hence the ratio of *AB* to *AD* is necessarily known. Therefore since one of the lines — namely *AB* — is given, by Th. 6 above the other line — namely *AD* — will be known, and thus the remaining side of the parallelogram has been determined. Q.E.D.

Another, more practical proof follows. Since side *AB* is given, let the known quantity *H* measure *AB* according to the number *K*, and let each of the lines *AG* and *AE*, measured off on the sides of our parallelogram, be equal to measure *H*. Then, when *EN* is drawn parallel to side *AB*, and *GF* parallel to side *AD*, the area *AF* is a square by [Euclid] I.29, I.34, and the definition of a square. This area measures parallelogram *AC* according to the known number *M*. Since [the area of] the parallelogram was given known, then this number [*M*] is divided by number *K* to get number *L*. Because *AB* is to *H*, or *AG*, as number *K* is to unity (*H* measuring line *AB* according to number *K*), then by [Euclid] VI.1 parallelogram *AN* is to the small square *AF* as number *K* is to unity. Moreover, small square *AF* is to parallelogram *AC* as unity is to number *M*, because the parallelogram itself is measured by the little square according to the number *M*. Hence, by simultaneous solution, the ratio of parallelogram *AN* to parallelogram *AC* is as that of number *K* to number *M*. Furthermore, *AN* is to *AC* as *AE*, or *H*, is to line *AD* by [Euclid] VI.1. Hence *H* is to *AD* as number *K* is to number *M*. But *K* is to *M* as unity is to number *L* by the definition of division. Therefore, the ratio of *H* to line *AD* is as that of unity to number *L*. Thus measure *H* is contained in line *AD* as many times as unity is contained in number *L*. Therefore line *AD* is inferred known because the known measure *H* is contained in it *L* times. And so the other side of the parallelogram has been found.

The mechanics in brief. Divide the area of the parallelogram by the given side to find the unknown side. For example, if the area of parallelogram *AC* is given as 36 square feet with side *AB* as 4 linear feet, divide 36 by 4 to get 9. The unknown side *AD* is therefore 9 feet long.

Theorem 18

From the given ratio of the sides of a rectangular parallelogram with known area, the length of each of the sides can be found.

If the area of rectangular parallelogram *ABCD* is known and its sides *AB* and *AD* have a known ratio to each other, then each of these [sides] can be found.

When the first figure of the previous theorem is used again, the ratio of parallelogram *AC* to square *AP* will be as that of line *DA* to line *AQ* by [Euclid] VI.1. Moreover, the ratio of line *DA* to line *AQ* is given, because *AQ* is equal to line *AB*. Therefore the ratio of parallelogram *AC* to square *AP* is known. Since parallelogram *AC* was given, square *AP* is found by Th. 6 above.

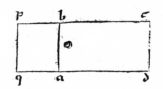

cognitum. inde quoqʒ per fecundã huius cõsta fua a b non ignorabitur: quæ quidẽ eft alterũ ex lateribus parallelogrami propofiti, datã autem tradidit hypothefis proportionem laterum dicti parallelogrami, ex latere igitur a b iã cognito fexta huius, reliquum latus a d fufcitabit notum. utrunqʒ ergo parallelogrami latus effecimus menfuratum, quod pollicebatur præfens theorema. ▼ Opus ita comparabis. Si proportio laterum data eft per denominationem, diuide numerum parallelogrami per denominatorem proportionis, & exibit numerus quadrati lateris confequentis, cuius radix quadrata latus ipfum confequens notificabit, poftea ad operationem fextæ huius aut præcedentis confugias, quæ reliquũ latus eliciet cognitum. Vt fi parallelogramum a c contineat 48 quadratos pedes, latus autẽ a d lateri a b triplum fuerit, ecce denominatorem proportionis 3. per quem diuido numerum parallelogrami 48, & exeunt 16. numerus qui debetur quadrato lateris a b confequentis, cuius radix quadrata 4. latus a b notum faciet. 4 autem triplicans, quoniã proportionem triplã elegimus; aut 48 diuifis per 4, exurget latus a d reliquum 12. Qd̃ fi proportio lateris ad latus data fuerit, non per denominationem, fed per fibi æqualem proportionem, ut fi diceretur, proportio lateris a d antecedentis ad latus a b confequentis eft, ut 5 ad 3; multiplicabis numerum parallelogrami dati per terminum confequentem, & productũ partieris in numerum antecedentem, exibit enim numerus affignandus quadrato lateris confequentis : cũ quo ut antehac ꝓcedendum erit. Vt fi parallelogramũ a c fuerit 60, latus autem a d ad latus a b fe habeat, ut 5 ad 3, multiplicabo 60 per 3. producũtur 180. quæ diuifa per 5, eliciunt 36. quadratum fcilicet lateris a b, cuius radix quadrata 6, ipfum latus a b notificabit. reliqua autem per operationẽ præcedẽtis abfoluent.

XIX.

Si quatuor quantitatum proportionalium tres quælibet datæ fuerint, & quarta reliqua innotefcet.

Sint quatuor quãtitates a b c d proportionales, quarum tres notæ fint quæcunqʒ, dico, cꝗ quarta reliqua nota ueniet. Quãuis autẽ ipfa ignota quãtitas nunc primum, nunc fecundum, interdũ uero reliqua foleat occupare loca, tamen ne operis uarietas, quæ neceffario hanc mutationem confequitur, lectorẽ perturbet; placuit femper ignotæ quãtitati poftremũ deputare locum. Præfens igitur theorema facile confirmabimus, fi prius quo pacto quãtitas ignota, quocunqʒ nobis offeraꝉ loco, poftrema fiat docebimus. Conftat autem huiufmodi quatuor quãtitatum ꝓportionalitas ex duabus proportiõibus, quarũ unius ambo termini funt cogniti, & illã faciemus primã, fecundæ autem ꝓportionis unus duntaxat notus eft terminus, qui fi fuerit antecedens, iã ordinate funt quatuor illæ quantitates, ut uolumus, habebit enim ignota quartum locum. Si uero confequens fecundæ proportionis notum fuerit, conuertemus ambas proportiões, & trãferetur ignota quãtitas ad poftremũ locũ. Nunc ad confirmationem proportionis defcendamus. Quatuor quãtitatum proportionaliũ a b c d, tres primæ funt notæ fecundum tres numeros f g & h, fimili ordine pofitos

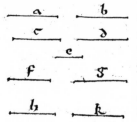

Thereupon, by Th. 2 above, its side *AB* is also known, and indeed, this is one of the sides of the given parallelogram. Furthermore, the hypothesis gave the ratio of the sides of the given parallelogram. Hence, with side *AB* found, the other side *AD* becomes known by Th. 6 above. Thus, both sides of the parallelogram are now measured. Q.E.D.

The mechanics: If the ratio of the sides is given by designation, divide the area of the parallelogram by the designator of the ratio and the square of the consequent* side emerges; the square root of this number identifies the consequent side itself. Then refer to the mechanics of Th. 6 above or of the preceding theorem, either of which may be used to determine the other side. For instance, if parallelogram *AC* contains 48 square feet and side *AD* is triple side *AB*, then the designator of the ratio is 3. Divide 48, the area of the parallelogram, by 3 to get 16. This number must be the square of the consequent side *AB*, and its square root is 4, the length of side *AB*. Whether you triple 4, since we chose the triple ratio, or divide 48 by 4, the remaining side *AD* is found to be 12. But if the ratio of one side to the other is given not by a designation but rather by [another] ratio equal to it — for example, if the ratio of the antecedent side *AD* to the consequent side *AB* is as 5 to 3 — then you will multiply the area of the parallelogram by the consequent term and divide [their] product by the antecedent number; the result is the square of the consequent side. With that found, one proceeds as before. For example, if the area of parallelogram *AC* were 60 and side *AD* to side *AB* were as 5 to 3, then multiply 60 by 3 to produce 180, which, divided by 5, yields 36, the square of side

AB. Its square root, 6, identifies side *AB* itself. The other side is obtained as before.

Theorem 19

If, of four proportional quantities, any three are given, the fourth one that remains will become known.

If four quantities, *A*, *B*, *C*, and *D*, are proportional and any three are known, then the fourth, remaining [quantity] can be found. Moreover, this unknown quantity may sometimes occupy the first place or the second place or, at other times, [one of the] other [places]; nevertheless, the change in the mechanics which necessarily follows such alteration does not disturb the proof. It has always been acceptable to designate the last place for the unknown quantity. Therefore, we will easily prove the present theorem if we agree that the unknown quantity is the last term, rather than in any place. Furthermore, the proportion of four such quantities is established by two ratios, in one of which both terms are known; we make that one the first ratio. Furthermore, in the second ratio only one term is known. If the known term is the antecedent, the four quantities are already in the order we desire, for the unknown is in the fourth place. If, however, the consequent of the second ratio is known, invert both ratios, thus transferring the unknown quantity to the last place. Now let us proceed to the solution of that ratio.

Of the four proportional quantities, the first three are known according to three numbers *F*, *G*, and *H*, placed in an analogous order;

*Since two adjacent sides of the rectangle can be set as a ratio, the sides may be identified by the names of the terms of the ratio, viz., "antecedent" side and "consequent" side.

ħe pofitos,ita ꝗ quãtitas tertia c, fcilicet nota fit per menfuram e fecundũ nu=
merum h .reperiaturꝗ numerus k ,ad quem fe habeat h ,ficut f ad g . quod fiet,
fi productum ex h in g ,per numerum f partiemur,quemadmodum ex uigefima
feptimi elementorũ trahitur.erit autem numerus k ,fecundũ quem nota habebit̃,
poſtrema quãtitatũ p menfurã quidẽ e. Nã ex quinta huius ꝓportio a quãtitatis
ad b erit,ficut numeri f ad numerũ g. fed a ad b ficut c ad d ex hypothefi, & f
ad g ficut h ad k, quare c ad d ficut h ad ĸ, & cõuerfim d ad c tanꝗ k ad h ,
& c ad menfurã e,ficut h numerũ ad unitatẽ, ꝗ e mefuret c fecundũ numerũ h.p
æquã igitur ꝓportio d quãtitatis poſtremæ ad e menfurã,tanꝗ numerũ k ad uni
tatem.eſt itaꝗ e menfura in d ,quoties unitas in k numero. d ergo quãtitas no
ta redditur per menfurã e fecundũ numerũ k ,quod libuit explanare. ▼ Opus,
multiplica numerum fecundæ quãtitatis per numerum tertiæ,& productũ in nu=
merum primæ quãtitatis diuide,exibit enim numerus poſtremæ quãtitatis quæfi=
tus.Vt fi a fuerit 4. & b 9. c uero 12. multiplico 12 per 9 ,producũtur 108 ,quæ
diuifa per 4 ,eliciunt 27 numerũ uidelicet quantitatis poſtremæ.

XX.

In omni triangulo rectangulo, fi fuper uertice acuti anguli,fecundũ
quantitatem lateris maximi circulum defcripferis,erit latus ipfum acu
tum,fubtendens angulum finus rectus conterminalis fibi arcus dictũ
angulum refpicientis:lateri autem tertio finus complementi arcus di=
cti æqualis habebitur.

Sit triangulus rectanglus a b c ,angulum c rectum habens,& a acutum,fu
per cuius uertice a fecundum quãtitatem lateris maximi a b ,maximo fcilicet an
gulo oppofiti defcribatur circulus b e d ,cuius circũferentiæ occurrat latus a c
quoad fatis eſt prolongatũ in e puncto. Di=
co quòd latus b c angulo b a c oppofitum
eſt finus arcus b e dictum angulum fubten
dentis.Latus autem tertium,fcilicet a c ,æ=
quale eſt finui recto complementi arcus b e.
Extendatur enim latus b c occurrendo cir=
cumferentiæ circuli in puncto d . à punctis
autem a quidẽ centro circuli exeat femidi
ameter a k æquediſtans lateri b c. & à pun
cto b corda b h æquediſtans lateri a c. fe=
cabunt autem fe neceſſario duæ lineæ b h &
a k,angulis a b h & b a k acutis exiſtenti
bus,quod fiat in puncto g.Quia itaꝗ femidi
ameter a e cordam b d fecat orthogonali=

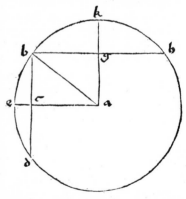

ter propter angulum a c b rectum,fecabit & ipfam per æqualia ex tertia tertij ele
mentoĸ.& arcum b d per æqualia ex 29.eiufdem. quemadmodum igitur tota li
nea b d per diffinitionem corda eſt arcus b d ,ita medietas eius,linea fcilicet b c
eſt finus dimidij arcus b e refpicientis angulũ b a e fiue b a c. quod aſſeruit pri
ma pars theorematis noſtri. Secundam deinceps partem ueram cõfiteberis, fi pri
us per 34.primi angulum a g b rectum eſſe didiceris,femidiameter enim a ĸ,&
cordam b h,& arcũ eius ex fupra memoratis medijs per æqua fcindet. quare per
diffinitionem linea recta b g finus erit arcus b ĸ,Eſt autem linea b g æqualis la

C 3   teri tri

so, for example, the third quantity C is known by the measure E according to the number H. Number K must be found such that H is to K as F is to G. This can be achieved if we divide the product of H times G by number F, as is shown in VII.20 of the *Elements*. Moreover, K is the number of times the last of the quantities [i.e., the unknown] will be known by measure E. Now by Th. 5 above, the ratio of quantity A to B is as that of number F to number G. But from the hypothesis, A to B is as C to D, and F to G is as H to K. Whence, C to D is as H to K, and inversely D to C is as K to H. C is to the measure E as number H is to unity because E measures C according to number H. By simultaneous solution, then, the ratio of the last quantity D to the measure E is as that of the number K to unity. Therefore measure E is in D as many times as unity is in number K. Quantity D therefore is known by measure E according to number K. Q.E.D.

The mechanics: Multiply the number of the second quantity by the number of the third and divide the product by the number of the first quantity. The result is the desired number of the last quantity. For example, if A is 4, B 9, and C 12, then multiply 12 by 9 to get 108 and divide this by 4 to obtain 27, clearly the number of the last quantity.

Theorem 20

In every right triangle, one of whose acute vertices becomes the center of a circle and whose [hypotenuse] its radius, the side subtending this acute angle is the right sine of the arc adjacent to that [side and] opposite the given angle, and the third side of the triangle is equal to the sine of the complement of the arc.*

If a right $\triangle$ ABC is given with C the right angle and A an acute angle, around the vertex of which a circle BED is described with the hypotenuse — that is, the side opposite

the largest angle — as radius, and if side AC is extended sufficiently to meet the circumference of the circle at point E, then side BC opposite $\angle$ BAC is the sine of arc BE subtending the given angle, and furthermore the third side AC is equal to the right sine of the complement of arc BE.

Extend side BC to meet the circumference of the circle at point D. From point A, the center of the circle, draw a radius AK parallel to side BC, and from point B draw chord BH parallel to side AC. The two lines BH and AK necessarily intersect, because angles ABH and BAK are acute, and this happens at point G. Therefore, since the radius AE intersects chord BD perpendicularly because of right $\angle$ ACB, [AE] bisects both chord [BD], according to [Euclid] III.3, and arc BD, according to [Euclid] III.29. Thus, just as by definition the entire line BD is the chord of arc BD, so also [by definition] its half, namely line BC, is the sine of half-arc BE opposite $\angle$ BAE, or BAC. And this is what the first part of the theorem asserts.

The second part is shown to be true if it is first understood that, by [Euclid] I.34, $\angle$ AGB is a right angle, for then radius AK bisects chord BH and its arc, as mentioned above. Hence, by definition, straight line BG is the sine of arc BK. Moreover, line BG is equal to

* To understand this and later proofs, the reader should recall that the "sine" of Regiomontanus' day differs slightly in its definition from the sine function of today. The sine, as used in this treatise, is a perpendicular dropped from one extremity of an arc of a circle to the diameter that has been drawn through the other extremity of the arc. The versed sine is the portion of the diameter between the foot of that perpendicular or sine and the arc. The sine of the complement of the arc is the sine of the difference between the arc and a quadrant; hence, if the arc is less than 90°, the complement of the arc is 90° minus the degrees of arc, but if the arc is greater than 90°, the complement is taken to be the degrees of arc minus 90°.

teri trianguli a b c. per 34.primi, ꝗ̄ ſupficies a g b c æquediſtātibus contineaͭ
lineis, angulus uero c a g, ſiue e a k rectus eſt per 29.primi, propter æquediſtan
tiam linearum b c ad a g, quare per ultimā ſexti arcus e k circumferentiæ ſuæ
quadrās proḃabitur. arcus itaꝗ̃ b k complementum ipſius arcus b e diffinietur,
cuius ſinus b g lateri a c æqualis nuperrime concludebatur. utranꝗ̃ igitur proportionis partem ſatis oſtendiſſe uidemur.

XXI.
Omnem angulum rectum notum eſſe oportet.

Vnus enim rectus angulus ad quatuor rectos notā habet proportionem, ſum
ma autem quatuor rectorum nota eſt, cum gradus unus ſcilicet 360. pars quatu-
or rectorum: qua tanꝗ̃ menſura famoſa utuntur uniuerſi Geometræ: ſecundū numerum notum 360.contineatur in quatuor rectis. quare per ſextā huius & unus
angulus rectus notus habebitur, quod erat lucubrādum. Quarta autē pars ex 360
eſt 90. Iuſto igitur computo 90 gradus angulo recto uēdicabimus. Miraberis for
ſitan, quo pacto diuerſi generis quātitates menſura gradualis metiatur. Dicimus
enim circumferentiā circuli uel arcum tot uel tot complecti gradus, item quatuor
rectos angulos, uel alium angulum quemcunꝗ̃ aliquot continere gradus. Quid
igitur uocabulo gradus ſignificare uelimus, paucis habeto. Menſura famoſa arcu
um eſt gradus circumferentialis ſcilicet 360.pars circumferentiæ circuli, menſura autem famoſa angulorum eſt gradus angularis, uidelicet 360.pars quatuor rectorum angulorum, id eſt, ſpacij plani, quod circa punctum quodlibet exiſtit. Ima
ginādo enim duas ſemidiametros circuli ſuper puncto quocunꝗ̃ tanꝗ̃ centro deſcripti: gradum circumferentiæ circuli dicti intercipientes, angulus quem ipſe ſe
midiametri ambiunt gradus uocabitur angularis, quonia angulus ille 360.in qua
tuor rectis, ſiue toto ſpacio centrum circuli ambiente continetur: ſicut & gradus
circumferentialis in tota circumferentia, huius enim anguli ad quator rectos, & il
lius arcus ad totā circumferentiā, eandem eſſe proportionem, ex ultima ſexti facile comprobabiͭ. ▼Trahimus poſtremo ex iam recitatis, ꝗ̄ cuilibet angulo & arcui ſe reſpicienti de circūferentia circuli ſuper uertice ipſius anguli deſcripti, unus
& idem ſeruit numerus. uerbi gratia, ſi angulum quemlibet 36 graduū ſtatuimus,
erit & arcus ſe reſpiciens 36 graduum, & econtra. quod quidem ex identitate numeri totius circumferentiæ circuli & quatuor rectorum, ultima ſexti ratiocinante
pendere dinoſcitur.

XXII.
Altero duorum acutorum, quos habet triangulus rectangulus, da-
to, reliquus non latebit.

Duo enim acuti anguli, quos habet triangulus rectangulus, per 32.primi, ua
lent unum rectum, ꝗ̄ tertius angulus ſit rectus, aggregatū itaꝗ̃ ex duobus dictis
acutis angulis notum eſt, quonia ex præmiſſa rectus angulus notus eſt, ſed & alter
acutorū ex hypotheſi datus eſt, quare per quartam huius reliquum cognoſcemus.
▼ Opus, numeri anguli acuti dati, ex numero unius recti minuas, & relinquetur quantitas alterius. Vt ſi angulus b fuerit 20, minuo 20 ex 90, relinquuntur
70.tantum igitur habebo angulum c reliquum.

Si duo

side [*AC*] of △ *ABC* by [Euclid] I.34 because the area *AGBC* is bounded by parallel lines. But ∠ *CAG*, or *EAK*, is a right angle by [Euclid] I.29 because line *BC* is parallel to *AG*. Hence by the last theorem of [Euclid] VI, arc *EK* is shown to be a quadrant of the circumference. Consequently arc *BK* is defined as the complement of arc *BE*, and the sine *BG* [of arc *BK*] was just shown to be equal to side *AC*. And thus both parts of the theorem* have been proven.

Theorem 21

Every right angle is necessarily known.

One right angle has a known ratio to four right angles. Furthermore, the sum of four right angles is known, because one degree — namely, a 360th part of four right angles, [such a part being] used as an accepted measure by geometry throughout the world — is contained in four right angles 360 times. Hence, by Th. 6 above, one right angle can be found, as was to be determined. A fourth part of 360 is 90. Thus, by accurate calculation, we may assert that there are 90° in a right angle.

You may perhaps wonder how degrees can measure quantities of diverse kinds, for we say that the circumference of a circle or an arc encompasses so many degrees [and] similarly that four right angles or any other angle contains several degrees. What we wish to indicate by the word "degree" can be expressed briefly.

The accepted measure of arcs is the circumferential degree, namely, a 360th part of the circumference of a circle. The accepted measure of angles is the angular degree, namely, a 360th part of four right angles — that is, of the planar space that exists around any point. For if it is imagined that two radii of a circle, described around any point as center, intercept a degree of the circumference of this circle, then that angle which the radii

enclose is called an angular degree, since that angle is contained 360 times in four right angles or the total space surrounding the center of the circle, just as a circumferential degree [is contained 360 times] in the total circumference. Then the fact that the ratio of this angle to four right angles is the same as the ratio of that arc to the total circumference is easily confirmed by the last theorem of [Euclid] VI.

Finally, we conclude from the above discourses that one and the same number describes an angle and the arc opposite it on the circumference of the circle that has been described around the vertex of that same angle. For example, if we take any angle of 36°, it and the arc determined by it will be of 36°, and the converse. Indeed, this is understood, by the reasoning of the last theorem of [Euclid] VI, as a consequence of the exact equality of the numerical values of the total circumference of the circle and of four right angles.

Theorem 22

If one of the two acute angles of a right triangle is given, the other can be found.

By [Euclid] I.32 the two acute angles of a right triangle have the value of one right angle. Because the third angle [of the given triangle] is a right angle, the sum of the two given acute angles is known. Since from the preceding [theorem] any right angle is known, and since from the hypothesis one of the acute angles is given, then, by Th. 4 above, the other acute angle is found.

The mechanics: Subtract the value of the given acute angle from the value of one right angle, and the quantity of the other [angle] remains. For example, if ∠ *B* is 20, then 90 minus 20 leaves 70, the size of the other ∠ *C*.

*In Latin, for *proportionis* read *propositionis*.

XXIII.

Si duo latera trianguli,rectum continentia angulum,fuerint æqua
lia,duo acuti anguli eis oppositi reddentur noti.

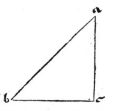

Duo latera a c,c b trianguli a b c rectanguli, re‑
ctum angulum c ambientia,sint æqualia.Dico,qp uterqp
angulorum b & c notus prodibit.Erunt enim per hypo‑
thesim & quintã primi duo anguli a & b æquales, cunqp
per 3 2.primi ipsi ualeant unum rectum. angulo c recto
existente,erit uterqp eorum medietas per 2 1 huius cogni‑
ti.quare per 6.huius uterqp eorum notus habebitur, quod
libuit explanare.   ▽ Opus. Quantitatem anguli recti
dimidiabis,& apparebit utriusqp angulorũ acutorum quãtitas.Verbi gratia.Re‑
cto angulo habente 90 gradus,dimidiabo 90 ,& habebo 45 pro medietate recti,
tantusqp pronunciabitur uterqp angulorum a & b.

XXIIII.

Si latus trianguli rectum subtendens angulum , alteri.duorum re‑
cto subiacentium fuerit duplum,angulus acutus ab eis contentus,reli‑
quo angulo acuto duplus enunciabitur. Vnde etiam utrunqp eorum
agnoscet Geometra.

Sit triangulus a b c,rectum angulum c habens
quem subtendat latus a b duplum lateri a c.Dico,
qp angulus b a c duplus erit angulo a b c.Extenda
tur enim a c usqp ad punctum d, donec c d habeat
æqlis lateri a c,ducta linea b d,erit itaqp a b linea
æqualis ipsi a d,qp utriusqp medietas sit a c.sed per
quartã primi duæ bases a b,b d triangulorum a b
c,& b c d sunt æquales, anguli quoqp a b c & d b
c æquales.totus igitur angulus a b d duplus est ad angulum a b c.est autem to
tus angulus a b d æqualis angulo b a d,siue b a c per quintã primi triangulo a
b d æquilatero existente,unde & angulus b a c duplus erit ad angulũ a b c,qd
oportuit demonstrare.   ▽ Ex hoc patebit corollarium. Quoniã enim in trian‑
gulo nostro angulus a duplus iam declaratus habeatur ad angulum b,id est,sicut
2 ad 1 ,erit coniunctim aggregatũ ex duobus angulis a & b ad angulum b,sicut
3 ad 1.Illud autem aggregatum æquipollet angulo recto, hypothesi & 3 2.primi
docentibus.proportio igitur anguli recti ad angulum b nota est,uidelicet sicut 3
ad 1.quare per sextã huius angulus b notus erit,recto per 2 1 huius noto existen‑
te.postremo etiã residuus ex recto angulus,scilicet a p 4 huius notus declarabĩ.

XXV.

Duobus trianguli cuiuscunqp cognitis angulis , tertium reliquum
datum iri.

Trianguli a b c,duo anguli a & b sint cogniti.Dico,qp & angulus c notus
emerget.Tres enim anguli a b c,duobus rectis æquantur.3 2.primi id confirmã‑
te.duo aũt recti sunt noti per 2 1 & 3 huius,quare & aggregatum ex tribus angu
lis tri‑

Theorem 23

If, in a [right] triangle, the two sides containing the right angle are equal, the two acute angles opposite the sides may be found.

If in a right △ *ABC* the two sides *AC* and *CB* enclosing the right ∠ *C* are equal, then each of the angles *B* and *C** can be found.

The two angles *A* and *B* are equal, by the hypothesis and [Euclid] I.5. Since by [Euclid] I.32 they have the value of one right angle when *C* is a right angle, then, by Th. 21 above, each of them will be half of that known angle. Therefore by Th. 6 above, each of the angles will be known.

The mechanics: Halve the quantity of a right angle, and the quantity of each of the acute angles appears. For example, since the right angle is 90°, take half of 90 and you will have 45 for half the right angle; each of the angles *A* and *B* will be declared [to be] that much.

Theorem 24

If the [hypotenuse] of a right triangle is double the length of one of the sides adjacent to the right angle, then the acute angle included by that side and the hypotenuse is double the other acute angle. Hence geometry also reveals each of the angles.

If right △ *ABC* is given with right ∠ *C* subtended by hypotenuse *AB* which is double side *AC*, then ∠ *BAC* is double ∠ *ABC*.

Extend *AC* all the way to point *D* until *CD* is equal to side *AC*. Draw line *BD*. Then *AB*

is equal to *AD* because half of each is *AC*. But by [Euclid] I.4 the two bases *AB* and *BD* of triangles *ABC* and *BCD* [respectively] are equal, and the angles *ABC* and *DBC* are also equal. Therefore the total ∠ *ABD* is double ∠ *ABC*. Moreover, the total ∠ *ABD* is equal to ∠ *BAD*, or *BAC*, because by [Euclid] I.5 △ *ABD* is equilateral. Therefore ∠ *BAC* is double ∠ *ABC*. Q.E.D.

From this a corollary arises. Since, in our triangle, ∠ *A* has already been determined to be double ∠ *B* — that is, as 2 to 1 — then by addition the sum of the two angles *A* and *B* to ∠ *B* is as 3 to 1. Furthermore, this sum is equal to a right angle, by the teachings of the hypothesis and [Euclid] I.32. Thus the ratio of a right angle to ∠ *B* is known to be as 3 to 1. Therefore, by Th. 6 above, ∠ *B* will be known because by Th. 21 above the right angle is known. And finally by Th. 4 above, the other angle is found.

Theorem 25

If two angles of any triangle are known, the third may be found.

If, in △ *ABC*, the two angles *A* and *B* are known, then ∠ *C* can be found.

The three angles *A*, *B*, and *C* equal two right angles, a fact confirmed by [Euclid] I.32. Now the [value of] two right angles is known by Ths. 21 and 3 above; hence the sum of the three angles

*For *C* read *A*.

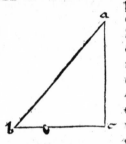

lis trianguli propoſiti notum habebitur, cunꝗ duos eorũ
datos ſubiecerit hypotheſis, p 4. huius tertius reliquus nõ
ignorabitur, quod libuit inferre.　◤Opus.Summã duo=
rum angulorum, qui dati ſunt ex quãtitate duorum recto=
rum minuas, & relinquetur tertij anguli quantitas deſide=
rata. Vt ſi angulus a fuerit 20, & angulus b 35 graduũ,
collectos 20 & 35 gradus, qui reddunt 55.ex 180 minuo.
relicti enim 125 gradus, angulo c adnumerabuntur.

X X V I.

Omnis trianguli rectanguli duobus lateribus cognitis, tertium ex
templo manifeſtari.

Triangulus a b c angulum c rectum habeat, cuius duo latera quælibet ſint
nota. Dico, ꝗ reliquum eius latus notum habebitur. Si enim duo latera rectũ con
tinentia angulum offerantur nota, erunt per primã huius quadrata eorum nota,
aggregatum quoꝗ ex eis per tertiã huius notum, quod æquipollet quadrato a b
per penultimã primi, unde ipſum notũ, & ideo per ſecun
dam huius coſta ſua, latus ſcilicet a b non ignorabitur.
Si uero alterũ eorum ſit datum cum latere rectum ſubten=
dente angulum, quadratũ minoris demptum ex quadra=
to maioris, per penultimã primi & quartã huius relinquet
quadratũ reliqui lateris notum, & ideo per ſecundã huius
coſta eius cognita orief, quæ fuere lucubrãda.　◤Opus
uulgare. Si latera rectũ ambientia angulum fuerint da
ta, quadrabis ea, quadrataꝗ congregabis, & collecti ex
eis radix quadrata quãtitatem lateris quæſiti manifeſta=
bit. Si uero alterũ eorum ſit datum cum latere rectum ſubtendente angulum, qua=
dratum minoris ex quadrato maioris demas, & relicti quadrata radix tertium la=
tus notificabit. Vt ſi latus a c fuerit 12, & b c 5, quadrabo 12, exurgunt 144. itẽ
quadrabo 5, ueniunt 25. colligo 144 & 25, fiunt 169. quorum radicem quadratam
inuenio 13. tantumꝗ fore didici latus a b. Sed ponatur latus a b 29, & latus a
c 20. duco 29 in ſe, ueniunt 841. ſimiliter 20 in ſe, faciunt 400. aufero 400 ex 841.
relinquũtur 441, quorum radicẽ quadratã 21 lateri b c deputabo. Ita in cæteris.

X X V I I.

Trianguli rectanguli duobus lateribus cognitis, omnes angulos
datum iri.

Si alterum datorum laterum recto opponatur angulo, ſatis eſt. ſi uero nõ, per
præcedentem ipſum addiſcemus, nam abſꝗ eo propoſitum attingendi nõ erit po=
teſtas. Sit itaꝗ triangulus a b c, angulum c rectum habens, cuius duo latera a b
& a c ſint data. Dico, ꝗ omnes anguli ipſius noti erunt. Super uertice enim an
guli acuti b, quem uidelicet latus ſubtendit datum tanꝗ centro, ſecundum quan=
titatem lateris b a circulo deſcripto, erit per 20 huius a c ſinus arcus ſibi conter=
minalis, qui reſpondet angulo a b c quem inquirimus. cũꝗ duæ lineæ a b & a c
inter ſe datæ ſint ex hypotheſi per menſurã ueterem, a b autem ſemidiameter cir=
culi deſcripti data ſit per menſurã noua, quæ quidem eſt una partium ſinus totius,
erit & a c nota per eandem menſurã docente ſeptima huius, dum igitur a b eſt
ſinus to=

of the given triangle will be known. Since the hypothesis gave two of these angles, the third will be found by Th. 4 above. Q.E.D.

The mechanics: Subtract the sum of the two given angles from the quantity of two right angles, and the desired quantity of the third angle remains. For example, if ∠ A is 20° and ∠ B is 35°, then subtract their sum, which is 55, from 180. The 125° left are the value of ∠ C.

Theorem 26

If two sides of a right triangle are known, the third is directly apparent.

If △ ABC has a right ∠ C [and] any two of its sides are known, then the third side can be found.

Now if the two given sides include the right angle, then, by Th. 1 above, their squares will be known. The sum of these [squares] will also be known by Th. 3 above, this [sum] being equivalent to the square of AB by the penultimate theorem of [Euclid] I. Hence [the square] itself is known, and therefore, by Th. 2 above, its root — namely, side AB — will be found.

But if one of the given sides subtends the right angle, then, when the square of the smaller [side] is subtracted from the square of the larger [side], by the penultimate theorem of [Euclid] I and Th. 4 above, the square of the third side is found. Therefore the side itself is found by Th. 2 above. Q.E.D.

The mechanics: If the two given sides include the right angle, then square the sides and add the squares. The square root of this sum is the length of the sought side. However,

if one of the given sides subtends the right angle, then subtract the square of the smaller [side] from the square of the larger, and the square root of the remainder will identify the third side. For instance, if side AC is 12 and BC 5, then square 12 to produce 144. Similarly, square 5 to find 25. Add 144 and 25 to make 169, of which the square root is found to be 13, the length of side AB. On the other hand, let side AB be 29 and side AC 20. Then 29 squared is 841 and similarly 20 squared is 400; 400 subtracted from 841 leaves 441, of which the square root is 21, the length of side BC. And so on.

Theorem 27

When two sides of a right triangle are known, all the angles can be found.

If one of the given sides is opposite the right angle, that is sufficient; if not, however, we will find it, also, by the preceding theorem, for without it, it will not be possible to handle the theorem. Thus, if △ ABC is given with C a right angle and sides AB and AC known, then all the angles can be found.

When a circle is described with ∠ B, which the given side [AC] subtends, as center and side BA as radius, then, by Th. 20 above, AC will be the sine of its adjacent arc, which is opposite the angle, ABC, that we seek. Now since from the hypothesis the two lines AB and AC are expressed in the same terms through the old measure, and furthermore since AB, the radius of the described circle, is known by the new measure — that of one of the parts of the whole sine — then by the teaching of Th. 7 above, AC will be known by that same measure. Since AB is

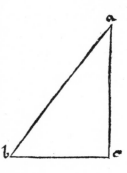

ſinus totus uel ſinus quadrantis, erit a c ſinus notus, &
per tabulã ſinus,qua neglecta,hoc in propſito nihil effice
re poſſumus,arcũ eius addiſcemus, cognito autẽ arcu di
cti ſinus,dat & angulus quẽ reſpicit arcus ille, nã arcus
ipſe & angulus ſecundũ eundẽ numerũ mẽſurant,quẽad-
modũ tota circũferẽtia & q̃tuor recti anguli ſecundũ eun
dẽ numerũ cõiter mẽſurãtur,q̃d in 2 1 huius cõmemoraui
mus,per 22 itaq̃ huius reliquũ acutum angulum b a c
cognoſcemus.Rectum autem angulum a c b, 2 1 huius
notum demonſtrabat.Vniuerſos igitur trianguli noſtri
angulos reddidimus notos, quod decuit explanare .
▼ Opus.Numerũ lateris rectum ſubtendentis angulum
conſtitue primum,& numerum lateris reſpicientis angu
lum quæſitum pro ſecundo ponas,numerũ uero ſinus totius tertium . Multiplica
igitur ſecundum per tertium,& productum diuide per primum,exibit enim ſinus
arcus reſpicientis angulum quæſitum,cui per tabulã ſinus arcum ſuum elicias,cu
ius etiã numerus angulum quæſitum manifeſtabit.hunc ſi ex anguli recti quanti
tate dempſeris,relictum numerabis ſecundum angulũ acutum. Vt ſi a b fuerit 20
a c 12,& b c 16,ſinus autem totus quemadmodũ in tabula noſtra ſuppoſuimus
60000,multiplicabo 1 2 per 60000,producuntur 720000,quæ diuido per 20,exe
unt 36000,huius ſinus arcum tabula præbet gradus 36,& minuta 52 ferè. tantũ
igitur pronunciabimus angulum a b c,qui tandem ſublatus ex 90,relinquet 53
gradus & 8 minuta ferè,& tantus habebitur angulus reliquus acutus.

XXVIII.

Data proportione duorum laterum trianguli rectanguli, angulos
eius percontari.

Aut enim alterum duorũ laterum opponitur recto angulo,aut non. Si primũ
ſit latus a b recto angulo a c b oppoſitum,cuius proportio ad latus a c ſit no-
ta.Dico,ꝙ anguli huius trianguli innoteſcent.Eſt enim a c ſinus arcus anguli a
b c per huius,dum a b eſt ſemidiameter circuli ſcilicet ſinus totus, proportio ergo ſinus totius ad ſinum anguli a b c nota eſt,hinc ſinus ille notificabitur, & tan
dẽ angulus a b c non latebit.Si uero proportio duorum laterum b c & a c data
fuerit,erit proportio quadratorũ notum data,& coniunctim proportio aggregati ex quadrato b c cum quadrato a c,hoc eſt quadrati a b propter angulum c
rectũ ad quadratum a c nota erit:unde & linearum proportio non ignorabit,
reliqua ut ante. ▼ Operatio.Si alterum duorum laterum recto angulo opponatur,multiplica terminũ minorem proportionis datæ per ſinum totum, & productum diuide per terminũ maiorem,exibit enim ſinus anguli,cuius latus breuius
opponitur.Si uero duorum laterum recto circumſtantiũ data fuerit proportio,
duc utrunq̃ terminorũ in ſe,& collecti ex productis radicẽ accipe quadratã, ipſa
enim erit terminus lateri,quod rectum ſubtendit angulum accomodandus , perduceris ergo ad iter priſtinũ.Vt ſi proportio a b ad a c fuerit ſicut 9 ad 7. multi
plico 7 in ſinum rectum totum 60000,fiunt 420000 . quæ diuido per 9,exeunt
46667 ferè.arcus autem reſpondens huic ſinui recto eſt gr.5 1 . minuta 3 ferè, &
tantus habebitur angulus a b c.Sed ponatur proportio a c ad c b ſicut 1 2 ad
5.duco 1 2 in ſe,fiunt 144.item 5 in ſe,reddunt 2 5 .hæc coniungo,faciunt 169. ho
rum radix eſt 13,attribuenda lateri a b,ſic proportio a b ad a c erit ut 13 ad 1 2.

D unde

the whole sine, or [in other words] the sine of the quadrant, sine *AC* therefore becomes known. Through the table of sines, by whose neglect we can accomplish nothing in this theorem, we also ascertain the arc. Moreover, when the arc for this sine is known, the angle opposite this arc is also given, for the arc itself and the angle are measured according to the same number, just as the total circumference and the four right angles are conjointly measured according to the same number, as was mentioned in Th. 21 above. Thus we know the other acute ∠ *BAC* by Th. 22 above. Th. 21 above showed the right ∠ *ABC* to be known. Therefore all the angles of our triangle have been found. Q.E.D.

The mechanics: Take the value of the side subtending the right angle as the first number, and take the value of the side opposite the desired angle for the second number, while the value of the whole sine is the third number. Then multiply the second by the third and divide the product by the first, for the sine of the arc opposite the desired angle will result. From the table of sines you may determine that arc, whose value equals the desired angle. If you subtract this [angle] from the value of a right angle, the number that remains is the second acute angle. For example, if *AB* is 20, *AC* 12, and *BC* 16 and the whole sine found in our table is 60000, then multiply 60000 by 12 to get 720000, which, divided by 20, leaves 36000. For this sine the table gives an arc of approximately 36° 52 min. This amount is ∠ *ABC*, which, subtracted from 90, finally yields about 53° 8 min., the size of the other acute angle.

Theorem 28

When the ratio of two sides of a right triangle is given, its angles can be ascertained.

One of the two sides is opposite the right angle or else none is. First, if side *AB*, whose ratio to side *AC* is known, is opposite right ∠ *ACB*, then the angles of this triangle become known.

AC is the sine of the arc of ∠ *ABC* by Th. 20 above, provided that *AB* is the radius — that is, the whole sine — of the circle. Therefore, the ratio of the whole sine to the sine of ∠ *ABC* is known. Hence the latter sine can be found, and finally ∠ *ABC* will be known. However, if the ratio of the two sides *BC* and *AC* is given, then the ratio of their squares will be known. And by addition the ratio of the sum of the square of *BC* plus the square of *AC* — that sum being the square of *AB* because of right ∠ *C* — to the square of *AC* will be known. Hence the ratio of the lines is known. And what remains is as before.

The mechanics: If one of the two sides is opposite the right angle, multiply the smaller term of the given ratio by the whole sine and divide the product by the larger term. This will yield the sine of the angle that is opposite the shorter side. However, if the ratio of the two sides which include the right angle is given, square each of the terms and take the square root of the sum of the squares. This [square root] will be the appropriate term for the side which subtends the right angle. Then continue in the previous way. For example, if the ratio of *AB* to *AC* is as 9 to 7, multiply 7 by 60000, the whole right sine, to get 420000. This, divided by 9, yields about 46667. The arc corresponding to this right sine is about 51° 3 min., and that amount is ∠ *ABC*. But if the ratio of *AC* to *CB* is given as 12 to 5, then square 12 to get 144, and similarly square 5 to find 25. These, when added, make 169, of which the root is 13, this being assigned to side *AB*. Thus the ratio of *AB* to *AC* will be as that of 13 to 12.

unde ut prius angulo a b c cognoſcendo uía parata eſt.

XXIX.

Altero duorum acutorum angulorum , quos habet triangulus re⸗
ctangulus ,cognito, cum uno eius latere & angulos cunctos & latera
metiemur.

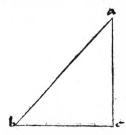

Trianguli a b c angulum c rectum habẽtis , angu
lus b ſit cognitus cũ latere uno quocũcꝫ(uerbi gratia)a c.
Dico, ꝙ omnes eius anguli cum lateribus omnibus inno
teſcent. Anguli profecto cognoſcentur ex 21 & 22 huíus.
reſtat igitur inuenire latera. Per 20 autem huíus & tabu⸗
lam ſinus hypotheſi iuuante, erit utruncꝫ laterum a c &
b c cognitum, ut a b eſt ſinus totus, duo itacꝫ latera quæ
libet trianguli propoſiti datam inuicem habebunt ,ppor⸗
tionem, cuncꝫ ex hypotheſi unũ eorum datum ſit per mẽ
ſurã nouam,erunt per 6 aut 7 huíus reliqua data,quod li⸗
buit attingere. ▼ Opus pulchrũ & perutile.Sinum arcus anguli dati,& eius cõ
plementi addiſcas,habebiscꝫ tria latera nota per menſuram ueterem,quæ eſt pars
una ſinus totíus,nam latus rectum ſubtendens angulum eſt ſinus totus . Si igitur
latus,quod rectum ſubtendit angulum,fuerit datum per menſuram nouam, pone
ſinum totum pro primo,& ſinum arcus anguli,cui opponitur latus quæſitum,pro
ſecundo,numerũ autem nouæ dationis tertiũ.multiplicatocꝫ ſecundo per tertium
productum diuide per primum,& exibit numerus lateris quæſiti.Si uero alterum
duorum laterum recto ſubiacentium detur,uolendo menſurare latus rectum ſub⸗
tendẽs angulum,pone ſinum arcus anguli,cui opponitur ipſum latus datum pro
primo,& ſinum totum pro ſecundo,numerum autem dationis nouæ tertium,ab⸗
ſolutocꝫ opere uulgari quatuor numerorũ proportionaliũ ad metam perduceris
cupitam. Quod ſi reliquũ latus recto ſubſtratũ angulo inueſtigaueris, pone ſinũ
arcus anguli,cui opponitur,latus datum pro primo,& ſinum complementi eius
pro ſecundo,numerũ uero dationis nouæ tertium,reliqua ut antehac executurus.
In exemplo.Detur angulus a b c 36 graduũ,& latus a b 20 pedũ,ſubtraho 36 à
90,manebunt 54 gradus,qui determinant quantitatem anguli b a c.inuenio au
tem lineam a c 35267 ex tabula ſinus,b c uero 48541,dum a b eſt ſinus totus
60000.Multiplico igitur 35267 per 20, producuntur 705340, quæ diuiſa per
60000,eliciũt 11 $\frac{45}{60}$ ferè.latus itacꝫ a c habebit pedes 11 & $\frac{45}{60}$,id eſt tres quartas
pedis unius.Similiter multiplico 48541 per 20,producuntur 970820,quæ diui⸗
do per 60000,exeunt 16 & 11 minuta ferè, tantumcꝫ latus b c pronunciabitur.
Quòd ſi ponatur latus a c 20 ,reliquis ut antea manentibus,erit iterum a c la⸗
tus 35267,& b c 48541,dum a b eſt ſinus quadrantis 60000. Ad inueniendũ
igitur latus a b,multiplico 60000 per 20, producuntur 1200000, quæ diuido
per 35267,exeunt 34 & 2 minuta ferè,habebit itacꝫ latus a b pedes 34 & 2 minu
ta ferè.Sed libeat menſurare latus b c,multiplicabo 48541 per 20,producuntur
970820 ,quæ diuiſa per 35267 ,eliciunt 27 & 32 minuta ferè.latus igitur b c 27
& 32 minuta ferè pedis unius complectiſꝰ . ▼ Hic parumper auſculta, quãdoquí
dem opus noſtrũ Triangulorũ Aſtronomiæ ſeruit plurimũ,quæ fractionibus nõ
tam uulgaribus cꝫ Phyſicis utiſꝰ,quo pacto fractiones uulgares in phyſicas cõmu
tentur,non erit ſilentio prætereundum.Omni nancꝫ integrorũ abſoluta diuiſiõe
ſi aliquid de numero diuiſo:quod neceſſario minus inuenietur diuiſore : relictum
fuerit,

Hence, as before, the way is prepared for finding ∠ *ABC*.

Theorem 29

When one of the two acute angles and one side of a right triangle are known, all the angles and sides may be found.

If in right △ *ABC*, with *C* the right angle, ∠ *B* is known together with any one side — say, *AC* — then all its angles and sides may be found.

Indeed, the angles are known from Ths. 21 and 22 above. Therefore all that remains is to find the sides. Now with Th. 20 above and with the table of sines assisting the hypothesis, each of the sides *AC* and *BC* will be known since *AB* is the whole sine. Thus any two sides of the given triangle will have a known ratio to one another. And since from the hypothesis one [of the sides] is given by a new measure, then by Ths. 6 and 7 above the others will be found. Q.E.D.

The mechanics are precise and very useful. Ascertain both the sine of the arc of the given angle and that of its complement, and you will have the three sides in terms of the old measure. This [measure] is some portion of the whole sine, for the side subtending the right angle is the whole sine. If, therefore, the side which subtends the right angle is given by the new measure [i.e., length], take the whole sine for the first [number] and the sine of the arc of the angle opposite the desired side for the second [number], the value of the new datum being the third [number]. When the second has been multiplied by the third, divide the product by the first to obtain the numerical value of the desired side. However, if one of the two sides adjacent to the right angle is given, then, in order to measure the side subtending the right angle immediately, take the sine of the arc of the angle opposite the given side for the first [number] and the whole sine for the second [number], the value of the new datum being the third. Having

solved, by the usual method, for the fourth proportional number, you will continue [as before] to the desired end. But if [one of the sides enclosing the right angle is given and] you seek the other side beneath the right angle, then take the sine of the arc of the angle opposite the given side for the first [number] and the sine of the complement [of this arc] for the second [number], the value of the new datum being the third [number]. Then the rest is executed as before.

For instance, let ∠ *ABC* be given as 36° and side *AB* as 20 feet. Subtract 36 from 90 to leave 54°, the size of ∠ *BAC*. Moreover, from the table of sines it is found that line *AC* is 35267 while *BC* is 48541, when *AB*, the whole sine, is 60000. Therefore multiplying 35267 by 20 yields 705340, which, divided by 60000, leaves about 11 45/60. Thus side *AC* will have 11 feet and 45/60 — that is, three-fourths of one foot. Similarly multiply 48541 by 20, giving 970820, which, divided by 60000, leaves about 16 [feet] and 11 min., the length of side *BC*. But if side *AC* is taken as 20 and everything else stays the same, *AC* will be 35267 and *BC* 48541, when *AB*, the sine of the quadrant, is 60000. Thus to find side *AB*, multiply 60000 by 20, producing 1200000, which is divided by 35267, leaving about 34 and 2 min. Therefore side *AB* will be about 34 feet and 2 min. But if it is agreed to measure side *BC*, multiply 48541 by 20, producing 970820, which, divided by 35267, leaves about 27 and 32 min. Therefore side *BC* is about 27 [feet] and 32 min. of one foot.

Pay attention to this for a little while. Because our predominant work of the *Triangles* serves astronomy, which does not use the common fractions so much as those of physics, the method by which common fractions are converted into physical fractions will not be passed by in silence. For in every complete division of integers, if any of the dividend is left which is found unavoidably smaller than the divisor

fuerit,diuifor autem numerus alius cp̃ 60 fuerit,fractio uulgaris habebitur, cuius
numerator quidem eſt ipſe numerus relictus,denominator autem numerus diui-
for.uolendo igitur eã minutiã reddere phyſicam,cum phyſicas minutias ſexage
narius numerus denominare ſoleat:multiplicabimus numeratorem minutiæ uul
garis per 60 ,productũ uero per denominatorem minutiæ uulgaris partiendo,ex
íbit numerator minutiæ phyſicæ æquipollentis dictæ fractioni uulgari.Quòd ſi
iterũ fuerit reſiduũ facta diuiſione iam dicta,ipſum per 60 extendemus,& pductũ
per diuiſorem priſtinũ partiemur,exibit enim fractio phyſica ſecundi ordinis,hu
iuſmodi demũ ſi repetieris opus , minutia phyſica tertij ordinis emerget.Primas
aũt minutias,quæ ſtatim poſt diuiſionenintegrorũ eliciuntur,minuta prima uul
gus appellat Aſtronomorũ,ſecundas uero ſecunda,tertias tertia,& ita de cæteris
ex ordine ſuo.Huic autem præcepto ne iuſto amplius digreſſi uideamur, maiorũ
exempla non ſubiecimus,cum & inuentu facilia,& in pleriſcp̃ locis alijs ſatis aper
ta dinoſcantur.

<h2 style="text-align:center">X X X.</h2>

Si trianguli rectanguli datus fuerit alter acutorum angulorum ,
proportiones laterum inter ſe,quamuis latera ipſa non dentur, notas
habebimus.

Sit triangulus rectangulus a b c , angulum c rectum
habens,cuius angulus b acutus ſit datus.Dico,cp̃ proportio
nes quorumlibet duorũ laterum notæ depromentur.Erit ẽ
per 22 huius & angulus a cognitus,quare per tabulã ſinus
uigeſima huius dirigente utruncp̃ laterũ a c & b c notum
dabitur,ut a b eſt ſinus totus,tribus itacp̃ lateribus tres notos aſſignabimus numeros,per diffinitionem igitͭ notæ proportionis theorema uerum enũciaſſe confiteberis. ℣ Opus
facile.Quære ſinum arcus anguli dati, & ſinũ eius complementi,habebiscp̃ duos numeros duobus lateribus rectum ambientibus deputan-
dos,pro latere autem tertio ſinum quadrantis conſtitues,quo facto , ſcies proportionem quorũlibet duorum laterum eſſe ut proportionem numerorũ ſibi reſpon-
dentium.Vt ſi angulus b fuerit 36,quæro ſinum 36 graduũ.qui eſt 35267.ſinũcp̃
complementi eius ſcilicet 54.graduum,qui eſt 48541.pro latere autem a b poño
60000.ſcio itacp̃ pportionẽ a b ad a c eſſe ut 60000 ad 35267,& ita de cæteris.

<h2 style="text-align:center">X X X I.</h2>

Si ſupra baſim trianguli,cui perpendicularis à uertice anguli cuiuſ-
cuncp̃ ducenda inſidebit,alter duorum angulorum obtuſus fuerit,ex-
tra triangulum ipſa cadet.ſi uero rectus coincidet cum latere ſibi con-
terminali,quod recto ſubſternitur angulo , utercp̃ autem acutus eam
intra triangulum conſiſtere iubebit.

Supra baſim b c trianguli a b c alter duorum angulorum b & c, uerbi gratia angulus b ſit obtuſus.Dico,cp̃ perpendicularis à uertice anguli a demittenda
ad baſim b c cadet extra triangulum a b c.& ſi angulus b rectus fuerit, coincidet cum latere a b.ſi uero utercp̃ angulorũ b & c acutum ſe offeret,ipſa perpendicularis triangulũ non egredietur,uerum in eo cõſiſtere coget.Aduerſario enim

D 2   contra-

and if the divisor is a number other than 60, you will have a common fraction whose numerator is the remaining number itself and whose denominator* is the divisor. In order to make this a physical fraction, since the number 60 customarily designates the physical fractions, we will multiply the numerator of the common fraction by 60, and division of that product by the denominator of the common fraction gives the numerator of the physical fraction that is equivalent to the given common fraction. But if there were a remainder again in the completed division just discussed, we will multiply that remainder by 60 and divide the product by the original divisor; then a physical fraction of the second order will be found. Finally, if you repeat such an operation, a physical fraction of the third order results. People of astronomy designate the first fractions, which are obtained immediately after division of the integers, as the prime minutes, the second [fractions] as the seconds, the third [fractions] as the thirds, and so on in this order.† Now, lest we seem to have digressed more than is necessary on this rule, we have submitted no examples of the higher orders since they are quite easily found and would be readily distinguished in most other situations.

Theorem 30

If one of the acute angles of a right triangle is given, the ratios of the sides can be found although the [lengths of the] sides themselves are not known.

If right $\triangle ABC$ has C as the right angle and acute $\angle B$ is given, then the ratios of any two of the sides can be found.

$\angle A$ will be found by Th. 22 above. Therefore by the table of sines and the directions in Th. 20 above, each of the sides AC and BC will be known as long as AB is the whole sine. Thus three known numerical values may be assigned to the three sides. Then by the definition of a known ratio, the theorem has indeed been proven.

The work is simple: Find the sine of the arc of the given angle and that of its complement, and you will have two numerical values for the two sides that include the right angle. For the third side, find the sine of the

quadrant. When this is done, you know that the ratio of any two sides is as the ratio of their corresponding [sine] values. For example, if $\angle B$ is 36[°], find the sine of 36°, namely 35267, and the sine of its complement, 54°, namely 48541. Furthermore, take 60000 for side AB. Therefore the ratio of AB to AC is as 60000 to 35267. And so on.

Theorem 31

If one of the two angles on the base of a triangle is obtuse, then a perpendicular drawn from the vertex angle to the base will fall outside the triangle. But if [one of the base angles] is a right angle, the perpendicular will coincide with the side adjacent to the right angle. If both [base angles] are acute, the perpendicular must remain within the triangle.

If one of the two angles B and C on the base BC of $\triangle ABC$ is obtuse — for example, $\angle B$ — then the perpendicular drawn from the vertex of $\angle A$ to base BC will fall outside $\triangle ABC$. If $\angle B$ is a right angle, [the perpendicular] will coincide with side AB. But if each of the angles B and C is acute, the perpendicular will not leave the triangle but will stay within it.

*Here the old and the modern uses of "denominator" coincide, but it should be borne in mind, as this discourse continues, that 60 is the "designator" (as well as the modern "denominator") of the "physical" fraction. The reader should also remember that decimal fractions, as such, were not grasped until about a century after Regiomontanus wrote his treatise; hence, astronomers used the cumbersome sexagesimal fractions — fractions whose denominators are some power of sixty (e.g., $\frac{1}{60}$, $\frac{1}{3600}$, $\frac{1}{216000}$). Of course the sexagesimal number system is still used in circular measurement and time-telling, so this notation is not totally unfamiliar to the modern reader.

†Ptolemy, who divided the diameter of a circle into 120 small pieces on the basis of the Babylonian division of circumference into 360 units, also divided each of the 120 segments into sixtieths (*partes minutae primae*) and each sixtieth again into sixtieths (*partes minutae secundae*). Regiomontanus' terminology is similar, while twentieth-century terminology has abbreviated these to "primes" or "minutes" and "seconds." Incidentally, the reader should note Regiomontanus' usage of *minutum, -i*, to indicate the minute or prime ($\frac{1}{60}$) whereas *minutia, -ae*, indicates small part or fraction in general.

contrarium primæ partis affirmanti concludemus impossibile hoc pacto. Perpendicularis non cadens extra triangulum necessario intra triangulū mane bit, aut cum altero laterum sibi conterminaliū coin cidet. Cadat primo sententia quidem aduersarij in tra triangulum, sitcę a d, erit itacę angulus a d c re ctus per diffinitionem, & per 16 primi: cum sit extrin secus ad triangulū a d b: maior angulo a b d in trinseco sibi opposito, quem obtusum statuimus, rectum igitur maiorem obtuso confitebitur falsigraphus, quod est impossibile. Quòd si dixerit eam coincidere al teri duorum laterum a b & a c, coincidat prius lateri a b, erit itacę angulus a b c rectus, quem hypothesis assentiēte etiā aduersario obtusum præbuit, quod est in cōueniēs. si uero coincidat lateri a c, erit angulus a c b rectus, angulo itacę b ob tuso existente per hypothesim, non erunt duo dicti anguli minores duobus rectis, quod cum repugnet decimæseptimæ primi impossibile perhibebitur, destructis au tem impossibilibus illis, inferemus aperte primæ partis ueritatem. Similiter ratā faciemus secundā partē. Nam siue proteruiendo dicatur eam cadere intra triangu lum, ut est a d, siue extra, qualis est a e, sequitur ex diffinitione perpendicularis angulum extrinsecū æqualem esse intrinseco sibi opposito, quod cum esse nequeat 16 primi prohibente, constat ipsam intra triangulū non cadere, necę extra eum, & ideo coincidere lateri apud quod rectus est angulus, reliquo enim sibi cōtermi nali lateri non potest coincidere, sic enim acutus angulus rectus haberetur, quod acuti anguli diffinitio non permittit. Tertiæ tandem parti certitudinem compa rabimus haud difficilius. Si enim dixeris eam coincidere cum altero duorū laterū, erit idem angulus & rectus per diffinitionem, & acutus per hypothesim, quod est impossibile. si uero extra triangulū cadere putabis eam, sit ipsa a e. erit igit per 16 primi angulus a c b extrinsecus ad triangulū a e b, quē acutū p̄buit, hypothesis maior angulo a e c recto per diffinitione, quod est impossibile. quo interempto, relinquetur ueritas tertiæ partis, sicę totum theorema nostrum ueritatem enunci asse monstrauimus.

XXXII.

In omni triangulo ex notis tribus constante lateribus, si una trium perpendiculariū data fuerit, reliquæ duæ non latebunt.

Perpendiculares intelligo ductas à uerticibus angulorū uersus latera ipsis an gulis opposita. Sit triangulus a b c, cuius tria latera sint data, producāturcę tres perpendiculares a d, b e, & c g uersus latera subtendentia angulos, à quorū uer ticibus ducuntur. non refert autem siue omnes ipsæ intra triangulū cadat, siue ali qua earum extra, nisi cę latus uersus quā ducitur perpendicularis extra triangulū cadens prolongetur, donec suæ occurrat perpendiculari. Tres autem perpendicu

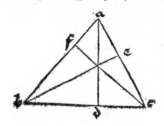

lares illæ in eodem puncto se intersecabunt, quod alio in loco demonstratum tradidimus. Sit itacę data p̄pendicularis a d, dico, cę reliquæ duæ da buntur notæ. Tendamus primum ad inquisitio nem perpendicularis b e. Cum igitur duo trian guli a d c & b e c sint æquianguli, cę angulum c communem habeāt, & utruncę angulorū apud puncta

When the contrary of the first part is asserted, we shall conclude [that] contrary to be impossible. If the perpendicular does not fall outside the triangle, it must either remain within the triangle or coincide with one of the sides adjacent to itself. First, take the contrary position that it falls within the triangle. Let *AD* be [the perpendicular]. Then ∠ *ADC* will be a right angle by definition. And by [Euclid] I.16 since [∠ *ADC*] is an exterior angle of △ *ADB*, it must be greater than interior ∠ *ABD*, opposite it, which is obtuse. This reveals the erroneous statement that a right angle is greater than an obtuse angle, which is impossible. Now let [*AD*] be said to coincide with one of the two sides *AB* and *AC*: First let it coincide with side *AB*. Then ∠ *ABC* would be a right angle, [but] to the contrary, the hypothesis stated that [*ABC*] was obtuse. This is a contradiction. If, however, [the perpendicular] were to coincide with side *AC*, ∠ *ACB* would be a right angle. Therefore, because ∠ *B* is obtuse by the hypothesis, the two mentioned angles will not be less than two right angles. But since that is inconsistent with [Euclid] I.17, it is asserted to be impossible. Moreover, having eliminated these impossibilities, we may readily infer the truth of the first part of the theorem.

We will confirm the second part in a similar manner. For, as before, whether [the perpendicular] is said to fall within the triangle, as *AD* does, or outside, as *AE* does, it follows from the definition of the perpendicular that the exterior angle is equal to the interior angle opposite it. But since, by the prohibition of [Euclid] I.16, this cannot be, it is established that [the perpendicular] falls neither within nor outside the triangle. Therefore it must coincide with the side next to the right angle. It cannot coincide with the other side adjacent to itself, for then an acute angle would have to be a right angle, which

the definition of an acute angle does not permit.

We prove the third part with no difficulty whatsoever. For if [the perpendicular] were said to coincide with one of the two sides, the angle would be a right angle by definition and an acute angle by hypothesis, which is impossible. However, if it were considered to fall outside the triangle, let it be *AE*. Then by [Euclid] I.16 ∠ *ACB*, which is exterior to △ *AEB** [and] which the hypothesis stated to be acute, will be greater than ∠ *AEC*, a right angle by definition. This is impossible. This having been disposed of, the truth of the third part remains. Q.E.D.

Theorem 32

If three sides of any triangle are known, and if one of the three perpendiculars is given, the other two can be found.

I understand the perpendiculars to be drawn from the vertices of the angles to the sides opposite the [respective] angles. Let △ *ABC* be given with its three sides known, and let three perpendiculars, *AD*, *BE*, and *CG*, be extended to the sides subtending the angles from whose vertices the [respective] perpendiculars are drawn. It makes no difference whether all of them fall within the triangle or some of them fall outside it, except that the side to which a perpendicular falling outside the triangle is drawn will be extended until it meets its perpendicular. Furthermore, the three perpendiculars will intersect each other at a single point, as has been proved in another [treatise]. Then, if perpendicular *AD* is given, the other two [perpendiculars] can be found.

First, consider perpendicular *BE*. The two triangles *ADC* and *BEC* are equal-angled, since they have ∠ *C* in common and each of the angles at points

*For *AEB* read *AEC*.

punẽta d & e rectum per diffinitionẽ anguli recti, & ideo per 3 2 primi reliquos duos angulos æquales, erit per quartã sexti proportio a c lateris ad latus c b ; sicut a d perpendicularis ad ppendicularem b e .tres autem primas harum quatuor quantitatũ proportionaliũ notas obtulit hypothesis, quare per 1 9 huius quarta uidelicet perpendicularis b e scita concludetur. Non aliter ad reliquã perpendicularem mensurandã perducemur, quod expectabas ostendendũ. Vniuersaliter autẽ ex hoc trahimus, cp eandem proportionẽ habeant quælibet duæ perpendiculares quã sibi conterminalia latera continentia angulum, cui ipsæ perpendiculares opponunt.   ▼ Opus huius ab ope, q̃d 1 9 huius tradidit generale, nõ discrepabit.

<h2 style="text-align:center">XXXIII.</h2>

Omnis trianguli æquilateri tres angulos conuincemus esse notos, unde quemlibet eorum acutum esse constabit.

Habet nanq̃ omnis triangulus æquilaterus tres æquales angulos per quintã primi, qui per 3 2 eiusdem duobus rectis æquipollent, unde quilibet eorum tertia pars duorum rectorũ angulorum habebitur. Summã autem duorum rectorum ex 2 1 . & 3 . huius notam addiscemus, quare per sextam huius quilibet eorũ datus ex urget, quod cupiebas explanandũ. Cum autem summa duorum rectorũ tripla sit ad unumquenq̃ eorum, dupla uero ad rectum unum, erit per 1 0 . quinti unusquisq̃ eorum minor recto, & ideo per diffinitionem acutus, quod pollicebat̃ corollariũ.

<h2 style="text-align:center">XXXIIII.</h2>

Omnis trianguli æquilateri latus perpendiculari suæ potentialiter sesquitertium esse. Vnde ex latero noto perpendicularis & econtra noticiam consequemur.

Ab angulo a trianguli æquilateri a b c descẽdat per pendicularis a d ad basim b c cadens per corollariũ præmissæ & 3 0 . huius intra triangulum a b c. Dico, cp latus trianguli a b c potentialiter sesquitertiũ sit ppendiculari a d. Ipsa enim perpendicularis basim in æquas partitur sectiones, penultima primi & cõmunibus animi conceptionibus id concludentibus. b c igitur aut a c sibi æqualis, dupla est ad lineam d c, quare per quartã secundi, aut 1 8 sexti quadratũ a c quadruplũ erit quadrato d c. penultima autẽ primi quadratũ a c duobus quadratis linearum a d & d c æquiualere docuit. Duo igitur huiusmodi quadrata coniũcta quadruplũ efficiunt quadrato d c, habent itaq̃ proportionem ad quadratũ d c sicut 4 ad 1 . quare diuisim pportio quadrati a d ad quadratum d c erit sicut 3 ad 1 , triplã uidelicet. erat autem quadratũ a c ad quadratum d c quadrupla. maioris itaq̃ harum pportionũ denominator est 4 , minoris uero 3 , quare per corollariũ 1 2 huius quadrati a c ad quadratũ a d sicut 4 ad 3 sesquitertia scilicet concludetur. potentia igitur lateris a c, quã uocant quadratũ eius potentiæ perpendicularis a d sesquitertia conuincitur, quod intendebat propositio. Quod autem corollariũ huius pollicebatur, sexta & prima huius enitent. Nam sexta huius ex pportione iam data quadrati a c, per primam huius noti ad quadratũ a d, ipsum quadratũ a d suscitabit notum, cuius demum costa ppendicularis scilicet a d per secũdam huius emerget cognita, similiter ex a d perpendiculari data latus a c notum explicabimus. Poteris etiam, si libeat, ex latere a c

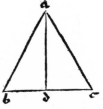

D 3   noto

D and *E* is a right angle by the definition of a right angle, and therefore by [Euclid] I.32 the other two angles are equal. Then by [Euclid] VI.4 the ratio of side *AC* to side *CB* will be as that of perpendicular *AD* to perpendicular *BE*. Now the hypothesis gave the first three of these four proportional quantities. Hence, by Th. 19 above, the fourth [term] — perpendicular *BE* — can be found. We will determine the value of the other perpendicular no differently. Q.E.D.

From this we may draw a general conclusion that any two perpendiculars have the same ratio as that of the adjacent sides forming the angle opposite these perpendiculars. The mechanics are the same as those for Th. 19 above.

Theorem 33

The three angles of every equilateral triangle can be proven known; whence it is agreed that any one of [the angles] is acute.

Every equilateral triangle has three equal angles by [Euclid] I.5, and by [Euclid] I.32 they are equivalent to two right angles. Hence, any one of them will be a third of two right angles. We already know the sum of two right angles from Ths. 21 and 3 above. Therefore, by Th. 6 above, any one of them can be found. Q.E.D.

Moreover, since the sum of two right angles is thrice any one of these [angles] but twice one right angle, each of these [three angles] will be less than a right angle by [Euclid] V.10, and thus by definition [each will be] an acute [angle]. This is offered as a corollary.

Theorem 34

In every equilateral triangle the [second] power of a side is four-thirds of the [second] power of its perpendicular. Hence, if the side is known, the perpendicular can be found, and vice versa.

In equilateral △ *ABC*, drop a perpendicular *AD* from ∠ *A* to base *BC*. By the preceding theorem and Th. 30[31]* above, this [perpendicular] falls within the triangle. Then the [second] power of a side of △ *ABC*

is four-thirds of the [second] power of perpendicular *AD*.

Now the penultimate theorem of [Euclid] I and the axioms show that the perpendicular divides the base into two equal parts. Therefore *BC*, or its equal *AC*, is twice line *DC*. Thus by [Euclid] II.4 or VI.18 the square of *AC* is four times the square of *DC*. Furthermore, the penultimate [theorem] of [Euclid] I shows the square of *AC* to be equal to the square of line *AD* plus the square of line *DC*. Therefore the sum of these two squares is four times the square of *DC*. Hence the ratio of their sum to the square of *DC* is as that of 4 to 1. Then by division, the ratio of the square of *AD* to the square of *DC* is as that of 3 to 1 — namely, threefold. Moreover, the square of *AC* was fourfold the square of *DC*. Then the designator of this larger ratio is four while that of the smaller [ratio] is three. Therefore, by the corollary of Th. 12 above, [the ratio of] the square of *AC* to the square of *AD* is as that of 4 to 3 — namely, four-thirds. Hence the [second] power (which [is another name for] the square) of side *AC* is proven to be four-thirds of the [second] power of its perpendicular *AD*, as the proposition indicated.

What the corollary of this promised Ths. 6 and 1 above will show. For since the ratio of the square of *AC*, which is known by Th. 1 above, to the square of *AD* is already found [to be $\frac{4}{3}$], the square of *AD* itself will be known by Th. 6 above. Then its root — namely, the perpendicular *AD* — will become known by Th. 2 above. Similarly when the perpendicular *AD* is given, side *AC* may be found. Also, if it is desired, since side *AC*

*The reference numbers to theorems subsequent to Th. 25 should be one number greater, e.g., 28 should read 29, 34 should read 35, etc. The frequent occurrence of this error might suggest that Regiomontanus inserted a theorem, possibly Th. 26, belatedly. Henceforth, when necessary, the correct theorem number will appear in brackets; however, in several places the numbers had already been corrected prior to publication of the Latin text.

notó per hypothefim corollarij præfentis,& angulo a c b per præcedentem co=
gnito 28 huius dirigente perpendiculare a d menfurare,& uiceuerfa ex perpen=
diculari data latus reddere cognitum.   ▼ Operatio.Dimidiu latus notum in fe
ductum triplabis,triplatícǫ radicem extrahes quadratã,ut libet ǫ ꝓpinquiſſime,
quã aſcribes perpendiculari a d .Aut ex quadrato lateris cogniti quarta fui par=
tem demas,& relícti radicem elicias quadratã ꝓpinquã,ut placet,quæ ꝑpendicu=
larem a d notificabit.In exemplo.Sit latus a c 12.& medietas eius 6.quæ ducta
in fe reddunt 36.hæc triplata faciunt 108.quorũ radix quadrata ꝓpinqua eſt 10
$\frac{391}{1000}$.tantam igitur feré ꝑpendiculare a d prædicabis.Aut multiplicatis 12. in fe
fiunt 144.quorum quarta pars 36.dempta ex ipfis 144. relinquet 108 . quadratũ
uidelicet perpendicularis a d,cum quo ut antehac operabimur.Qđ fi ex ꝑpen=
diculari data libeat elicere latus,ꝑpendicularem in fe multiplica, productóǫ ter=
tiam fui partem adijcias,& refultabit numerus quadrato lateris deputandus,cuius
radix ꝓpinqua latus ipfum manifeſtabit. Vt fi perpendicularis fuerit 6. multipli
co 6 in fe,fiunt 36.quorũ tertia pars eſt 12.hæc adiecta,congregabit 48. quorum
radix quadrata ad ꝓpinquũ eſt 6$\frac{115}{125}$.& tantum habitur ad propinquum latus a
c .Iuberis autẽ quærere radicem quadratã ad ꝓpinquũ,nam fi latus ipfum pofue=
ris notum fecundũ numerum quemcunǫ,impoſſibile eſt perpendicularem præci=
fe notam fieri,latus enim & ꝑpendicularis potentia tantũ cõmunicantes funt,quo
niam quadrata fua funt in ꝓportione numerorũ non quadratorum, quemadmo=
dum in præfenti didicimus theoremate.fimiliter fi perpẽdicularis fupponatur no=
ta,non erit unǫ latus trianguli æquilateri cognitum præcife, Nunc ad triangu=
los æquicrures defcendamus .

XXXV.

Quǫcunǫ angulorum trianguli æquicruris cognito ,reliquos da=
tum iri.Cum quo declarabitur, utrunǫ angulorum,quos bafis fuften
tat, acutum effe.

Trianguli a b c,duo latera a b & a c æqualia habentis, angulus b fit co=
gnitus.Dico, ǫ reliqui duo non uenient.Eſt enim per quintã primi angulus c æ=
qualis angulo b,ꝓpter latera a b & a c æqualia,quare & ipfe notus,& ideo per
25.huius tertius reliquus angulus b a c innotefcet.Non aliter ꝑcedemus,fi quis
angulum c notum obtulerit.Quđ fi angulus a detur,erit per 4. huius refidrum
duorum rectorũ cognitum,quod eſt aggregatũ ex duobus angulis b & c æquali
bus,& ideo per fextã huiufmodi uterǫ eorũ datus cũ fit medietas huius aggrega
ti.Vnicus igitur angulus quilibet trianguli æquicrurij cognitus focios fufcitabit
fuos,quod expectabatur oftendendũ.Cum autem duo recti ad unum rectũ fe ha=

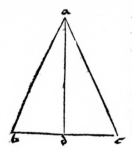

beant,ficut aggregatum ex duobus angulis b & c ad
utrunǫ eorum,eſt enim utrobiǫ dupla proportio , duo
autem recti maiores funt duobus angulis b & c colle=
ctis per 17. primi,erit quoǫ per 14.quinti unus rectus
maior utroǫ angulorũ b & c .per diffinitionem igitur
anguli acuti uterǫ eorum acutus habebitur,quod enun
ciabat corollariũ. ▼ Operationem accipe.Si alter du
orum æqualium,qui bafi infident,angulorũ notus præ=
beatur,erit & reliquus fecundum eundem numerum no
tus,hunc numerũ duplabis,& duplum fubtrahes à nu=
mero

is known by the hypothesis of the present corollary, and since ∠ ACB is known by the preceding theorem, it is possible to measure perpendicular AD by the directions of Th. 28[29] above. Conversely, from a given perpendicular, the side can be found.

The mechanics: Triple the square of half the known side and take the nearest possible square root of that tripled product. The [result] is the value of perpendicular AD. Or from the square of the known side subtract a fourth of [the square] itself, and of the remainder take the closest possible square root. This will identify perpendicular AD. For example, let side AC be 12, and half of that is 6. Square 6 to get 36, and triple [36] to make 108. The closest square root [of 108] is 10 ^{391}/$_{1000}$. This, then, is approximately the [length of] perpendicular AD. Alternatively, 12 squared is 144, of which one-fourth is 36. Subtract 36 from 144 to leave 108, the square of perpendicular AD; from here on, proceed as before. But if it is desired to find a side when a perpendicular is given, square the perpendicular and to this product add one-third of itself. The resulting number is the square of the side, and the nearest square root will be the side itself. For example, if the perpendicular is 6, square 6 to make 36, of which one-third is 12. When these are added together, the sum is 48, whose square root is approximately 6 ^{116}/$_{125}$. This is the approximate [length] of side AC. You are told to seek [only] the closest square root, because, if the side itself were given according to some [whole] number, it would be impossible to find the perpendicular exactly. For the side and the perpendicular have [only] their [second] powers sharing in this value [⅓] since their squares are in a ratio of nonsquare numbers, as we determined in the present theorem. Similarly, if the perpendicular is

taken as the known, it will not be possible to know the exact side of the equilateral triangle. Now we shall proceed to isosceles triangles.

Theorem 35

If any one of the angles of an isosceles triangle is known, the other angles will be found. In addition, each of the base angles will be shown to be acute.

If in △ ABC, having two equal, known sides AB and AC, ∠ B is known, then the other two angles will become known.*

By [Euclid] I.5 ∠ C is equal to ∠ B, because sides AB and AC are equal. With this known, then, the third ∠ BAC is found by Th. 25 above. We would proceed in the same way if ∠ C were the given angle. But if ∠ A were given, then by Th. 4 above [when ∠ A is subtracted] from two right angles, the remainder will be the sum of the two equal angles B and C. And so by Th. 6 above, each of them is found [to be] half of this sum. Therefore any one angle of any isosceles triangle reveals its supplements. Q.E.D.

Moreover, as two right angles are to one right angle, so also is the sum of the two angles B and C to either one of them, for each ratio is twofold. Furthermore, two right angles are greater than the sum of the two angles B and C by [Euclid] I.17. And by [Euclid] V.14 one right angle will be greater than either ∠ B or C. Therefore, by the definition of an acute angle, each of them is acute, as the corollary states.

The mechanics: If one of the two equal base angles is given, the other will be known [to be] the same value. Then double this value and subtract the double from the value

*In the Latin text, for *non venient* read *nota venient*.

mero duorum rectorum, qui enim relinquitur numerus, angulo, quem basis respi‑
cit, deputabitur. Si uero angulus, quem duo æqualia ambiunt latera, notus offera‑
tur, numerum eius à numero duorum rectorum minues, nam relictu numeri medi
etas utrunque æqualiũ angulorum patefaciet. In exemplo: Sit angulus b 30 gradu
um, erit autem & angulus c sibi æqualis 30. hi collecti, reddunt 60. qui sublati ex
180. numero duorum rectorum usitato, relinquũt 120. & tantus erit angulus a.
Sed ponatur angulus a 150. minuo 150. ex 180. ualore scilicet duorum rectorũ
relinquuntur 30. quorũ medietas 15 utrunque angulorũ b & c cognitũ efficit.

XXXVI.

**Perpendicularem duobus trianguli æquicrurij notis lateribus con
terminalem, cui basis nota per medium diuisa substernitur, faciliter
indagare.**

Sit triangulus æquicrurius a b c, cuius basis b c data
sit, lateraque a b & a c cognita, à quorũ cõmuni puncto a
ad basim b c descendat perpendicularis a d, cadens intra
triangulum, quemadmodũ ex præcedenti & 30. huius con‑
cluditur. Dico, ꝗ ipsa ꝑpendicularis a d nota ueniet. Quo
niam enim uterque angulorũ supra basim apud punctum d
rectus est, erit per penultimã primi tam quadratũ lineæ a b
æquale duobus quadratis linearum a d & d b, ꝗ quadra‑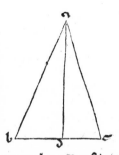
tum a c duobus quadratis linearum a d & d b. sunt autẽ
duo quadrata linearum a b & a c æqualia, ꝓpter costas suas æquales, quare & ag
gregatum ex duobus quadratis a d & d b æquale aggregato ex duobus quadra
tis a d & d c. dempto igitur cõmuni quadrato a d, relinquetur quadratum b d
æquale quadrato d c, unde & linea d b æqualis lineæ d c concluditur. cunque to‑
ta b c nota sit ex hypothesi, erit per sextam huius utraque linearum d b & d c no
ta, sunt enim eius medietates. quadratum itaque lineæ d c per primã huius notum
habebitur, sed & quadratum a c per eandem iuuante hypothesi notum est, à quo
cum superet quadratum d c in quadrato a d per penultimã primi, si quadratum
d c notum abstraxeris, relinquetur per quartã huius quadratum perpendicularis
a d notum, & ideo per secundã huius ipsa ꝑpendicularis a d cognita, quod placu
it attingere.   ▽ Opus. Quadratum dimidiæ basis ex quadrato lateris minuas,
relicti enim radix quadrata ꝑpendiculare manifestabit. Vt si basis b c fuerit 10.
& utrunque laterum a b & a c 13. quadrabo medietatem basis, quæ est 5. exurgũt
25. item quadrabo numerum lateris scilicet 13. producuntur 169. à quibus postꝗ
25 dempsero, relinquent 144. pro quadrato perpendicularis a d, quorum radice
quadratã 12 perpendicularis sibi uendicabit.

XXXVII.

**Quales habeat angulos æquicrurius triangulus, ex cognitis lateri‑
bus & basi faciliter indagare.**

Qualitatem anguli dicimus rectitudinem acutiem & obtusitatem. Duobus
ad hoc iudicijs perducemur, quorũ unũ accipit ex ꝑpendiculari ad basim demissa
& basi, aliũd uero ex latere & basi. Nam si ꝑpendicularis dimidiæ basi fuerit æqua
lis, angulus, cui basis opponitur, rectus erit, si uero minor fuerit medietate basis ob
tusus, & si maior acutus, quorũ demonstrationem breuiter afferemus. Sit enim in
triangu‑

of two right angles, for the remainder is the value of the angle opposite the base. However, if the angle included by the two equal sides is given as the known, subtract its value from the value of two right angles. Then half of the remainder will be [the value of] each of the two equal angles. For example, let ∠ B equal 30°; then ∠ C will also equal 30. These are added, yielding 60, [and 60] subtracted from 180, which is the accepted value of two right angles, leaves 120 as the value of ∠ A. But if ∠ A were 150, subtract 150 from 180 — namely, the value of two right angles — and 30 remains. Half of 30 is 15, which is the value of each of the angles B and C.

Theorem 36

The perpendicular which is conterminous with the two known sides of an isosceles triangle and which bisects the known base may be easily found.

If isosceles △ ABC is given with base BC and sides AB and AC known, and if the perpendicular AD, which falls within the triangle as was proved in the preceding theorem and in Th. 30[31] above, is drawn to base BC from the point A shared by the two sides, then perpendicular AD itself can be found.

Since each of the angles on the base at point D is a right angle, then by the penultimate theorem of [Euclid] I the square of line AB is equal to the [sum of] the squares of lines AD and DB, and the square of AC is equal to [the sum of] the squares of lines AD and DB. Furthermore, the squares of lines AB and AC are equal because the [lines themselves] are equal. Therefore the sum of the two squares of AD and DB is equal to the sum of the two squares of AD and DC. Then, when the square of AD, common [to both sides of the equation], is subtracted, the square of BD is left equal to the square of DC. Hence line DB is equal to line DC. Now since the entire line BC is known by the hy-

pothesis, each of the lines DB and DC will be found by Th. 6 above, for they are half of [BC]. Therefore the square of line DC will be found by Th. 1 above. In addition, the square of line AC is found by the same theorem with the assistance of the hypothesis. Because the square of AC exceeds the square of DC by the square of AD from the penultimate theorem of [Euclid] I, then, if the known square of DC is subtracted [from the square of AC], the square of perpendicular AD is left known by Th. 4 above. Therefore by Th. 2 above the perpendicular AD itself is known. Q.E.D.

The mechanics: Subtract the square of half the base from the square of the side, and take the square root of the remainder to find [the length of] the perpendicular. For example, if the base BC is 10 and each of the sides AB and AC is 13, then square half the base, which is 5, to get 25. Similarly square the value of the side — namely 13 — to get 169. When 25 is subtracted from [169], 144 is left for the square of the perpendicular AD. Its square root is 12, the [length of] the perpendicular itself.

Theorem 37

The types of angles in an isosceles triangle may be easily found when the sides and base are known.

We describe an angle's type as right, acute, or obtuse. We judge this [type] on the basis of two criteria, of which one is learned from the perpendicular drawn to the base and the base, while the other is found from a side and the base. For if the perpendicular is equal to half the base, the angle opposite the base is a right angle. However, if [the perpendicular] is less than half the base, [the opposite angle] is obtuse, and if [the perpendicular] is greater than half the base, then [the opposite angle] is acute. We shall prove these briefly.

triangulo æquicrurio a b c perpēdicularis a d æqualis medietati basis d c, erit
itacʒ ex proceffu 23 huius bis affumpto utercʒ angulorum d a c, & d a b medie
tas recti, totus igitur angulus b a c rectus habebitur. Si autem a d perpendicu‑
laris minor fuerit medietate basis b c, prolongetur d a in e, donec d e fiet æqua

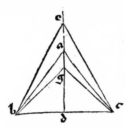

lis lineæ d c, ductis lineis e b & e c nouū æquicruriū tri
angulum claudentibus per 4 primi, cuius angulus b e c
ex recitato proceffu rectus declarabit.unde & per 21 pri‑
mi angulus b a c maior eo, & ideo obtufus enunciabitur.
Sed fi perpendicularis a d maior fuerit medietate basis d
c, abfcindatur ex ea d g æqualis d c, extensifcʒ lineis b
g & g c, probabitur ut prius angulus b g c rectus, qui p
21 primi maior est angulo b a c, angulū itacʒ b a c mi‑
norem effe recto, & ideo acutum nemo dubitabit.   ▼ Ali
ud uero indicium apparebit hoc pacto. Si latus medietati basis potentialiter du‑
plum occurrat, rectus prædicabitur, quem basis fubtendit angulus, fi uero poten‑
tialiter minus cʒ duplum obtufus, & fi maius cʒ duplum acutus, quod fic cōstabit.
Nā fi quadratū lateris, ʠd fuā dicimus potētia, duplū fuerit quadrato dimidiæ ba
fis, cū ipfum æqpolleat p penultimā primi duobus quadratis ipfius fcilicet ppendi
cularis & dimidiæ basis, planū erit quadratū ppedicularis æquari quadrato dimi‑
diæ basis, & ideo ppediculare dimidiæ basi. ad prius igit explanata fi refugerimus
cōstabit angulū b a c effe rectum. Si uero quadratū lateris a b minus fuerit cʒ du
plū quadrati dimidiæ basis, erit & aggregatū ex duobus quadratis linearū a d &
d b per penultimā primi minus cʒ duplū quadrati b d. unde quadratū a d minus
oportebit effe quadrato b d, & ideo costam huius lineam fcilicet a d minorē co‑
sta illius b d, ex prædictis ergo angulum b a c obtufum effe non ignorabimus.
Quod fi quadratū a b maius fuerit duplo quadrati b d, cōcludemus ut nuncnūc
fecimus lineam a d longiorem linea b d. quamobrē ex fupra memoratis in pri‑
mo indicio angulus b a c acutus explorabitur. Qualis itacʒ fit angulus, quem ba
fis fubtendit, gemino monstrauimus indicio. utruncʒ autem reliquorū, quos dicta
fustentat basis, acutum effe docuit corollarium 34 huius.   ▼ Opus uero postea‑
cʒ ex præcedenti perpendicularem didiceris, ex proceffu huius facile comparabis.

XXXVIII.

Trianguli æquicrurij fiue latus quodcuncʒ dederis, fiue perpendi‑
cularem cum uno angulorum, & reliqua latera & perpendiculares
menfurabuntnr.

Sit triangulus æquicrurius a b c, cuius unum latus quodcuncʒ fit notū cum
uno angulorum eius, Dico, ꝗ reliqua duo latera nota fient cum perpendiculari‑

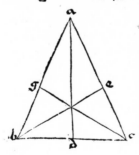

bus. Detur enim primo alterum duorū laterū & fit a c,
demiffacʒ perpendiculari a d ad basim b c, erit trian‑
gulus a d c rectangulus, cuius angulus c acutus ex co
rollario 34.huius notus habebitur, fiue per hypothefim
folam, fiue per hypothefim & 34.huius, quare per 28 hu
ius latere a c noto existente, tam linea a d perpendicu
laris, cʒ d c notæ occurrent, duplicata autem d c notæ
proueniet basis b c data. latus autem a b cum fit æqua
le lateri a c, nemini erit ignotum, Sic igitur & latera re
liqua

Let the perpendicular *AD* in isosceles △ *ABC* be equal to half the base *DC*. Then when the procedure of Th. 23 above is applied twice, each of the angles *DAC* and *DAB* will be half of a right angle. Therefore the whole ∠ *BAC* is a right angle. Now if the perpendicular *AD* were less than half the base *BC*, extend *DA* to *E* until *DE* is equal to line *DC*. When lines *EB* and *EC* are drawn, by [Euclid] I.4, a new isosceles triangle is formed, whose ∠ *BEC* is determined to be a right angle by the process [just] stated. Hence by [Euclid] I.21 ∠ *BAC* is greater than [∠ *BEC*] and therefore must be obtuse. But if perpendicular *AD* were greater than half the base, or *DC*, decrease [*AD* by] *DG*, which is equal to *DC*. [Then] extend lines *BG* and *GC*. As before, ∠ *BGC* is ascertained to be a right angle, which by [Euclid] I.21 is greater than ∠ *BAC*. Thus ∠ *BAC* is smaller than a right angle and therefore acute — which nobody will doubt.

However, the other criterion will be evident in this way: If the [second] power of the side is twice the [second] power of half the base, the angle which the base subtends will be a right angle. However, if [the second power of the side] is less than twice [the second power of half the base, the angle] will be obtuse, and if [the second power of the side] is more than twice [the second power of half the base, the angle] is acute. This is established as follows: If the square (which we call the [second] power) of the side were twice the square of half the base, then, since by the penultimate theorem of [Euclid] I [the square of the side] equals the [sum of] the squares of the perpendicular and half the base, it will be obvious that the square of the perpendicular is equal to the square of half the base. Therefore the perpendicular [is equal] to half the base. Thus, if we refer to the previous explanation, it is established that ∠ *BAC* is a right angle. But if the square of side *AB* were less than twice

the square of half the base, then, by the penultimate theorem of [Euclid] I, the sum of the squares of lines *AD* and *DB* will be less than twice the square of *BD*. Hence the square of *AD* must be less than the square of *BD*, and therefore the square root of it — namely, line *AD* — must be less than the square root of the other, or line *BD*. Consequently, from the aforementioned [facts], we know ∠ *BAC* to be obtuse. But if the square of *AB* were greater than twice the square of *BD*, we conclude now, as we did before, that line *AD* is longer than line *BD*. Whereupon by reference to the first criterion above, ∠ *BAC* is found to be acute. Therefore we have demonstrated by two criteria [how to determine] what type of angle the base subtends. Moreover, the corollary of Th. 34[35] above showed that each of the base angles is acute.

The mechanics are easily determined from the procedure above after the perpendicular has been found from the preceding theorem.

Theorem 38

If, in an isosceles triangle, either a side or a perpendicular is given together with one of the angles, the other sides and perpendiculars can be measured.

If isosceles △ *ABC* is given with any one side and one of its angles known, then the other two sides [and] the perpendiculars can be found.

First, assume that one of the two sides — say, *AC* — is given. When perpendicular *AD* is drawn to base *BC*, △ *ADC* will be a right triangle, whose ∠ *C* [is] acute, by the corollary of Th. 34[35] above, [and] is known either by the hypothesis alone or by the hypothesis and Th. 34[35] above. Therefore, by Th. 28[29] above and because side *AC* is known, so perpendicular *AD* and [segment] *DC* are also known. Furthermore, doubling *DC* gives the base *BC*. And since side *AB* is equal to side *AC*, it will [also] be known. And thus we have measured the remaining sides

liqua & perpendicularem unam menſi ſumus. Qdˀ ſi detur baſis b c cum aliquo
angulorum, erit & eius medietas d c data, habebitᶜꝗ triangulus a ꝺc rectangulus notum latus d c, & angulum c acutum cognitum, quare per 29 huius reliqua
eius duo latera non ignorabuntur, quorū unum eſt perpendicularis a d quæſita,
reliquū uero etiam triangulo æquicrurio propoſito commune eſt. Perpendicularē
autem b e aut c g baſi conterminalē, ex 3 2 huius perpendiculari a d nota exiſtente, faciliter addiſcemus. Poſtremo ex perpendiculari nobis data, cum angulo
quocūꝗ reliqua ſcibilia depromemus. Deꞇ eͫ primo ꝑpendicularis a d baſi inſi-
ſtens, qua intra triangulum cadet, ut ſupra confirmauimus, oportet autē & angulum c acutum eſſe notum, ſiue per hypotheſin ſolam, ſiue per hypotheſim & 3 5
huius. Triangulus itaꝗ a d c rectangulus, cum &
latus a d notum habeat, & angulum c datum, re
liqua ſua latera a c & d c, per 29 huius adducet
cognita, cunꝗ d c ſit medietas baſis, tota quoꝗ ba
ſis b c trianguli ꝓpoſiti non erit ignota. eſt autē
a b æqualis ipſi a c, notæ igitur uenerunt omnes
lineæ trianguli ꝓpoſiti, perpendicularem autem
b e aut c g, 3 2. huius afferet cognitam. Sed ſi deꞇ

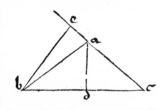

angulus quicunꝗ cum altera perpendiculariū b e & c g, habebit triangulus b e
c latus b e cognitum cum angulo acuto c, & ideo per 29 huius b c linea dabiꞇ
cum eius medietate d c. Iterum ergo triangulus a d c rectangulus, notum latus
d c habens cum angulo c, latus ſuū a d, perpendicularem ſcilicet trianguli æqui
cruriꝗ ꝓpoſiti cum latere a c per 29 huius manifeſtabit. Vna igitur ex ꝰeis memoratis quæcunꝗ cum unico angulo quicquid in triangulo æquicruriꝗ inquiri
ſolet, apertum efficiet, quod libuit abſoluere.   ꝰ Opus autē huius, ne diutius æ-
quo detinearis, miſſum facimus, quod quidem haud difficulter colligemus, ſi ad 29
& 3 2 huius confugerimus.

XXXIX.

Lateribus trianguli æquicruriꝗ cum baſi cognitis, omnes ipſius an
gulos manifeſtare.

Triangulus a b c æquicrurius duo latera a b & a c no
ta habeat cum baſi b c. Dico, ꝗ omnes eius anguli noti fiͤt.
Demiſſa enim ad d baſim ꝑpendiculari a d, erit d c nota,
cum ſit medietas baſis, ut ſupra cͦmemorauimus. duo igitur
latera a c & d c trianguli, a d c rectanguli nota ſunt, quare per 27 huius angulus eius c, qui & triangulo a b c cͦmu
nis eſt, notus comprehͤditur. unde & per 3 5 huius reliqui anguli trianguli a b c propoſiti non latebunt, quod cenſebam
demonſtratū iri.   ꝰ Operationes autem 27 & 3 5 huius, ſi
commiſceas opus theorematis præſentis facile conflabis.

X L.

Si perpendicularem trianguli æquicruriꝗ datam habueris, ex baſi
nota latus, aut econtra ex latere noto baſim elicies.

Sit triãgulus æquicrurius a b c, cuius altera perpendiculariū a d & b e, uel
g data ſit. Dico, ꝗ ſi etiam baſis b c nota fuerit, latus a c cognitū erit, & econ-
E   tra, ſi

and one perpendicular.

But if the base *BC* is given together with one of the angles, half [the base], *DC*, will also be given, and right △ *ADC* will have a known side *DC* and a known acute ∠ *C*. Therefore, by Th. 29 above, its other two sides will be found, of which one is the desired perpendicular *AD* while the other is [the side] shared by the given isosceles triangle. Moreover, since perpendicular *AD* is known, by Th. 32 above we can easily determine perpendicular *BE*, or *CG*, adjacent to the base.

Finally, if the perpendicular is given together with any one angle, the remaining parts can be ascertained. First, let the perpendicular *AD*, which falls within the triangle, as we established above, and rests on the base, be given. ∠ *C* must be acute, either by the hypothesis alone or by the hypothesis and Th. 35 above. Therefore [in] right △ *ADC*, since side *AD* is known and ∠ *C* is given, the other sides *AC* and *DC* can be found by Th. 29 above. And since *DC* is half the base, so the whole base *BC* of the given triangle will also be known. Furthermore, *AB* is equal to *AC*. Thus all the lines of the given triangle have been found, and in addition Th. 32 above will give perpendicular *BE*, or *CG*.

But if any angle is given together with one of the perpendiculars *BE* and *CG*, △ *BEC* will have side *BE* known along with acute ∠ *C*, and therefore by Th. 29 above line *BC*, with its half *DC*, will be found. Hence, again, right △ *ADC* has side *DC* known with ∠ *C*. Its side *AD* — namely, the perpendicular of the given isosceles triangle — as well as side *AC* will be ascertained by Th. 29 above. Thus,

in an isosceles triangle, any one of the mentioned lines together with only one angle will reveal whatever may be desired.

The mechanics will be by-passed lest the reader be detained, for they are hardly different from those of Ths. 29 and 32 above.

Theorem 39

When the sides and base of an isosceles triangle are known, all its angles may be found.

If isosceles △ *ABC* has two sides *AB* and *AC* known together with base *BC*, then all its angles can be found.

When perpendicular *AD* is drawn to *D* on the base, *DC* will be known since it is half the base, as we mentioned above. Therefore the two sides *AC* and *DC* of right △ *ADC* are known. Then by Th. 27 above its ∠ *C*, which is also shared by △ *ABC*, will be known. Hence, by Th. 35 above, the other angles of the given △ *ABC* will be found, which was promised.

The mechanics of this theorem are easily derived from a combination of the mechanics of Ths. 27 and 35 above.

Theorem 40

If the perpendicular of an isosceles triangle is given, then either the side can be found when the base is known or the base can be found when the side is known.

When *ABC* is an isosceles triangle, one of whose perpendiculars, *AD*, *BE*, or *G** is given, then if the base *BC* is known, side *AC* may be found, or if

*For *G* read *CG*.

34

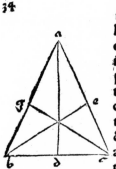

tra,ſi latus a c uel a b notum fuerit,baſis ipſa non ignora-
bitur.Sit enim primo perpendicularis a d nota cum baſi b
c,erit itacɜ & d c medietas baſis cognita,quare per 26 huius a c nota dabitur.Si uero a c latus offeratur notum,erit
per allegatam 26 huius linea d c nota,quæ cum ſit medietas baſis,duplata baſim, ipſam conſtituet.Sit deinceps ppen
dicularis b e uel c g nota cum baſi b c ,ſiue intra ſiue extra
triangulum cadat.duobus itacɜ triangulis rectangulis b e c
& a d c,angulusꝰc cõmunis erit,quare per 32 primi æquianguli concludentur,& ideo per quartam ſexti erit proportio e c ad c d,ſicut b c ad c a .tres autem primæ harum li
nearum proportionaliũ notæ ſunt, e c quidem ex hypotheſi & 26 huius, b c ex
hypotheſi,& d c medietas ipſius b c. quare per 19
huius a c quarta nota ueniet,ſcilicet latus trianguli
æquicrurij quæſitum.Si uero perpendicularẽ b e da
tam habueris cũ latere a b ,erit per 26 huius a e no-
ta.eſt autem ex diffinitione æquicrurij trianguli a c
æqualis ipſi a b ,quare & a c nota, & per tertiã aut
quartã huius e c cognita.duobus itacɜ lateribus b e
& e c trianguli b e c rectanguli notis exiſtentibus,

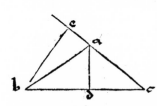

erit per 26 huius linea b c nota,baſis ſcilicet trianguli noſtri. Quo autem pacto
perpendicularis una reliquam ſuſcitare ſoleat,ſuperius explanatũ eſt. ꟼ In ope
ratione b ius non morabimur,quam quidem ex operibus præallegatarum ꝓpo-
ſitionum faciliter conflabimus.

XLI.

Vno duntaxat angulo trianguli æquicrurij cognito, utruncɜ latus
ad baſim & perpendiculares,notas habebit proportiones.

Sit triangulus æquicrurius a b c,unum habens notum angulũ quemcuncɜ,
cuius duæ perpendiculares ſint a d & b e .Dico,ꝙ proportio a c lateris ad baſim b c ,& ambas perpendiculares nota fiet.Erit em triangulus a d c rectangu
lus,notũ habẽs c angulũ acutũ ex hypotheſi ſola, aut ex hypotheſi & 35 huius,qua

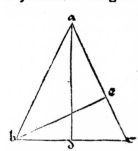

re per 30 huius proportio a c ad a d perpendicularem
nota erit,ſed & eiuſdem a c ad lineã d c ex eadẽ 30 ꝓ
portio non ignorabitur.cuncɜ proportio d c ad c b ſit
nota,eſt enim ut medietatis ad totum,erit per 12 huius
proportio a c ad baſim b c cognita.Sic quo pacto no
tæ fiant proportiones lateris a c ad perpendicularẽ a d
& baſim b c iam explanauimus.Rurſus triangulo a b
e rectangulo angulũ a notum habente, aut per hypotheſim ſolam,aut per hypotheſim & 35 huius, erit ꝑ 30
huius proportio a b lateris ad perpendicularem b e no
ta.quicquid autem de altero laterum a b & a c prædicamus,& de reliquo cum
ſint æqualia enunciatum intelligemus.uerum igitur eſt,quod theorema ꝓpoſuit.
ꟼ Operari aũt,ſi uoles,propoſitiones theorema noſtrũ confirmantes repetito.

In tertio

side *AC* or [side] *AB* is known, the base may be found.

First, let the perpendicular *AD* be known along with base *BC*. Then half the base, *DC*, will be known. Therefore, by Th. 26 above, *AC* is found. But if side *AC* is given, then, by the argument of Th. 26 above, line *DC* will be known. Since [*DC*] is half the base, its double constitutes the base itself. Next, let the perpendicular *BE*, or *CG*, be known together with base *BC*. [The perpendicular] may fall either within or without the triangle. Then ∠ *C* is common to the two right triangles *BEC* and *ADC*. Hence by [Euclid] I.32 [the triangles] are found to be equal-angled. Thus by [Euclid] VI.4 the ratio of *EC* to *CD* will be as that of *BC* to *CA*. Moreover, the first three of these proportional lines are known: *EC* by the hypothesis and Th. 26 above, *BC* by the hypothesis, and *DC* as half of *BC*. Hence, by Th. 19 above, the fourth [term], *AC*, will become known — namely, the desired side of the isosceles triangle. But if perpendicular *BE* were given together with side *AB*, *AE* will be found by Th. 26 above. Furthermore, from the definition of an isosceles triangle, *AC* is equal to *AB*, and consequently *AC* is known. By Th. 3 or 4 above *EC* is known. Therefore, because the two sides *BE* and *EC* of right △ *BEC* are known, by Th. 26 above line *BC* will be found — namely, the base of our triangle. Furthermore, how one perpendicular may reveal the other has already been explained.

No delay will be made over the mechanics of this theorem since they may be easily constructed from the mechanics of the foregoing theorem.

Theorem 41

If only one angle of an isosceles triangle is given, the ratios [of] each side to the base and the perpendiculars may be known.

If isosceles △ *ABC*, with perpendiculars *AD* and *BE*, has any one angle known, then the ratio of side *AC* to the base *BC* and to both perpendiculars can be found.

Triangle *ADC* will be a right triangle with acute ∠ *C* known by the hypothesis alone or by the hypothesis and Th. 35 above. Therefore, by Th. 30 above, the ratio of *AC* to the perpendicular *AD* will be found, and furthermore, the ratio of this same *AC* to line *DC* will be found from the same Th. 30 above. Since the ratio of *DC* to *CB* is known, for it is as the half to the whole, then by Th. 12 above the ratio of *AC* to the base *BC* is known. Thus we have now explained how the ratios of side *AC* to the perpendicular *AD* and to the base *BC* can be found. On the other hand, when △ *ABE* has ∠ *A* known either by the hypothesis alone or by the hypothesis and Th. 35 above, then by Th. 30 above the ratio of side *AB* to the perpendicular *BE* will be found. No matter which one of the sides, *AB* and *AC*, we find, we will know the other because they are equal. Therefore, what the theorem proposed is true.

If you wish, you may repeat the theorems that confirm this proposition.

In tertio demum triangulorum genere ludendum censeo.

Tria superius trianguloru diffiniuimus genera, quoru primo ab æqualitate laterum sumenti originem modicum accidit uarietatis, in eis quas Mathematici scrutantur rebus. in secundo autem magis uaria est scibiliu inquisitio, φ ipsum ab unitate simplicitateςφ laterum recedat. tertium uero genus, quo amplius à primo distat genere, eo difficilius se offert. In primo præterea genere, omnes anguli, non quidem ad spontaneam positionem tuam, sed necessitate cogniti sunt, unuquenςφ enim eorum tertiam partem duoru rectoτum demonstrauimus, uno autem latere quolibet dato, reliqua duo non latebunt, φ ipsa dato lateri sint æqualia. Secundu uero trianguloru genus angulos suos non præbet cognitos, nisi aliquid ex angu= lis suis aut lineis præcognitu habeatur, quod & in lateribus mensurandis euenire compertum est. Tertij autem generis tanta tamςφ uaria est intricatio, ut non satis sit unum angulu cum uno latere præsciuisse ad reliqua cognoscenda, aut duo eius latera tantu, uerum ut latera & angulos metiamur uniuersos, aut tria latera præ= scienda sunt, ad tres angulos reperiendos, aut duo latera cum uno angulo, aut duo anguli cum uno latere. duobus enim angulis datis, tametsi tertius extemplo no= tus reddatur per 25 huius, non tamen latera nota ueniunt, uerum proportiones ipsorum lateru duntaxat, quemadmodu infra docebimus, notas fieri oportet. Po= stremo in his triangulis absςφ notitia duorum casuu aut alterius eorum, quos per= pendicularis ex ipsa basi separat, nihil efficiet Geometra, quæ quidem perpendicu= laris diuersimode cadere solita, nunc intra nunc extra triangulum, ut supra tetigi= mus, multiuariam ignotorum faciet inquisitionem. Principio igitur exploran= dum arbitror, quales sint uniuersi anguli, quos habet propositus nobis triangulus trium datoru laterum, unde perpendicularis quælibet quo pacto casura sit, dirigē= te 32. huius callebimus, cuius demum perpendicularis quantitatem metiri non prorsus inutile uidebitur.

XLII.

Triangulus notorum trium laterum, qualem quemuis angulum habeat percontari.

Sit triangulus a b c trium inæqualiu, & notorum inter se laterum, cuius qua= les sint anguli experiudum est. Faciamus primo periculu de angulo a. dico autē, φ si quadratu lateris b c, ipsum angulu a subtendētis, æquale fuerit duobus qua= dratis laterum a b & a c, quæ dictum ambiunt angulu, rectus erit angulus ille. si uero minus illis quadratis, erit acutus, & si maius, obtusus, quæ sic constabunt. Si enim quadratum b c æquale reperiatur duobus quadratis a b & a c, erit per ultimam primi angulus a rectus. certum est igitur primu indicium. Si uero qua= dratum b c minus fuerit quadratis a b & a c, non potest angulus a esse rectus, neςφ obtusus, eo nanςφ recto existente, quadratu b c duobus quadratis a b & a c per penultimam primi æquabitur, quod est cōtra= riu posito. Sed non obtusus. sic enim per 12. secun= di quadratu b c maius duobus quadratis a b & a c habetur, quod cum repugnet posito, relinquit angulum a esse acutum, & ita secundum firmaui= mus indiciu. Quòd si quadratu b c maius fuerit quadratis a b & a c, nō poterit angulus a esse re

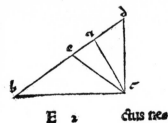

A digression on the third type of triangle.

The three types of triangles have already been defined. Of these, the [one chosen] as a beginning was the [equilateral, because] there is very little diversity in those things which mathematicians investigate. However, in the second of these, the search for information is more diversified, because [this type] departs from oneness and simplicity of [its] sides. The third type, however, which* is much different from the first type, appears more difficult than [the second]. Besides, in the first type all the angles are known, not through a given premise, but by necessity, for each of them has been shown to be a third of two right angles. Moreover, when any one of the sides is given, the others are found because they are equal to the given side. However, the second type of triangle does not let its angles be found unless something about its angles or lines has already been given. This [information] permits the sides to be measured. The variability of the third type, however, is so complex that it is not enough to know one angle with one side or to know only two sides to find the others. Indeed, in order to measure the sides and all the angles, the three sides determining the angles, or two sides with one angle, or two angles with one side must be known. For if two of the angles are given, although the third can be found immediately by Th. 25 above, the sides still cannot be found. Indeed, only the ratios of the sides to each other will necessarily be known, as shall be developed below. Finally, in these triangles, without knowledge of the two segments, or one of them, into which the perpendicular divides the base, geometry can accomplish nothing; indeed, the perpendicular [to the base] may fall in various ways, now inside and now outside the triangle, as we mentioned above. This makes a very diversified search for the unknowns.

Thus, to begin the investigation, one must first determine which kinds of angles a postulated triangle of three given sides has.

Hence, from the directions of Th. 32[31] above, we will know in what way any one perpendicular may have fallen. Finally the perpendicular [of the triangle] will be seen to measure a quite useful quantity.

Theorem 42

When the three sides of a triangle are known, the type of any angle may be determined.

If $\triangle ABC$ is given with the three sides unequal and known in terms of each other, then the types of its angles may be found.

First, let us prove [this] for $\angle A$. If the square of side BC, subtending $\angle A$, is equal to the [sum of the] squares of sides AB and AC, which include the mentioned angle [A], then this angle will be a right angle. However, if this square is less [than the sum of the squares of the other sides], the angle will be acute, and if [this square] is greater [than the sum of the squares of the other sides], the angle will be obtuse. These [statements] may be established thus: If the square of BC is equal to [the sum of] the squares of AB and AC, then by the last theorem of [Euclid] I [angle] A will be a right angle. Thus the first criterion is proven.

However, if the square of BC is less than [the sum of] the squares of AB and AC, $\angle A$ cannot be a right angle, nor can it be obtuse. For if it were a right angle, the square of BC would be equal to [the sum of] the squares of AB and AC by the penultimate theorem of [Euclid] I, which is contrary to what is given. Nor is it obtuse, for then by [Euclid] II.12 the square of BC would be greater than [the sum of] the squares of AB and AC. But since this is inconsistent with what was given, [the only possibility] left is that $\angle A$ is acute. And thus the second criterion is substantiated.

But if the square of BC is larger than [the sum of] the squares of AB and AC, $\angle A$ cannot be

*In the Latin text, for *quo* read *quod*.

ctus neqʒ acutus,nam ſi alterum illorum dixeris,erit quadratum b c, aut æquale
duobus quadratis a b & á c per penultimam primi,aut minus eis per 13.ſecun-
di,neutrū aūt horum cum poſito ſtabit, cui igitur dubiū erit angulū a obtuſum
eſſe.Ad reliquos demū angulos quales ſint,ſimili pducemur examine.   ¶Ope-
rationem ex proceſſu iam facto non potes non comprehendere.In exemplo.Sit la
tus a b 7,latus b c 9,& latus a c 12,uolo ſcire qualis ſit angulus a,quadrabo
ſingula latera,quadratū de 7 eſt 49.quadratū de 9 eſt 81. quadratū uero de 12 eſt
144,colligo duo quadrata 49 & 144.reſultant 193.cum itacʒ quadratū de 9,quod
eſt 81 ſit, minus cʒ 193.pronuncio angūlū a eſſe acutū.Ita in cæteris.

XLIII.

Datis tribus lateribus trianguli,duos caſus, quos perpendicularis
à puncto angulari ad baſim deſcendens,ex ipſa diſtinguit baſi,com-
perire.

Perpendicularis intra triangulū manens,duos caſus profecto diſtinguet ex
ipſa baſi.quæ uero cū altero laterum coincidit,unum duntaxat caſum habebit.p-
pendiculari autem extra triangulum cadente,caſus huiuſmodi non ſunt portiones baſis,uerum baſis eſt pars alterius eorum.Sanè igitur intelligenda erat diffini
tio caſus ab initio poſita.uocabulo enim baſis,baſim ſimpliciter dictam,& baſim
quantū oportet prolongatam ſignificauimus.Cognitio autem caſuū dictorū,aut
alterius eorum,neceſſaria eſt ad perpendicularem trianguli tria latera nota haben
tis cognoſcendam, per quam denicʒ perpendicularem anguli quæri ſolent.Cum
autem d his,quæ in triangulis rectangulis quæri ſolent,ſuperiori loco ſatis dixiſ
ſe uideamur,ad triangulos non rectangulos præcepta futura ſonabunt potiſſimū,
licet quædam ad rectangulos etiam applicari poſſit.Ex puncto igitur a triangu
li a b c tria latera habentis nota uerſus lineam b c procedat perpendicularis a
d,diſtinguens ex baſi duos caſus b d & d c,dico q̓ illi duo caſus noti uenient.
Qua nancʒ lege ſiue intra ſiue extra triangulū cadat huiuſmodi perpendicularis

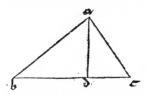

præcedens & 31.huius indicabunt. Cadat itacʒ primo intra triangulum duobus angulis b & c acutis
exiſtentibus.argumento igitur 13.ſecundi quadratū
lateris a b ſuperabit à duobus quadratis linearū a c
& c b,in eo,quod ſit ex b c in c d bis.cūcʒ tam qua
dratum a b notum ſit ex hypotheſi & prima huius,
cʒ aggregatum ex quadratis a c & b c ex hypothe-
ſi,prima & tertia huius,erit per quartam huius,quod ſit ex b c in c d bis notum,
& eius dimidiū,quod ſit ex b c in c d notum.eſt autem latus b c notum ex hypotheſi,quare per 17.huius linea c d nota ueniet, alter uidelicet duorū caſuum,
quo dempto ex linea b c nota p hypotheſim,reliquus caſus b d innoteſcet.Qd́
ſi alter angulorum b & c obtuſum ſe præbeat,perpendicularis a d trianguli are
am tranſiliet,ad partem quidem anguli obtuſi,qui uerbi gratia ſit c, erit igitur p
12.ſecundi quadratum lateris a b maius duobus quadratis linearū a c & c b,in
eo,quod ſit ex b c in c d bis. Ex prius igitur adductis

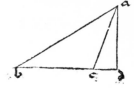

locis(ut breuis ſim)exceſſus ille notus erit,ſcilicet, quod
ſit ex b c in c d bis,& eius dimidiū quod ex b c in c d.
cuncʒ b c nota ſit ex hypotheſi,erit per 17. huius & c d
nota,ſic minorem duorum caſuū notum eniſi ſumus, cui
ſi baſim

a right [angle] or an acute [angle]. For if it is said to be either of these, the square of *BC* will be either equal to [the sum of] the squares of *AB* and *AC* by the penultimate theorem of [Euclid] I or less than them by [Euclid] II.13. Neither of these [alternatives] is supported by what is given. Therefore, owing to this, there will be no* doubt that ∠ *A* is obtuse. Finally, by a similar consideration, we may find the types of the other angles.

The mechanics are easily understood from the procedure already done. For example, let side *AB* be 7, side *BC* 9, and side *AC* 12. One wants to know of what type ∠ *A* is. Each side is to be squared: The square of 7 is 49; the square of 9 is 81; and the square of 12 is 144. The sum of the two squares, 49 and 144, is 193. Therefore since the square of 9 [the length of the side opposite ∠ *A*], which is 81, is less than 193, ∠ *A* must be acute. And so on.

Theorem 43

If three sides of a triangle are given, the two segments into which the perpendicular, drawn from the vertex to the base, divides the base may be found.

When the perpendicular remains within the triangle, it divides the base into two segments. But when [the perpendicular] coincides with one of the sides, there will be only one segment. Furthermore, when the perpendicular falls outside the triangle, these segments are not portions of the base but rather the base is part of one of [the segments]. Indeed, then, an understandable definition of "segment" was given at the beginning [of this book]. By the word "base" we indicate both the base as it is meant simply and the base as its length is necessarily extended. Knowledge of the given segments or segment is necessary for identification of the perpendicular in a triangle that has three known sides. Finally, through this perpendicular the angles may be sought. Since enough has been said above about [finding the unknowns] in right triangles, the subsequent rules will be concerned

chiefly with nonright triangles. One can apply any of them also to right triangles.

Thus, from point *A* to line *BC* of △ *ABC*, which has three known sides, let the perpendicular *AD* be drawn, dividing the base into two segments *BD* and *DC*. Then these two segments may be found. The preceding theorem and Th. 31 above indicate by which rule — whether inside or outside the triangle — this perpendicular falls. First of all, then, it falls within the triangle if the two angles *B* and *C* are acute. Thus, by the argument of [Euclid] II.13, the square of side *AB* will be exceeded by [the sum of] the squares of the lines *AC* and *CB* by twice the product of *BC* times *CD*.† And as the square of *AB* is known from the hypothesis and Th. 1 above, so is the sum of the squares of *AC* and *BC* known from the hypothesis and Ths. 1 and 3 above. Therefore, by Th. 4 above, twice the product of *BC* and *CD* will be known, and its half, the product of *BC* and *CD*, is known. Moreover, side *BC* is known by hypothesis; whence by Th. 17 above, line *CD* — namely, one of the two segments — becomes known. When [line *CD*] is subtracted from line *BC*, which is known by the hypothesis, the other segment *BD* is also known.

But if one of the angles *B* and *C* were given obtuse, the perpendicular *AD* of the triangle would lie outside the triangle on the side of the obtuse angle, which in [our] example will be *C*. Then by [Euclid] II.12 the square of side *AB* will be greater than [the sum of] the squares of lines *AC* and *CB* by twice the product of *BC* and *CD*. From the previously cited [theorems] (to be brief), the excess — namely, twice the product of *BC* and *CD* — will be found, as will its half, the product of *BC* times *CD*. Since *BC* is known by the hypothesis, *CD* also will be found by Th. 17 above. Thus the shorter of the two segments is known, and if

*In the Latin text, for *erit* read *non erit*.
†*Quod fit ex . . . in . . .* is yet another expression for a multiplicative product.

ſi baſim b c notam adieceris,reſultabit caſus maior b d notus, quæ fuere lucubranda. ⸪ Operatio uaria proceſſum huiuſmodi conſequitur.Nam ſi uterⳃ an
gulorum,quos baſis ſuſtinet,acutus fuerit,deme quadratū lateris uñ eorum op=
poſiti ex aggregato duorū quadratorū reliqui ſcilicet lateris & ipſius baſis, quod=
ⳃ relinquitur, dimidiatū in baſim partiaris,exibit enim caſus,qui eſt apud an=
gulum acutum prædictū,quem ex baſi minuendo,reliquū habebis caſum . In ex=
emplo.Ponat quis mihi latus a b 20,pedum,b c 21.& a c 13,monitu præcedē
tis utrunⳃ angulorū b & c acutum eſſe coniicio.quadrabo igitur a b,fiunt 400.
quadratū autem a c,quod eſt 169.coniungam quadrato b c,quod eſt 441.& re=
ſultabunt 610.à quibus demo quadratū a b,manent 210.quorum dimidiū 105.
diuido per 21.exeūt 5.& tantus eſt caſus d c,aufero 5 ex 21 numero baſis,manēt
16 pro caſu reliquo.Quòd ſi alter angulorū prædictorū obtuſus extiterit,à qua=
drato lateris obtuſum ſubtendentis angulum ſubtrahe aggregatū quadratorum
reliqui lateris & baſis ipſius,quodⳃ remanebit,dimidiatū per baſim partire, exi=
bit enim caſus minor,cui poſteaⳃ baſim adiiciemus,emerget caſus maior. Ponaɟ
in exemplo a b 51.b c 38.& a c 25.erit igitur angulus c obtuſus, quadrabo a
b,ueniunt 2601.quadrabo b c,exurgunt 1444. item quadratum a c eſt 625.
colligo duo quadrata b c & c a,reſultant 2069.quæ dempta ex quadrato a b,re
linquunt 532.quorum dimidiū 266 diuido per b c ſcilicet 38.exeunt 7.& tanta
eſt linea c d,caſus uidelicet minor,cui adiungo baſim 38.congregantur 45 pro
caſu maiore.

XLIIII.
Quod præcedens docuit,alio tramite inueſtigare.

Trianguli a b c tria latera ſupponantur nota, a b quidem longius latere ā
c,perpendicularis autem a d cadat intra triangulū ſuper baſim b c,erit itaⳃ ca
ſus b d maior caſu d c ex hypotheſi,penultima primi & cōmuni animi conce=
ptione,ſit igitur e d æqualis ipſi d c ducta linea a e.Cum autem per penultimā
primi quadratū a b æquipolleat duobus quadratis
a d & d b,quadratum uero a c duobus quadratis
a d & d c,quadrato perpendicularis a d communi
ablato,erit per cōmunem ſcientiam differentia qua=
dratorum b d & d c æqualis differentiæ quadrato=
rum a b & a c.duo autem quadrata linearum a b
& a c nota per hypotheſim,ex quarta huius notam

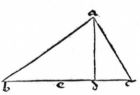

habebunt differentiam,quare & differentia quadratorū b d & d c non ignorabi
tūr,ſed per ſextam ſecundi quadratū b d æquatur ei, quod fit ex e b in b c cum
quadrato e d.differentia igitur quadratorum b d & e d ſiue d c eſt id,quod fit
ex e b in b c,erat autem hæc differentia nota,quare quod fit ex e b in totam b
c notum declarabitur.cunⳃ ipſam b c notam attulerit hypotheſis,erit per 17 hu
ius ipſa b e nota,quam baſi b c cognita dementes,linea e c relinquiemus cogni
tam & eius medietatem d c,quæ eſt caſus minor.huic cum ſit æqualis lineæ e d,
ſi adiecerimus b e prius notam,caſum maiorem b d ſcitum reddemus . Quòd ſi
perpendicularis aream trianguli egrediatur,continuata baſi b c,donec concurret cum perpendiculari in puncto d,ponatur ipſi c d æqualis d e.priſtino igitur
fretus argumento,confiteberis differentiam quadratorū a b & a c æqualem eſſe
differentiæ quadratorū b d & d c,quam quidem differentiam hypotheſis,prima
& quarta huius latere non ſinunt,quadratū autem b d ſuperat quadratum c d,

E 3 per

the known base *BC* is added to it, the longer segment *BD* becomes known. This was promised.

The various mechanics of the procedure follow: If each of the base angles is acute, subtract the square of the side opposite one of [the angles] from the sum of the squares of the other side and the base. [Halve] whatever is left and divide [that] half by the base, for the result will be the segment which is next to the previously mentioned acute angle. Subtracting this [segment] from the base will leave the other segment. For example, let side *AB* be given as 20 feet, *BC* as 21, and *AC* as 13. By the advice of the preceding [theorem], each of the angles *B* and *C* is inferred to be acute. Then square *AB* to make 400. Furthermore, add the square of *AC*, which is 169, to the square of *BC*, which is 441, and the result is 610. From this, subtract the square of *AB*, leaving 210, of which half is 105. Divide 105 by 21, obtaining 5, and that amount is segment *DC*. Subtract 5 from 21, the value of the base, leaving 16 for the other segment. But if one of the aforementioned angles is obtuse, then, from the square of the side subtending the obtuse angle, subtract the sum of the squares of the other side and the base. [Take half of] whatever remains and divide [that] half by the base, for the result will be the shorter segment. Finally to that [segment] add the base to determine the longer segment. For example, let *AB* be given as 51, *BC* as 38, and *AC* as 25. Then ∠ *C* will be obtuse. Square *AB*, making 2601; square *BC*, yielding 1444. Similarly the square of *AC* is 625. Add the squares of *BC* and *CA*, resulting in 2069. This, subtracted from the square of *AB*, leaves 532, of which half is 266. Then 266 divided by *BC* — namely 38 — leaves 7, and that amount is line *CD*, the shorter segment. To that add the base, 38, and the sum is 45 for the longer segment.

Theorem 44

Another proof for the previous theorem.

If the three sides of △ *ABC* are given with *AB* longer than side *AC*, and if perpendicular *AD* falls within the triangle on base *BC*, then segment *BD* will be greater than segment *DC* from the hypothesis, the penultimate theorem of [Euclid] I, and the axioms. Therefore let line *AE* be drawn such that *ED* is equal to *DC*. Since, by the penultimate theorem of [Euclid] I, the square of *AB* is equal to [the sum of] the squares of *AD* and *DB*, and since the square of *AC* is equal to [the sum of] the squares of *AD* and *DC*, then, when the square of the common perpendicular *AD* is subtracted, by axiom the difference of the squares of *BD* and *DC* will be equal to the difference of the squares of *AB* and *AC*. Because the two squares of the lines *AB* and *AC* are known by the hypothesis, from Th. 4 above their difference may be found. Therefore the difference of the squares of *BD* and *DC* is also known. But by [Euclid] II.6 the square of *BD* is equal to the product of *EB* and *BC* plus the square of *ED*. Thus the difference of the squares of *BD* and *ED*, or *DC*, is the product of *EB* and *BC*. Furthermore, this difference was known. Hence, the product of *EB* and the whole [base] *BC* is known. And since the hypothesis gave *BC*, by Th. 17 above *BE* itself may be found. Subtracting this from the known base *BC*, we find line *EC* and its half *DC*, which is the shorter segment. Since line *ED* is equal to this [segment], then if we add [it to] *BE*, found earlier, we cause the longer segment *BD* to be known.

But if the perpendicular falls outside the triangle, then extend base *BC* until it meets the perpendicular at point *D*. Let *DE* be taken equal to *CD*. Then using the previous reasoning, you will show the difference of the squares of *AB* and *AC* to be equal to the difference of the squares of *BD* and *DC*. The hypothesis and Ths. 1 and 4 above do not allow this difference to be hidden. Moreover, by [Euclid] II.6 the square of *BD* exceeds the square of *CD*

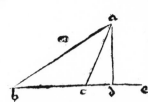

per 6.ſecundi in eo,quod ſit ex c b in totã b e . q̃d igitur ſit ex c b in totam b e ,notum habebitur,eſt autem ipſa b c data per hypotheſim,quare per 17. huius tota b e cognoſcetur,ex qua ſi dempſeris baſim b c datam,reſidua c e nota uidebitur, cũ eius medietate c d.caſum itaq̃ c d minorem eniſi ſumus,cui ſi baſim adiunxeris datam, caſus maior b

d menſuratus emerget,quod decuit explanare. ▼ Operatio.Subtrahe quadratum lateris minoris a quadrato maioris,relicto’q̃ diuiſo per baſim, quod exibit ab ipſa baſi minuas,ſi perpendicularis intra triangulũ ceciderit,aut ab eo quod ex ibit baſim demas,ſi extra ceciderit,eius autẽ quod relinquitur dimidiũ,ꝓ caſu minori teneto,cui ſi id,quod ex diuiſione elicitum eſt,ſociaueris,caſum maiorem con gregabis,perpendiculari quidem intra triangulũ cadente.ſed ſi extra ceciderit,ca ſum minorem cum baſi ſummabis,reſultabit enim caſus maior quæſitus.In exem plo.Sit latus a b 20.a c 13.&baſis b c 21.quadratũ de 13.eſt 169.quadratũ de 20.eſt 400,ſubtraho 169.â 400.manent 231.quæ diuido per 21.exeunt 11.hæc ſublata ex 21.relinquũt 10.medietas de 10.eſt 5 . tantumq̃ minorem pronuncto caſum,cui adiectis 11.colligo 16.pro caſu maiori,hæc dum perpẽdicularis intra. Sed extra offeratur latus a b 51.a c 25.&baſis b c 38.quæ quidẽ perpẽdicularẽ extra triangulum cadere ſignificant.ſubtraho quadratum de 25.quod eſt 625 . ex quadrato de 51.ſcilicet 2601.manent 1976.quæ diuiſa per 38.eliciunt 52.â quibus demo 38.manent 14.quorum medietatem ſcilicet 7 . caſui minori deputabo. colligo 7 & 38.reſultant 45.tantũ igiͧ maiorem enunciabo caſum. ▼ Quod’ſub differentia laterũ & congerie cõmuni continetur,æquum eſt ei,quod ſub differentia caſuum atq̃ congerie eorum,ſcilicet ipſa baſi continet̃ rectangulũ.

XLV.

Huiuſmodi caſus per alia media numerare.

Sit triangulus a b c non rectangulus trium notorum laterum inæqualium, â cuius puncto angulari a demittatur perpendicularis a d ſupra baſim b c tri anguli ſuperficiem non tranſgrediens,quæ ex baſi b c duos ſeparet caſus b d & d c,quorum alterũ altero maiorem eſſe in præcedenti concluſimus, propter inæqualitatem laterum a b & a c,hos caſus menſurandos præſtola mur. Super pun

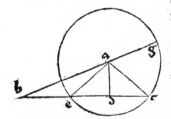

cto a tanq̃ centro ſecundum quantitatem lateris minoris a c deſcribo circulũ h e c,cuius circum ferentia neceſſario ſecabit & baſim b c,&latus a b,q̃ a c linea longior ſit perpendiculari a d,bre uior autẽ latere a b ,ſecet itaq̃ baſim b c in puncto e,quo cum centro circuli copulabo per lineã e a,lineam uero a b ſecet in puncto h . extendã deinceps b a ultra centrum circuli,donec occur ret circumferentiæ eius in puncto g,erit igitur li

nea e d per tertiam tertij æqualis d c caſui minori,& ideo linea b e eſt differen tia caſuum.Quoniã autem à puncto b extra circulum ſignato, duæ lineæ b g & b c productæ circulũ ſecent,erit per 35 tertij,quod ſit ex g b in b h æquale ei,q̃d ex c b in b e,ſed quod ſit ex g b in b h,notum eſt per 16 huius,linea enim g b nota eſt cum ſit æqualis duobus trianguli lateribus a b & a c per hypotheſim da tis,ſed & b h differentia ſcilicet duorum laterum ſcita eſt per hypotheſim & quartam huius

tam huius

by the product of *CB* times all of *BE*. Therefore the product of *CB* times the whole of *BE* will be known. Furthermore, *BC* is given by the hypothesis. Thus by Th. 17 above the whole line *BE* may be found. If you subtract the given base *BC* from this, the remainder *CE* will be known together with its half *CD*. Hence the shorter segment *CD* has been found. If the known base is added to this [segment], the longer segment *BD* results. Q.E.D.

The mechanics: Subtract the square of the shorter side from the square of the longer [side], and after the remainder is divided by the base, subtract the result from the base itself if the perpendicular falls within the triangle. But if the perpendicular falls outside the triangle, subtract the base from the result [of that division]. Furthermore [in either case] take half of this remainder to obtain the shorter segment. If to that [shorter segment] you add the quotient found from the division, you will have the longer segment, provided the perpendicular falls within the triangle. But if [the perpendicular] falls outside, add the shorter segment to the base, for the desired longer segment will result. For example, let side *AB* be 20, *AC* 13, and the base *BC* 21. The square of 13 is 169, and the square of 20 is 400. Subtract 169 from 400, leaving 231, which, divided by 21, leaves 11. This is subtracted from 21, and 10 is left. Half of 10 is 5, and this amount is the shorter segment. When 11 is added to that, 16 is found for the longer segment, provided the perpendicular is within [the triangle]. But if [the perpendicular] is outside, let side *AB* be 51, *AC* 25, and the base *BC* 38. These [values] do indeed indicate that the perpendicular falls outside the triangle. Subtract the square of 25, which is 625, from the square of 51 — namely 2601 — leaving 1976. This, divided by 38, gives 52, from which 38 is subtracted, leaving 14. Half of 14 — namely 7 — is the shorter segment. Adding 7 and 38 gives 45, and that amount is declared to be the longer segment.

[In other words,] the product of the difference of the sides times their sum is equal to the product* of the difference of the segments times their sum — namely, the base.

Theorem 45

Segments of this type may be calculated in other ways.

Let △ *ABC*, which is not a right triangle, have three unequal, known sides. From the vertex of ∠ *A*, let a perpendicular *AD* be drawn to the base *BC* without going beyond the area of the triangle. [The perpendicular] divides the base *BC* into two segments *BD* and *DC*, one of which is longer than the other, as we concluded in the preceding theorem, because of the inequality of sides *AB* and *AC*. We expect to measure these segments.

Using point *A* as the center and the length of the shorter side *AC* [as the radius], draw circle *HEC*. Its circumference will necessarily intersect both the base *BC* and the side *AB* because line *AC* is longer than the perpendicular *AD* but shorter than side *AB*. Therefore [the circle] intersects base *BC* at point *E*, which is to be joined to the center of the circle by line *EA*. In addition, [the circle] intersects line *AB* at point *H*.† Then let line *BA* be extended beyond the center of the circle until it meets the circumference at point *G*. Therefore, by [Euclid] III.3, line *ED* is equal to the shorter segment *DC*, and thus line *BE* is the difference of the segments. Furthermore, because the two lines *BG* and *BC*, drawn from point *B* marked outside the circle, intersect the circle, then by [Euclid] III.35 the product of *GB* and *BH* is equal to the product of *CB* and *BE*. But the product of *GB* and *BH* may be found by Th. 16 above, for line *GB* is known since it is equal to the two sides of the triangle, *AB* and *AC*, which were given by the hypothesis; and furthermore, *BH* — namely, the difference of the two sides — is known by the hypothesis and Th. 4 above.

Rectangulum here, as observed in note, p. 51, does not refer to any geometric form with one or more right angles; rather, just as the area of a rectangle is the product of its right-angled dimensions, so the term "rectangle" was sometimes used in the past to indicate the product of two factors. For example, "the rectangle which is contained under *A* and *B*" means "the product of *A* and *B*."

†In the figure, for line *bag* read *bhag*.

tam huius.quare & quod fub c b & b notum habebitur, cunqz lineam b e notã
fubiecerit hypothefis,erit per 17 huius linea b e nota,differentia fcilicet cafuum,
qua dempta ex bafi b c nota,relinquetur linea e c cognita, cuius medietas d c
eft cafus minor.Item cafui dicto minori lineam b e notam adijcias, & prodibit
cafus maior.Si uero perpendicularis extra triangulum ceciderit defcripto circulo
fuper capite ipfius perpendicularis fecundũ quantitatẽ lateris minoris cõtinueť
latus longius ultra centrum circuli,donec obuiabit circumferentiæ circuli in pun
cto g,quẽadmodum fupra fecimus.Extendatur deniqz bafis,ut in ea refidere pof-
fit perpendicularis demiffa, cõueniantqz
perpendicularis ipfa & bafis prolongata
in puncto d.non tamen ibi fiftat, fed pro
cedat ꝗ́ufꝗz offendet circuli circũferentiã
in puncto e, ducta femidiametro a e.pri
ftino igiť freti fyllogifmo declarabimus
quod fit ex e b in b c notum. cũqz ex hy
pothefi notam habeamus bafim b c, erit
per 17.huius linea e b (fumma uidelicet
duoꝝ cafuum) nota: dempta ergo bafi b

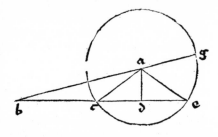

c nota per hypothefim ex b e iã inuenta, refidua c e nõ erit ignota. & eius medie
tas c d,cafus fcilicet minor.Item fi bafim b c cafui minori iam noto adiunxerimus,cafus maior b d notus conflabitur, quæ fuere depromẽda. ▽ Operatio.Ag
gregatum ex duobus lateribus per differentiam eoꝝ multiplica,producto'qz p ba
fim diuifo,quod exibit à bafi fubtrahas perpendiculari intra triangulẽcadente,
aut bafim ex eo quod diuifione facta elicitur minuas,refidui enim dimidiũ erit ca
fus minor quæfitus,cui fi id quod ex diuifione elicitũ eft,addideris perpendicula
ri intra cadente,aut ei bafim adieceris,fi extra ceciderit perpẽdicularis,cafum ma
iorem conftitues.Verbi gratia.Sit latus a b 25. a c 17.& bafis b c 28. congre-
gabo 25 & 17,refultabunt 42. differentia duoꝝ laterũ eft 8.multiplico igitur 42
per 8,producuntur 336.quæ diuido per 28.numerum fcilicet bafis,exeunt 12. fub
traho 12.ex 28.manent 16,quoꝝ medietas eft 8.Cafus ergo minor erit 8. Itẽ addo 12.ad 8.ueniunt 20.& tantus habebitur cafus maior.In hoc exemplo oportuit perpendicularem cadere intra triangulum. Sed offeratur mihi latus a b maius
20,a c 13.& bafis 11.quo fit ut perpendicularis a d triangulũ egrediatur.Sum-
ma duoꝝ laterum eft 33.differentia eoꝝ eft 7.multiplico 33. per 7. producuntur
231.quæ diuido per 11.numerum fcilicet bafis,exeunt 21.à quibus minuo 11.ma
nent 10.medietas de 10 eft 5.tantusqz numerabitur cafus minor.Item congrega-
bo numerum bafis 11.cum cafu minori 5.redduntur 16.pro cafu maiori. Tripli-
cem igitur huiufmodi cafus metiendi artem abfoluimus.Nunc quid utilitatis ipfi
afferant paucis lucubrabimus.

X L V I.

Perpendicularem à quouis puncto angulari ad oppofitum fibi latus
protenfam ex notis tribus trianguli lateribus reddere menfuratam.

Sit triangulus a b c,ex cuius puncto angulari a
defcendat perpendicularis a d ad bafim b c,fi in
tra triangulum ceciderit,aut ipfi bafi quantũ opor
tet continuate occurrens,fi extra triangulum profi
liet. Di-

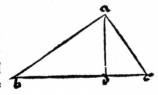

Therefore the product of *CB* and *B*** will also be known. And since the hypothesis gave line *BC*, by Th. 17 above line *BE* may be found — namely, the difference of the segments. When this difference is subtracted from the known base *BC*, line *EC* is left known, and its half *DC* is the shorter segment. Similarly, add known line *BE* to the known shorter segment, and the longer segment will be found.

However, if the perpendicular were to fall outside the triangle, then after the circle has been drawn with the top [*A*] of the perpendicular itself as center and with the value of the shorter side [as the radius], the longer side is extended beyond the center of the circle until it meets the circumference of the circle at point [*G*], just as we [drew it] above. Next, the base is extended so that the perpendicular, as drawn, can rest on it, and the perpendicular itself and the extended base meet at point *D*. However, [the base extended] does not stop there but proceeds until it hits the circumference of the circle at point *E*, [from which] radius *AE* is drawn. Therefore, using the previous reasoning, we will state that the product of *EB* times *BC* is known. And since by hypothesis the base *BC* is known, line *EB* (obviously the sum of the two segments) may be found by Th. 17 above. Then when the base *BC*, known by hypothesis, is subtracted from *BE* just found, the remainder *CE* is found and also its half *CD* — namely, the shorter segment. Similarly if the base *BC* is added to the now-known shorter segment, the longer segment *BD* is obtained. Q.E.D.

The mechanics: Multiply the sum of the two sides by their difference, and after the product is divided by the base, subtract the quotient from the base if the perpendicular falls within the triangle; otherwise subtract the base from the quotient. Half of the remainder will be the desired shorter

segment. If the quotient from the division is added to this [shorter segment] when the perpendicular falls within the triangle, or if the base is added to [the shorter segment] when the perpendicular falls outside, the longer segment is found. For example, if side *AB* is 25, *AC* 17, and the base *BC* 28, then add 25 to 17 to get 42. The difference of the two sides is 8. Then multiply 42 by 8, yielding 336, which, divided by 28 — namely, the value of the base — leaves 12. Subtract 12 from 28, leaving 16, of which half is 8. Therefore the shorter segment will be 8. Similarly, add 12 to 8, making 20, and this amount will be the longer segment. In this example the perpendicular necessarily fell within the triangle. But let the longer side *AB* be given as 20, *AC* as 13, and the base as 11, such that the perpendicular falls outside the triangle. The sum of the two sides is 33 and their difference is 7. Multiply 33 by 7, giving 231, which is divided by 11 — namely, the value of the base — and 21 results. From 21 subtract 11, leaving 10. Half of 10 is 5, and this amount will be the value of the shorter segment. Similarly, add the value of the base, 11, to the shorter segment, 5, and 16 is found for the longer segment.

Thus we have described three ways of measuring these segments. Now by means of a few theorems we will study how useful these are.

Theorem 46

When the three sides of a triangle are known, a perpendicular drawn from any vertex to the side opposite it can be measured.

When △ *ABC* is given, from whose vertex *A* perpendicular *AD* is drawn to the base *BC* if [the perpendicular] falls within the triangle or to meet the base extended as much as necessary if [the perpendicular] falls outside the triangle,

*For *B* read *BE*.

liet.Dico,ꝙ ipſa perpendicularis a d nota ueniet.Nam huiuſmodi perpendicula
ri deſcendente concludentur duo trianguli rectanguli, latus commune habentes,
ipſam ſcilicet perpendicularem,quorum ſiniſter ſiniſtrū trianguli propoſiti latus,
& caſum ſiniſtrum pro lateribus duobus reliquis accipit,dexter autē latus dextrū
cum caſu dextro,quemadmodū in figura apparet.per 26.igitur huius,quæ ex pe‑
nultima primi pendet,perpendicularis a d nota ueniet,latere a b noto exiſtente
per hypotheſim,caſu autem b d per quamlibet trium præmiſſarum.Idem efficies
ſi triangulo a d c rectangulo uſus fueris.   ꝟ Opus breue.Quadratum alterius
duorū caſuum ex quadrato lateris ſibi coterminalis minue,relicti enim radix qua
drata perpendicularem quæſitam manifeſtabit.

<h2 style="text-align:center">XLVII.</h2>

Si quis trianguli tria latera habuerit cognita, trium eius angulo‑
rum addiſcet quantitates.

De triangulis non rectangulis ſermo futurus habebitur,de rectangulis enim
ſuperius ſatis dixiſſe uidemur.Sint itaꝗ trianguli a b c tria latera nota, dico,ꝙ
tres eius anguli non latebunt.Demittatur enim ex puncto a ad baſim b c ꝑpen
dicularis a d,quæ cadat,ne intra triangulum an extra ſuperiores docuerunt. Ca

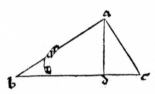

dat primo intra triangulum,erit autem ipſa nota ꝑ
præmiſſam.triangulus igī a d c rectangulus duo
latera a d & a c nota habens,per 27.huius angu‑
los ſuos acutos manifeſtabit . Similiter triangulus
rectangulus a b d notos habebit acutos angulos.
duobus autem angulis b & c cognitis,qui triangu
lis dictis rectangulis & propoſito noſtro triangulo
a b c cōmunes ſunt,tertius angulus b a c per 25.huius non ignorabitur.omnes
ergo angulos trianguli a b c notos effecimus. Cadat demum perpendicularis a
d extra triangulum,argumentis igitur priſtinis duo anguli a b d & a c d noti

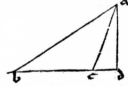

declarabuntur.cunꝗ ex 13 primi duo anguli a c d & a
c b duos rectos notos,per 21 & 3 huius ualeant,erit per
4 huius & angulus a c b cognitus, unde per 25 . huius
angulus b a c non ignorabitur.Poteris autem hæc bre‑
uius concludere,ſi loco perpendicularis duos caſus acce
peris.Nam propter duo latera a b & b d nota ex hypo
theſi,& aliqua trium concluſionū,quas de caſibus numerandis tradidimus, per 27
huius notus erit angulus b,ſimiliter propter latus a c notū ex hypotheſi , & ca‑
ſum d c ſuperius numeratum,angulus c patefiet,duo autem anguli b & c co‑
gniti,ſocium ſuum angulum a per 25 huius ſuſcitabunt.   ꝟ Ne autem æquo di
utius obfundaris,operationem duabus ex rebus colliges , nam perpendicularem
ex præmiſſa,aut utrunꝗ duorum caſuum ex aliqua trium præcedentium numera
bis.angulos autem triangulorum partialium rectangulorum,qui & triangulo ꝓ‑
poſito cōmunes habentur,27 huius non ſinet ignotos,qui tandem 25 huius diri‑
gente,tertium angulum elicient menſuratum.

<h2 style="text-align:center">XLVIII.</h2>

Duobus trianguli notis angulis, laterum proportiones inuicem
cognitum iri.

Trian‑

then perpendicular *AD* itself will become known.

The drawing of this perpendicular creates two right triangles having a common side — namely, the perpendicular itself. Of these [triangles], the one on the left has the left side of the given triangle and the left-hand segment as its two other sides; furthermore, the [triangle] on the right has the right side with the right-hand segment, just as it appears in the figure. Then by Th. 26 above, which is based on the penultimate theorem of [Euclid] I, perpendicular *AD* will become known because side *AB* is known by hypothesis and also because segment *BD* may be found by any one of the three preceding [theorems]. You would accomplish the same thing if you were to use right △ *ADC*.

The mechanics briefly: Subtract the square of one of the two segments from the square of the side adjacent to it. The square root of the remainder is the desired perpendicular.

Theorem 47

If three sides of any triangle are known, the sizes of its three angles may also be found.

The following discourse will concern non-right triangles, for we seem to have said enough above about right triangles. Therefore, if three sides of △ *ABC* are known, then its three angles may be found. Let perpendicular *AD* be dropped from point *A* to the base *BC*. Whether it falls inside or outside [the triangle] the above [theorems] have shown.

First, if the perpendicular falls within the triangle, [the perpendicular] will be known by the preceding [theorem]. Therefore since right △ *ADC* has two known sides *AD* and *AC*, then by Th. 27 above its acute angles will be found. Similarly, right △ *ABD* will have known acute angles. Because the two

angles *B* and *C*, which are common to the mentioned right triangles and our given △ *ABC*, are known, then by Th. 25 above the third ∠ *BAC* will be found. Therefore we have determined all the angles of the triangle.

Finally, if the perpendicular *AD* falls outside the triangle, then by the preceding arguments the two angles *ABD* and *ACD* will be declared known. And since from [Euclid] I.13 the two angles *ACD* and *ACB* have the value of two right angles, known by Ths. 3 and 21 above, then by Th. 4 above ∠ *ACB* will be known. Hence by Th. 25 above, ∠ *BAC* may be found.

You can conclude this even more briefly if you consider the two segments in place of the perpendicular. Because the two sides *AB* and *BD* are known from the hypothesis and one of the three conclusions that we drew concerning determination [of] the segments, then, by Th. 27 above, ∠ *B* will be known. Similarly, because side *AC* is given by the hypothesis and segment *DC* is measured as before, ∠ *C* will be revealed. Furthermore, with the two angles *B* and *C* known, their supplementary ∠ *A* will be found.

Not to prolong this matter more than [is] reasonable, you will derive the mechanics from two things, for you will calculate either the perpendicular, from the preceding [theorem], or each of the two segments, from one of the three foregoing [theorems]. Furthermore, Th. 27 above will determine the angles of the right-triangle portions, and these [angles] are common to the given triangle also. Finally, by the directions of Th. 25 above, these [angles] will make the third angle known.

Theorem 48

If two angles of a triangle are known, the ratios of the sides to one another can be found.

Trianguli a b c duos angulos ſupponamus da-
tos.Dico,ꝗ quorumlibet duorum laterum proportio
nota uidebitur.A communi enim termino duorum la
terum,quorum proportionem eniti uoles,demitte perpendicularem uerſus reliquũ latus, quæ cadat ne intra
an extra triangulum ex angulis,quos baſis ſuſtentat,fa
cile doceberis,illud autem proceſſum noſtrũ non uariabit,erit enim trianguli par
tialis a b d rectanguli angulus b notus ex hypotheſi,iuuãte 25 huius,ſi opus fue
rit,quare per 30 huius proportio a b ad a d nota accipietur. ſimili argumento
proportio a c lateris ad eandem perpendicularem nota
declarabitur.utroꝗ igitur laterum a b & a c ad ppendicularem a d proportionem habente,erit per 12.huius
eorum inter ſe proportio nota.Similiter procedemus cir
ca duo latera quæcunꝗ elegerimus,quod erat abſoluen-
dum.   ꝕ Operationi autem immorari non eſt conſiliũ
ipſam enim ex operibus allegatorum theorematum facile comparabimus.

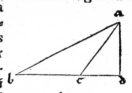

XLIX.

Si duo latera trianguli data notum ambiant angulum,reliquos angulos reſiduumꝗ latus dimetiri.

Sint duo latera a b & b c trianguli a b c,nota cum angulo quem ambiunt
a b c.Dico,ꝗ latus a c notum erit cum duobus angulis reliquis.Demittã enim
à uertice anguli ignoti perpẽdicularem ad latus ſibi oppoſitum, quæ uerbi gratia
ſit a d .nondum autem ex hypotheſi noſtra ſcirè poterimus,cadat ne perpendicu
laris illa intra triangulum an extra,hoc enim non ſtatim conſequitur noticiã anguli,quem duo data ambiunt latera,nihilominus metam attingemus cupitam,&
qua lege perpendicularis ipſa incedat explorabimus.Cum igitur triangulus a
b d rectangulus angulum b acutum habeat datum ex
hypotheſi cum latere a b,erit per 29 huius utraꝗ linea
rum a d & b d cognita reſpectu lateris a b .ſi itaꝗ b
d iam inuẽta per ſyllogiſmũ minor reperietur baſi b c
nota per hypotheſim,perpendicularem intra triangulum cadere nemo dubitabit.ſi uero maior fuerit baſi b
c ,cadet extra,& ſi æqualis, coincidet perpendicularis a d cum latere a c, eritꝗ
ppter hoc triangulus a b c rectangulus.Sit ergo b d caſus primo minor baſi b
c data,quo ablato ex b c nota,relinquetur per 4 huius linea d c cognita. cunꝗ
iam pridem a d perpendicularem notam concluſerimus,habebit triangulus a d
c rectangulus duo latera a d & d c nota,quare per 26 huius latus eius a c notũ
elicietur,quod & triangulo noſtro cõmune eſt,ſed & angulus eius acutus a c d ar
gumento 27 huius inuenietur.duobus autem angulis b & c cognitis,tertius angulus a per 25 huius latere non poterit.Quòd ſi linea
b d maior occurrat baſi b c,dempta ipſa baſi nota p
hypotheſim ex linea b d inuenta per argumentationem,manebit c d nota,deinde ut prius linea a c nota prodibit cum angulo a c d ,quem ſi ex duobus rectis abſtuleris,relinquetur per 13 primi & 4 huius angulus a c b cognitus,tandemꝗ ex 25 huius angulus

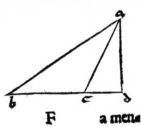

F   a mẽſu

If two angles of △ *ABC* are given, then the ratio of any two sides will be known.

From the point of intersection of the two sides whose ratio is sought, drop a perpendicular to the third side. Whether this [perpendicular] falls inside or outside the triangle can easily be determined from the base angles. Furthermore, that will not change our procedure, for ∠ *B* of right-triangle portion *ABD* will be known by the hypothesis, with the assistance of Th. 25 above if necessary. Therefore by Th. 30 above the ratio of *AB* to *AD* will be found. By a similar argument the ratio of side *AC* to the perpendicular is declared known. Thus because each of the sides *AB* and *AC* has a ratio to the perpendicular *AD*, then by Th. 12 above their ratio to each other is known. We proceed similarly for any two sides we choose. This was to be proven.

The mechanics: No need to tarry here since the mechanics are obvious from the cited theorems.

Theorem 49

If two sides of a triangle and their included angle are given, the other angles and the other side can be found.

If two sides *AB* and *BC* of △ *ABC* are given with their included ∠ *ABC* known, then side *AC* will be known along with the two other angles.

Drop a perpendicular — *AD*, for example — from the vertex of unknown angle [*A*] to the opposite side. We cannot yet know from our hypothesis whether that perpendicular falls within the triangle or outside, for this does not follow immediately from the identity of the angle included by the two given sides. Nonetheless, the desired goal can be reached, and in which way the perpendicular falls can be investigated. Then, since right △ *ABD* has acute ∠ *B* given from the hypothesis together with side *AB*, each of the lines *AD* and *BD* is known, by Th. 29 above, in the same terms as line *AB*. Thus if *BD*, already found by deduction, is discovered to be less than the base *BC*, known by hypothesis, then no one will doubt that the perpendicular falls within the triangle. However, if [*BD*] is longer than the base *BC*, [the perpendicular] falls outside; and if [*BD*] is equal to [*BC*], the perpendicular *AD* coincides with side *AC*, and because of this, △ *ABC* will be a right triangle.

Therefore, first let the segment *BD* be shorter than the given base *BC*. When [*BD*] is subtracted from known *BC*, line *DC* may be found by Th. 4 above. And because we determined perpendicular *AD* already, right △ *ADC* will have two known sides, *AD* and *DC*. Hence, by Th. 26 above, its side *AC* will be found. This [side] is common to our triangle also, and its acute ∠ *ACD* may be found by the reasoning of Th. 27 above. With the two angles *B* and *C* known, the third ∠ *A* can be found by Th. 25 above.

But if line *BD* is longer than the base *BC*, subtract the base, known by hypothesis, from line *BD*, found by deduction, and *CD* is left known. Thereupon, as before, line *AC* is found together with ∠ *ACD*. If this [angle] is subtracted from two right angles, then, by [Euclid] I.13 and Th. 4 above, ∠ *ACB* is left known. And finally from Th. 25 above ∠

a mensuratus emerget,quæ fuere lucubranda. ▼ Operationis uero tenorē mis=
sum facimus,qui ad allegata theoremata confugienti ultro se ingeret.

L.

Si alterum ex duobus notis lateribus trianguli, angulo obtuso da=
to opponatur,& latus & angulos reliquos non ignorabit Geometra.

Duo latera a b & a c trianguli a b c nota sint,quorum alterum scilicet a b
opponatur angulo a c b obtuso dato.Dico,ꝙ & latus b c cognitum ueniet cū
duobus angulis a & b.Ex termino enām communi datorum laterū descendat p̄=
pendicularis a d,concurrens cum basi b c,quantum oportet prolongata in pun
cto d,ipsam enim extra triangulum cadere cogit 31 huius.Triangulus itaꝗ a c

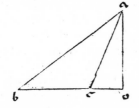

d angulum a c d notum habebit,ipse enim a c d an
gulus cum angulo a c b per hypothesim noto duo=
bus rectis æquiualent.cūꝗ latus a c trianguli recta̅n
guli prædicti notū sit ex hypothesi,erit per 29 huius
utraꝗ linearum a d & d c nota respectu lineæ a c.
item triangulus a b d rectangulus duo latera a b &
a d nota habens, a b quidem per hypothesim, a d ue
ro per argumentationē iam factam,ex 26 huius & 27

latus suū b d notum afferet cum angulo b,quare dempta c d prius cognita ex
tota b d iam nota,relinquetur basis b c nō ignota.duo autem anguli b & c tri
anguli a b c,tertium angulū a per 25 huius excitabunt.Verum igitur enuncia
bat theorema nostrum. ▼ Operationem ex eis,quæ circa triangulos rectan=
gulos tꝛ ꝯdidimus,facile colligemus.

LI.

Trianguli duo latera data cum angulo acuto, cui alterum eorum
opponitur,ad latus & angulos reliquos cognoscendos nequaꝗ suffi=
cere.Verū qua lege perpēdicularis cadat,si callebimus,oīa patesient.

In hac re accusanda uenit infirmitas anguli acuti,qui nequit docere cadat ne
perpendicularis intra triangulum propositum an extra,quod obtusus angulus in
præmissa indicabat.Nam sit triangulus a b c,cuius duo latera a b quidem ma=

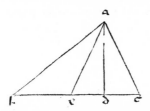

ius,& a c minus sint data cū angulo b acuto,sit=
ꝗ angulus c acutus non datus,& à puncto a de=
mittatur a d perpendicularis ad basim, quæ per
31 huius cadet intra triangulum.ex processu autē
45 huius,casus b d maior erit casu d c,abscinda
tur ergo ex b d linea e d æqualis ipsi d c,ducta
linea a e,quæ per quartā primi æqualis erit lineæ
a c.Quāuis itaꝗ latera a b & a c trianguli a b

c data sint,& æqualia duobus lateribus a b & a e trianguli a b e,angulus autē
b datus communis ambobus triangulis,tamen bases eorum uarie sunt & reliqui
anguli.Ad præcognitionem igitur duorū laterum & unius anguli acuti,cui alte
rū eorum opponitur,non ligatur noticia reliqui lateris & angulorū reliquorum,
quod pollicebatur theorema nostrum. ▼ Vt autem latus & angulos reliquos
agnoscamus,præsciendum est,qua lege perpendicularis à communi termino da=
torum laterum exorta cadat.Si enim intra triangulum ceciderit,triangulus a b d
                                                                    rectangulus

A is determined.

The mechanics: These can be by-passed by anyone who will refer to the cited theorems.

Theorem 50

If one of two known sides is opposite a given obtuse angle, both the [other] side and the remaining angles can be found geometrically.

If two sides *AB* and *AC* of △ *ABC* are known and one of them — namely *AB* — is opposite a given obtuse ∠ *ACB*, then both side *BC* and the two angles *A* and *B* can be found.

From the point of intersection of the given sides, let perpendicular *AD* be dropped, meeting base *BC*, extended as much as necessary, at point *D*, for by Th. 31 above [the perpendicular] is known to fall outside the triangle. Therefore △ *ACD* will have ∠ *ACD* known, for ∠ *ACD* plus ∠ *ACB*, known by the hypothesis, equals two right angles. And since side *AC* of the mentioned right triangle is known from the hypothesis, by Th. 29 above each of the lines *AD* and *DC* will be known in the same terms as line *AC*. Similarly, from Ths. 26 and 27 above, right △ *ABD*, having two known sides *AB* and *AD* — *AB* [known] from the hypothesis and *AD* [known] by the previous reasoning — will have its side *BD* known together with ∠ *B*. Therefore when *CD*, previously found, is subtracted from the already-known, whole [line] *BD*, the base *BC* is left known. Furthermore the two angles *B* and *C* of △ *ABC* reveal the third ∠ *A* by Th. 25 above. Q.E.D.

The mechanics for this are easily gathered from what has already been said about right triangles.

Theorem 51

When two sides of a triangle are given with an acute angle opposite one of these [sides], there is not enough [information given] to find the [other] side and the remaining angles. However, if we know which way the perpendicular falls, then all can be found.

The weakness of the acute angle must be blamed in this [theorem], for [it] cannot show whether the perpendicular falls inside the given triangle or outside, as the obtuse angle in the preceding theorem indicated. Now [in] △ *ABC*, whose two sides — *AB*, the longer, and *AC*, the shorter — are given together with acute ∠ *B*, let ∠ *C*, not given, be acute. From point *A* let a perpendicular *AD* be dropped to the base. By Th. 31 above [the perpendicular] falls within the triangle. Furthermore from the procedure of Th. 45 above, segment *BD* will be longer than segment *DC*. Thus on [line] *BD* mark off line *ED* equal to *DC*. Draw line *AE*, which by [Euclid] I.4 will be equal to line *AC*. Then although sides *AB* and *AC* of △ *ABC* are given and are equal to the two sides *AB* and *AE* of △ *ABE*, and although the given ∠ *B* is common to both triangles, nevertheless the bases and the remaining angles of these [two triangles] are different. Therefore the identity of the other side and the other angles is not [directly] connected with the foreknowledge of two sides and one acute angle that is opposite one of the [two sides]. This is what [the first statement] in our theorem promised.

For us to find the side and the other angles, it has to be known which way the perpendicular, drawn from the point of intersection of the given lines, falls. For if it falls within the triangle, right △ *ABD*

rectangulus habebit latus a b notum ex hypothesi cum angulo acuto b, quare p
29 huius tam perpendicularis a d nota uidebitur, ꝗ casus b d. Ex duobus autē
lateribus a d & a c notis, trianguli a d c rectanguli per 26 & 27 huius linea d
c inuenietur cum angulo c .duas igitur lineas b d & d c iam singulatim notos,
si congregabimus, tota basis b c per 3 huius mensurata ueniet. duo etiam anguli
b & c noti tertium angulum a 25. huius dirigente, cognitum elicient. Quòd si
perpendicularis a d extra triangulum ceciderit, oportuit angulum a c b esse ob
tusum, præmissam igitur consulendo, & basim b c & angulos reliquos trianguli
nobis propositi metiemur.   ⅴ Opus autem cum & facile sit, & ex superioribus
pendeat, missum facio.

<h3 style="text-align:center">LII.</h3>

**Si latus trianguli datum duos sustentet notos angulos, reliqua duo
latera non erunt ignota.**

Sit triangulus a b c, latus a b datum habens, cui insideant duo anguli a b c
& b a c noti. Dico, ꝗ duo eius latera reliqua fient cognita. Pró angulo autem ter
tio mensurando, non est terendus dies, quem 25. huius exemplo notū afferet. Ab
a termino lateris a b dati exoriaꝓ ppendicularis ad basim b c ignotā, quæ cadat
ne intra triangulum an extra anguli, quibus subiacet basis, per 31 . huius edoce=
bunt. Cadat igitur prius intra triangulnm, trian=
gulus itaꜗ a b d rectangulus angulum b acutū
habens datum cum latere a b per 29. huius, reli=
qua duo latera sua a d & d b cognita afferet re=
spectu lateris a b. item triangulus a d c rectangu
lus cum latus a d iam notum habeat, pari ratio=
ne sua latera a c & c d, manifestabit respectu per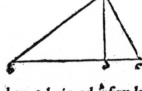
pendicularis a d notæ. cūꜗ perpendicularis a d
data sit ad latus a b, erit & utraꜗ linearum a c & a d per 8. huius ad ipsum latus
a b data. latus ergo a c trianguli propositi notum effecimus cū duobus casibus
b d & d c. quibus collectis b c basis cognita resultabit. Sed cadat perpendicula=
ris a d extra triangulum, eritꜗ per media prætacta
trianguli rectanguli a b d utrunꜗ latus a d & d b
notum respectu a b. item triangulo a c d rectangu=
lo angulum a c d notum habente propter duos an=
gulos b a c & a b c notos ex hypothesi, quibus ipse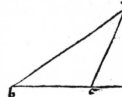
per 32. primi æquipollet, cū latere a d, per 29. huius,
reliqua duo sua latera a c & c d cognita habebunꝰ.
Sic ergo latus a c trianguli a b c propositi notū erit
cum duobus casibus b d & d c, quorum minor ex maiore demptus relinquet ba=
sim b c notā, quod uolebamus explanare.   ⅴ Operatio autem ex allegatis com
parabitur theorematibus .

<h3 style="text-align:center">LIII.</h3>

**Latus trianguli notum, quod alteri duorum datorum angulorum op
ponitur, reliqua suscitabit latera.**

In triangulo a b c latus a b notū angulo a c b dato opponatur, sitꜗ angu=
lus a b c datus. Dico ꝗ reliqua duo latera non erunt ignota. De angulo autē ter

F 2   tio

will have side *AB* known by hypothesis together with acute ∠ *B*. Then, by Th. 29
above, as perpendicular *AD* is found, so segment *BD* will be found. Furthermore, because
the two sides *AD* and *AC* are known, line *DC*
as well as ∠ *C* of right △ *ADC* will be found
through Ths. 26 and 27 above. Hence, if we
add the two lines *BD* and *DC*, now individually known, then, by Th. 3 above, the
total base *BC* will be measured. Also the two
known angles *B* and *C* reveal the third ∠ *A*
by the directions of Th. 25 above.

But if the perpendicular *AD* falls outside
the triangle, ∠ *ACB* is necessarily obtuse.
Therefore, by consulting the previous theorem, we can measure the base *BC* and the
other angles of our given triangle.

Since the mechanics are so simple and de-
pend on the previous work, I pass them by.

Theorem 52

If the given side of a triangle bears two
known angles, the other two sides can be
found.

If *ABC* is a triangle with a given side *AB*
on which two known angles *ABC* and *BAC*
rest, then the other two sides may be found.

The day is not to be wasted measuring the
third angle, which Th. 25 above makes known
forthwith.* From end *A* of given side *AB*,
let a perpendicular be drawn to unknown
base *BC*. By Th. 31 above the base angles
will show whether [the perpendicular] falls
within the triangle or outside.

As before, let it fall within the triangle.
Then △ *ABD* is a right triangle having acute
∠ *B* given along with side *AB*. By Th. 29
above, its other two sides *AD* and *DB* will
be known in terms of line *AB*. Similarly, since

right △ *ADC* has side *AD* already known,
by comparable reasoning its sides *AC* and *CD*
will be found in terms of the known perpendicular *AD*. And since perpendicular *AD* is
given in terms of side *AB*, by Th. 8 above
each of the lines *AC* and *AD*† will also be
given in terms of side *AB*. Hence side *AC* of
the given triangle becomes known from the
two segments *BD* and *DC*. When these are
added, the base *BC* will be found.

But if the perpendicular *AD* falls outside
the triangle, then by the aforementioned
method each of the sides *AD* and *DB* of right
△ *ABD* will be known in terms of *AB*. Similarly, in right △ *ACD*, which has side *AD*
[and] ∠ *ACD* known — [the latter] because
the two angles *BAC* and *ABC* are known
from the hypothesis and, by [Euclid] I.32,
[∠ *ACD*] is equal to [the sum of] those [angles
BAC and *ABC*] — by Th. 29 above its other
two sides *AC* and *CD* will be found. Thus
side *AC* of the given △ *ABC* will be known
together with the two segments *BD* and *DC*.
When the shorter of those [segments] is subtracted from the longer, the base *BC* is left
known. Q.E.D.

The mechanics are provided from [those
of] the cited theorems.

Theorem 53

If in a triangle the side opposite one of
two given angles is known, the other sides
can be found.

If in △ *ABC* side *AB*, opposite given ∠
ACB, is known, and if ∠ *ABC* is [also] given,
then the other two sides can be found.

Regarding the third angle,

*In the Latin text, for *exemplo* read *extemplo*.
†For *AD* read *CD*.

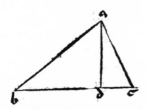

tio, quemadmodum in præcedenti, nihil ambigui nos uexabit. Lateri duobus datis subiacenti angulis insistat perpendicularis a d ex puncto angulari sibi opposito sumens originem, quæ qualiter cadet anguli dati 3 1. huius commonente exploratū reddent. Cadat igitur primo intra triangulū, ut si at triangulus a b d rectangulus, qui cū habeat latus a b notum cum angulo a b d acuto, habebit

& latus c d cognitum cum latere suo c d, 29. huius dirigente. Similiter triangulus a d c, cum & latus a d iam notum habeat, & angulū a c d ex hypothesi datū, duo latera sua reliqua a d & d c manifestabit. omnes igitur lineas b d, a d, a c & d c mensura una, per quā a b nota supponebatur, secundū numeros metietur notos, quare etiā linea b c ex duabus b d & d c notis resultās non erit incognita. Sed cadat perpendicularis a d extra triāgulum concurrens cum basi b c, quātum sat est

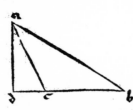

continuata. Cum itaqᵍ angulus a b c notus supponatur cum latere a b, erit per 29. huius triāgulo a b d rectangulo existente, utraqᵍ linearum a b & d b nota. Rursus in triāgulo a d c rectangulo, cum latus a d iā sit notum, angulus autem a c d propter uicinum suū a c b ex hypothesi notum 13. primi arguente, notus habeatur, erit per 29. huius utraqᵍ linearum a c & c d respectu lineæ perpendicularis a d, & ideo respectu a

b cognita, postqᵍ autem lineā c d ex d b minuemus, relinquetur basis b c cognita. reliqua igitur trianguli a b c latera, noticiæ nostræ subiecimus, quod expectabat ostendendum. ▼ In his autem postremis conclusionibus operationem missam facere libuit, ne sermone obtunderemus nimio, quā quidem operationem, si colligere uoles, ad 26. 27, & 29. huius refugias.

LIIII.

Trium laterum trianguli inter se datis proportionibus, omnes eius angulos mensurare.

Latus a b trianguli a b c, tā ad latus eius a c, qᵍ ad b c proportionem habeat datā. Dico, qᵒ omnes anguli eius noti fient. Pro libito enim ipsum latus a b in quᵗ

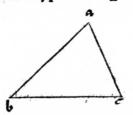

libet partes scindas, quarum una hoc in proposito mensura communis habebitur. cum itaqᵍ a b latus sit notum per mensurā huiusmodi, erunt per 6. huius reliqua duo latera nota per eandem mensurā, & ideo per 48. huius triāguli a b c omnes anguli patefient, quod erat ostendendum. ▼ Operatio autem huius, postqᵍ unum latus quodcunqᵍ tanqᵍ numerum notum pro libito posueris, & reliqua in

de latera per opus sextæ huius didiceris, ab operatione 48. huius nō discrepat.

LV.

Tribus angulis trianguli cuiuslibet datas, inter se proportiones habentibus, unusquisqᵍ eorum cognitus habebitur, latera quoqᵍ inter se proportiones accipient notas.

Sit trian

just as in the preceding theorem, no ambiguity will trouble us. Let perpendicular *AD*, originating at the vertex [*A*] opposite [*BC*], rest on the side [*BC*] with the two given angles. The perpendicular falls the way the given angles indicate, in accordance with the advice of Th. 31.

First, then, let it fall within the triangle, so that △ *ABD* is made a right triangle. Since [△ *ABD*] has side *AB* known together with acute ∠ *ABD*, side *CD** as well as side *CD*† will be known by the directions of Th. 29 above. Similarly, △ *ADC*, since it has side *AD* already known and ∠ *ACD* given by the hypothesis, will reveal its other sides *AD*‡ and *DC*. Therefore the one measure, by which *AB* was given known, measures all the lines, *BD*, *AD*, *AC*, and *DC*, a known number of times. Therefore line *BC* also becomes known from the two known [lines] *BD* and *DC*.

But [now] let the perpendicular *AD* fall to meet the base *BC*, extended as much as is necessary, outside the triangle. Since ∠ *ABC* is given known together with side *AB*, then by Th. 29 above, since △ *ABD* is a right triangle, each of the lines *AB*§ and *DB* will be known. Again, since side *AD* is now known and since ∠ *ACD* is also known because its supplement is known from the hypothesis by the reasoning of [Euclid] I.13, then by Th. 29 above each of the lines *AC* and *CD* in right △ *ADC* will be found in terms of perpendicular line *AD*. Therefore they will also be known in terms of line *AB*. After we subtract line *CD* from *DB*, the base *BC* is left known. Therefore we have determined the other sides of △ *ABC*, as was promised.

The mechanics in these last conclusions will be by-passed, lest this discussion be too prolonged. But if reference is needed, then consult Ths. 26, 27, and 29 above.

Theorem 54

If three sides of a triangle are in known ratios to each other, all the angles can be measured.

If in △ *ABC* side *AB* has a given ratio to side *AC* as well as to side *BC*, then all the angles may be found.

Side *AB* may be divided arbitrarily into any number of portions, one of which will be the common measure in this theorem. Therefore, since side *AB* must be known by this measure, then by Th. 6 above the other two sides will be known by this same measure, and thus by Th. 48 above all the angles of △ *ABC* will be found. Q.E.D.

The mechanics: After you have arbitrarily assigned a known value to any one side, then you can find the other sides by the mechanics of Th. 6 above. [The rest of the mechanics are] no different from the mechanics of Th. 48 above.

Theorem 55

If in any triangle the ratios of the three angles to one another are given, any one of [the angles] may be found, and the ratios of the sides to each other will also be known.

*For *CD* read *AD*.
†For *CD* read *BD*.
‡For *AD* read *AC*.
§For *AB* read *AD*.

Sit triangulus a b c, cuius tres anguli inter se proportiones habeāt datas. Dico, ꝙ quilibet eorum notus erit, proportionesq́ꝫ laterum inter se datas fieri oportebit. Quoniā enim anguli a ad angulum b data est proportio, sit ipsa ut numeri h ad numerum k (necesse em̄ est eā in numeris notis inueniri) & ꝓportio anguli b ad angulum c nota. tanq̄ꝫ numeri k ad numerum l. cum itaqꝫ proportio anguli a ad angulum b, sit sicut numeri h ad numerum k, erit coniunctim a & b angulorum ad angulum b sicut h & k numerorum ad k numerum, est autē b ad c sicut k ad k. per æquā igit̄ a b ad c tanq̄ꝫ h k ad l, & coniunctim a b c ad c sicut h k l ad l. aggregati itaqꝫ ex tribus angulis a b c ad angulum c est sicut aggregati ex tribus numeris h k l ad numerum l proportio. summa igitur trium angulorum a b c ad ipsum angulum c proportionē habet notā, summa autem huiusmodi per 32. primi duobus rectis æquipollet, quos oportet esse notos ex supradictis, unde & summa trium angulorum dictorum nota conuincitur, per sextam ergo huius angulus c notus declarabitur, qui cum ad reliquos angulos datas habeat proportiones, erunt & reliqui anguli ꝑ 6. huius noti. quare per 49. huius tria latera trianguli a b c proportiones inter se notas accipient, ambas ergo theorematis partes satis ostēdisse uidemur ◤ Operatio. Tribus angulis tres accomodabis numeros, ita ꝙ primus angulus ad secundum se habeat sicut numerus primus ad secundum, secundus uero angulus ad tertium, sicut secundus numerus ad tertium, quod facile fiet, si operationem 6. huius consulueris, quo facto, summabis dictos tres numeros, & collectum ex eis numerum pro primo statuas, numerum uero anguli, quem nosse desideras, pro secundo, & numerum duorum angulorum rectorum pro tertio. multiplicando itaqꝫ secundum per tertium, & diuidendo in primum, exibit quantitas anguli quæsiti. Vt si proportio anguli a ad angulum b fuerit sicut 10 ad 7. anguli autem b ad angulum c sicut 7 ad 3, colligo tres numeros 10. 7. & 3. fiunt 20. pro primo numero. libeat autem inuenire angulum a, cuius numerus est 10. quem pro secundo, numerus autem duorum rectorum usitatus est 180. multiplico igitur 180. per 10. producuntur 1800. quæ diuido per 20. exeunt nonaginta. angulum igitur a inuenio 90. gradus continentem, ut duo recti sunt 180. quare & ipse rectus habebitur. Repertis autem angulis ad operationem 49. huius confugiendum erit, ut proportiones laterum addiscamus.

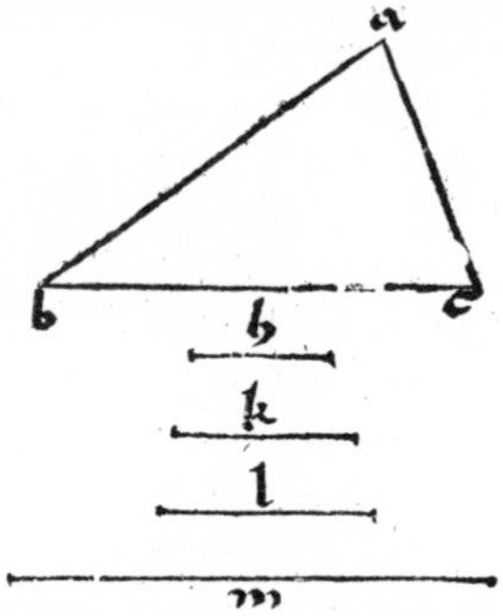

LVI.

Data proportione duorum laterum, unoq́ꝫ angulo cognito, reliqui duo anguli noti fiēt. Vnde utriusqꝫ dictorum laterum ad tertium proportio non latebit.

Sit proportio a b lateris ad b c latus trianguli a b c data, cum uno angulo quocunqꝫ. Dico, ꝙ reliqui duo anguli noti uenient, & proportio utriusqꝫ dictorum laterum ad tertium latus data erit. Diuiso enim ad libitum latere a b in quotlibet partes, una earum tanq̄ꝫ mensura communi utemur, quæ quoties in latere b c contineatur, sexta huius edocebit.

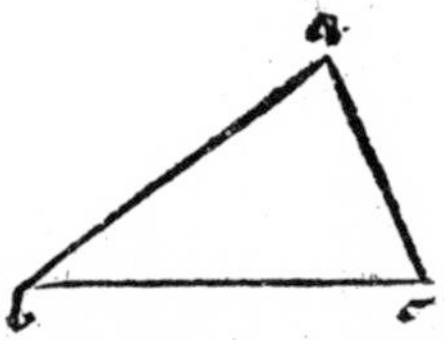

F 3 duo

If *ABC* is a triangle whose three angles have given ratios to each other, then any one of [the angles] will be known and the ratios of the sides to each other will necessarily be found.

Because the ratio of ∠ *A* to ∠ *B* is given, let it be [represented] as [the ratio] of number *H* to number *K* (for it is necessary that [the ratio] be expressed in known numbers). And let the ratio of ∠ *B* to ∠ *C* be known as [the ratio] of number *K* to number *L*. Then since the ratio of ∠ *A* to ∠ *B* is as that of number *H* to number *K*, by addition [the ratio] of angles *A* plus *B* to ∠ *B* will be as that of numbers *H* plus *K* to number *K*. Moreover, *B* is to *C* as *K* is to *K*.* By simultaneous solution, therefore, *A* plus *B* to *C* is as *H* plus *K* to *L*, and by addition *A* plus *B* plus *C* to *C* is as *H* plus *K* plus *L* to *L*. Thus the ratio of the sum of the three angles *A*, *B*, and *C* to ∠ *C* is as [the ratio] of the sum of the three numbers *H*, *K*, and *L* to number *L*. Hence [the ratio] of the sum of the three angles *A*, *B*, and *C* to ∠ *C* itself is a known ratio; moreover, by [Euclid] I.32 this sum is equal to two right angles, which must be known from what was said earlier. Hence the sum of the three mentioned angles is known conclusively, and then by Th. 6 above ∠ *C* may be found. Since [∠ *C*] has given ratios to the other angles, the other angles will also be found by Th. 6 above. Then by Th. 49 above the three sides of △ *ABC* will have known ratios to each other. And thus both parts of the theorem have been proven.

The mechanics: Assign three numbers to the three angles such that the first angle is to the second as the first number is to the second, and the second angle is to the third as the second number is to the third. This is easy if Th. 19 above is consulted. After this has been done, add the three mentioned numbers, and take the sum of these numbers for the first [term], the number of the angle which you wish to make known for the second [term], and the value of two right angles for the third [term]. By multiplication of the second term by the third and division of [that product] by the first term, the value of the desired angle will be found. For example, if the ratio of ∠ *A* to ∠ *B* is as 10 to 7 and [that] of ∠ *B* to ∠ *C* is as 7 to 3, add the three numbers, 10, 7, and 3, to get 20, which is the first [term]. Let it be desired to find ∠ *A*, whose number is 10, which [is taken] for the second term. Furthermore, the customary value of two right angles is 180. Then multiply 180 by 10, yielding 1800, which you divide by 20, and 90 results. Thus, ∠ *A* is found to be 90° when two right angles are [expressed as] 180. Therefore the angle is a right angle. When the angles have been found, refer to the mechanics of Th. 49 above to determine the ratios of the sides.

Theorem 56

When the ratio of two sides is given and any one angle is known, the other two angles may be found. Hence the ratio of each of the mentioned sides to the third will be disclosed.

If the ratio of side *AB* to side *BC* of △ *ABC* is given together with any one angle, then the other two angles may be found and the ratio of each of the given sides to the third side will be determined.

When side *AB* is arbitrarily divided into any number of portions, one of them is such that we may use it as a common measure. Th. 6 above will show how many times [this measure] is contained in side *BC*.

*For *K* read *L*.

duo igitur latera a b & b c inter se data habebuntur., cunq angulus unus qui
cunq per hypothesim datus sit, erit per 50.51.aut 52.huius reliquum latus cogni
tum, quare per diffinitionem laterum omniu inter se proportiones habebimus da
tas, sed & per easdem reliqui anguli mensurabunt, quod intendebamus cocludere.

¶ Operationem ex allegatis comparabimus locis, si prius alterum duorum la
terum tanq notum constituerimus.

LVII.

Datis proportionibus duorum angulorum, utriusq uidelicet seor
sum ad rectum angulum, unoq latere quolibet cognito, omnes angu
li cum reliquis lateribus cognoscentur.

Vtriusq duorum angulorum a & b ad rectum data sit proportio, sitq latus
a b aut aliud quodcunq cognitum.Dico, ꝙ omnes anguli trianguli a b c noti

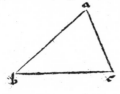

fient cum lateribus. Erit enim per sextam huius uterq
dictorum angulorum cognitus, recto per 21. huius no
to existente, quare per 53.& 54.huius, quod reliquu est,
absoluemus. ¶ Operationi autem nihil loci damus,
ꝙ ipsa ex supradictis facile decerpatur.

FINIS.

LIBER SECVNDVS
TRIANGVLORVM.

I.

In omni triangulo rectilineo proportio lateris ad latus est, tanq si
nus recti anguli alterum eorum respicientis, ad sinum rectum anguli
reliquum latus respicientis.

Sinum anguli, ut alibi uocamus, sinum arcus angulum ipsum subtendentis.
Sinus autem huiusmodi ad unam & eandem circuli semidiametrum, siue ad plu
res, æquales tamen, referri oportebit. Sit igitur triangulus a b g rectilineus. Di
co, ꝙ proportio lateris a b ad latus a g est ut sinus anguli a g b ad sinum angu

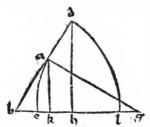

li a b g .item lateris a b ad b g tanq sinus an
guli a g b ad sinum anguli b a g . Si enim tri
angulus a b g fuerit rectangulus, ex 28 primi
huius comparabimus demonstrationem. Si uero
non fuerit rectangulus, duo tamen latera a b &
a g fuerint æqualia, erunt quoq duo anguli eis
oppositi, & ideo sinus eoru æquales, unde de ipsis
duobus lateribus propositionem nostram uerifi
cari constat.Quod si alterum altero longius exti

terit, sit uerbi gratia a g logius, dirigaturq b a usq ad d, donec tota b d æqua
lis habeatur lateri a g :deinde super duobus punctis b & g factis centris, descri
bi intel

Therefore the two sides *AB* and *BC* are given in terms of each other. And since some one angle is given by the hypothesis, then by Th. 50, 51, or 52 above the other side may be found. Hence by definition we will have the ratios of all the sides to each other. Furthermore, the other angles will be measured by these same ratios, as we intended to prove. The mechanics can be derived from the cited [theorems], if beforehand we take one of the two sides as a known [number].

Theorem 57

If the ratios of each of two angles to a right angle are individually given and any one side is known, all the angles and the other sides may be found.

If the ratio of each of two angles, *A* and *B*, to a right angle is given, and if side *AB* or any other side is known, then all the angles of △ *ABC* will become known together with the sides.

By Th. 6 above each of the mentioned angles will be found because [the value of] a right angle is known by Th. 21 above. Therefore by Ths. 53 and 54 above we will provide what is left. We give the mechanics no space here because they can easily be derived from what was said above.

The End [of Book One]

BOOK TWO OF THE TRIANGLES

Theorem 1

In every rectilinear triangle* the ratio of [one] side to [another] side is as that of the right sine of the angle opposite one of [the sides] to the right sine of the angle opposite the other side.

As we said elsewhere, the sine of an angle is the sine of the arc subtending that angle. Moreover, these sines must be related through one and the same radius of the circle or through several equal [radii]. Thus, if △ *ABG* is a rectilinear triangle, then the ratio of side *AB* to side *AG* is as that of the sine of ∠ *AGB* to the sine of ∠ *ABG*; similarly, that of side *AB* to *BG* is as that of the sine of ∠ *AGB* to the sine of ∠ *BAG*.

If △ *ABG* is a right triangle, we will provide the proof [directly] from Th. I.28 above. However, if it is not a right triangle yet the two sides *AB* and *AG* are equal, the two angles opposite [the sides] will also be equal and hence their sines will be equal. Thus from the two sides themselves it is established that our proposition is verified. But if one [of the two sides] is longer than the other — for example, if *AG* is longer — then *BA* is drawn all the way to *D*, until the whole [line] *BD* is equal to side *AG*. Then around the two points *B* and *G* as centers,

*Evidently Regiomontanus anticipated his subsequent theorems in Books III, IV, and V concerning spherical triangles and hence qualified his triangles here as rectilinear.

bi intelligantur duo circuli æquales secundum quantitates linearum b d & g a, quorum circumferentiæ occurrant basi trianguli in punctis l & e, ita, ut arcus d l quidem angulum d b l siue a b g, arcus autem a e angulum a g e siue a g b subtendat: ex duobus demum punctis a & d duæ perpendiculares a k & d h ba si incidant: palã autem q d h est sinus rectus anguli a b g, & a k sinus rectus anguli a g b. est autem per 4. sexti Euclidis proportio a b ad b d, & ideo ad a g sicut a k ad d h. quare certum est, quod asserebat propositio.

II.

Cognito aggregato ex duobus lateribus trianguli, cũ duobus an-gulis sibi oppositis, unumquodcp trianguli latus secernere.

Triangulus a b g congeriem duorum lateru a b & a g habeat datam, & utrunq angulorum a b g & a g b notum. Dico, q tria latera eius in uenientur. Erit enim ex præcedenti proportio a b lateris ad latus a g cognita propter angulos da-tos, & ideo cõiunctim proportio b a & a g ad a g dabitur: cunq congeries duorum laterum a b & d g sit nota per hypothesim, erit & latus a g cognitum, hinc & a b latus non latebit. ex hypothesi autem duo angu-lia b g & a g b noti, latere non sinunt angulum b a g, sic demum ex duobus an-gulis b a g & a g b cognitis cum latere a b, tertium quocp latus b g notum con cludemus. Potest autem aliter, quancp prolixius, idem absolui, si prius ex puncto a ad basim b g perpendicularem a d demiserimus: habebit enim triangulus partia lis a b d rectangulus, angulũ a b d acutũ cognitũ, quare p 31 primi huius propor tio a b ad a d nota prodibit: ex eisdem quocp medijs proportio a g ad a d non la-tebit, utrũcp ergo duarum linearum a b & a g ad perpendicularem a d proportio nota conclamabitur: hinc per 28 primi huius earum inter se proportio scita ueni-et, & ideo coniunctim aggregati ex eis ad utrancp earum proportio notificabitur quãobrem utracp earum nota prosiliet, hinc tandem linea b g, quemadmodũ in primo præcepimus, cognoscetur.

III.

In triangulo æquicruri, si unus angulus datus fuerit cum uno latere quocuncp, reliqua cognitum iri.

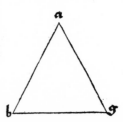

Quancp in primo sufficienter hanc rem explicasse ui-dear, libet tamen paulisper circa triangulos æquicrures, & deinde circa triangulos uarios immorari: q ea quæ superi us quærebantur breuiori tramite consequamur. Sit talis triangulus a b g, duo latera a b & a g habens æqualia, cuius unus angulus quicũcp sit datus cũ linea terali eius. Dico, q reliquæ lineæ eius notæ uenient. Erũt enim p 38 primi huius duo reliqui anguli cogniti. unde & per ante præmissam reliquæ lineæ faciliter notificabuntur. Sic abscp perpendiculari unde-cuncp ducta, propositum attingere didicimus. Quòd si unus angulus eius triangu li duntaxat fuerit datus, proportionem laterum non ignorabimus, erunt enim per 35 primi huius & reliqui anguli dati: hinc & per ante præmissam uerum esse, quod asserebam confiteberis.

Si quis

two equal circles are understood to be drawn with the lengths of lines BD and GA as radii [respectively]. The circumferences of these circles intersect the base of the triangle at points L and E, so that arc DL subtends ∠ DBL, or ABG, and arc AE subtends ∠ AGE, or AGB. Finally two perpendiculars AK and DH, from the two points A and D, fall upon the base. Now it is evident that DH is the right sine of ∠ ABG and AK is the right sine of ∠ AGB. Moreover, by VI.4 of Euclid, the ratio of AB to BD, and therefore to AG, is as that of AK to DH. Hence what the proposition asserts is certain.

Theorem 2

If the sum of two sides of a triangle is known together with the two angles opposite them, each side of the triangle may be found.

If [in] △ ABG the sum of the two sides AB and AG [is] given, and each of the angles ABG and AGB is known, then its three sides can be found.

By the preceding [theorem] the ratio of side AB to side AG is known because the angles are given. Hence by addition, the ratio of BA plus AG to AG is known. And since the sum of the two sides AB and DG* is known by hypothesis, side AG becomes known. Thereupon side AB will be found. Moreover, the two angles ABG and AGB, known from the hypothesis, reveal ∠ BAG. Thus, finally, since the two angles BAG and AGB are known along with side AB, we will also determine the third side BG.

Now, the same thing can be proved in another, though more lengthy, way, if first we were to drop perpendicular AD† from point A to base BG. The right-triangle portion ABD will have acute ∠ ABD known. Therefore by Th. I.31 above the ratio of AB

to AD may be found. By these same means the ratio of AG to AD becomes apparent. Then the ratio of each of the two lines AB and AG to the perpendicular AD will be determined. Hence by Th. I.28 above the ratio of these [lines] to one another becomes known, and therefore by addition the ratio of the sum of these to each of them [individually] will be identified [i.e., AB + AG:AB and AB + AG:AG], whereupon each of them is revealed. Hence, just as we showed in the first [proof above], line BG is finally known.

Theorem 3

In an isosceles triangle, if one angle as well as any one side is given, the rest can be found.

Although it would seem that this matter was explained sufficiently in the First Book, still it is desirable to dwell a while upon isosceles triangles and thereafter upon scalene triangles, so that we may follow by shorter paths those things which were sought above. If such a △ ABG has two equal sides AB and AG [and] any one angle along with the third side is given, then its other lines can be found.

By Th. I.38 above the other two angles will be known. Hence, from the theorem before the preceding [Th. 1 above], the other lines are also readily identified. Thus without any perpendiculars having been drawn, we have learned to handle the theorem.

But if one angle only of the triangle were given, we would [still] know [at least] the ratio of the sides, for by Th. I.35 above the other angles will be found; then from this and from the [theorem] before the preceding [Th. 1 above], you will acknowledge what I [just] set forth to be true.

*For DG read AG.
†In the figure, for line dg read bdg.

IIII.

Si quis trianguli uarij duos angulos seorsum dederit cū uno latere eius quoliẛet, reliqua latera faciliter metiemur.

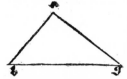

Det mihi quispiā duos angulos trianguli a b g tria latera inæqualia habentis, cum uno ipsoᵹ laterū, uerbi gratia a b. Dico, ꝙ reliqua duo latera accipiet cognita. Nam 3 2.primi elementoᵹ intercedēte tertius quoꝗ angulus innotesᛒet: cunꝗ per antepræmissam proportio si nus anguli a gˉb noti ad sinum anguli a b g noti sit uelut lateris a b ad latus a g. tresꝗ harum quantitatum notæ sint, cui nisi prorsus ignaro quarta quantitas, uidelicet latus a g non manifestabitur? Idem eodemꝗ modo lateris b g cognoscendi præceptum habebitur. Hoc pacto perpendicularem undecunꝗ etiam duxisse superuacaneum censebitur.

V.

Ex duobus lateribus trianguli datis cum angulo alteri eorū opposi to, reliquos angulos ac tertium latus notificare.

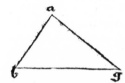

Sit talis triangulus a b g, duo latera a b & a g habēs cognita cum angulo a b g. Dico ꝙ reliqui duo anguli cum tertio latere suo innotescēt. Superiori enim freti syl logismo ex duobus lateribus a g & a b cognitis cum si nu recto anguli a b g noti per hypothesim, angulus a g b notus emerget. Hinc quoꝗ 3 2.primi ratiocinante tertius angulus b a g haud ignorabitur. Proportio autē sinus anguli a b g noti per hypothesim ad sinū anguli b a g noti per argumentationē est ut lateris a g ad latus g b. quare & latus g b non erit ignotum. Quāuis autem ex duobus lateribus a b & b g cognitis cū angulo b a g ab eis comprehenso aliter ꝗ in primo reliquos angulos cum tertio latere dimetiendi sit potestas, quemadmodū nunc recitabitur. Non tamē per hāc uiam operandum suadeo. erit enim propter angulum b a g notum congeries du orū anguloᵹ a b g & a g b cognita. cunꝗ proportio sinus unius eoᵹ ad sinū alte rius sit cognita: est enim sicut duoᵹ laterum a b & a g proportio data, fieret per tertij uterꝗ anguloᵹ a b g & a g b notus. Sed hæc uia nihil facilitatis addit, quare in tali proposito à   primi recedere non licebit.

VI.

Triangulus trium notorum angulorū lateribus suis proportiones uendicabit cognitas.

Nihil habet difficultatis hæc propositio, nisi 1.huius negligenter præterieris: nam quorumlibet duorum laterum ea erit proportio, quam habent sinus anguloᵹ rum eis oppositorum ordine uidelicet præpostero, ut superius traditum est.

VII.

Data perimetro trianguli cū duobus angulis eius, unumquodꝗ la tus seorsum cognoscere.

Congeries trium laterū trianguli a b g sit data cū duobus angulis eius a b g & a g b. Dico ꝙ omnia latera eius seorsum innotescent. Erūt em tres anguli eius noti,

Theorem 4

If in any scalene triangle two angles are given individually with any one of its sides, the other sides are easily measured.

If any two angles of △ *ABG*, having three unequal sides, are given together with one of its sides — for example, *AB* — then the other two sides can be found.

By the application of I.32 of the *Elements* the third angle also becomes known. And since by an earlier theorem [Th. 1 above] the ratio of the sine of known ∠ *AGB* to the sine of known ∠ *ABG* is as that of side *AB* to side *AG*, and since three of these quantities are known, then, except through utter ignorance of this, the fourth quantity — namely, side *AG* — may be found.* In the same way, the same will be found concerning side *BG*. By this method one may judge that it was unnecessary to draw any perpendicular.

Theorem 5

When two sides of a triangle are given together with the angle opposite one of them, the other angles and the third side may be determined.

If such a △ *ABG* has two sides *AB* and *AG* known along with ∠ *ABG*, then the other two angles and the third side may be found.

By reliance upon the earlier syllogism, ∠ *AGB* is determined from the two known sides *AG* and *AB* together with the right sine of known ∠ *ABG*. Hence, also, by the reasoning of [Euclid] I.32, the third ∠ *BAG* is by no means unknown. Furthermore, the ratio of the sine of ∠ *ABG*, known by hypothesis, to the sine of ∠ *BAG*, known by the proof, is as that of side *AG* to side *GB*. Therefore, side *GB* will be found.

Now, when two sides *AB* and *BG*† are known together with their included ∠ *BAG*, one can determine the other angles and the third side in a different way from that in the first [part], as will now be discussed. However, I do not recommend working by this method. Because ∠ *BAG* is known, the sum of the two angles *ABG* and *AGB* is known. And since the ratio of the sine of one of these to the sine of the other is known because the ratio of the two sides *AB* and *AG* is given, then by [Euclid] III.[not given] each of the angles *ABG* and *AGB* will be known. But this method adds nothing in the way of ease; therefore in [solving] such a premise one should not retreat from the first [method above or from Ths. I.50 and I.51 above].

Theorem 6

If the three angles of a triangle are known, the ratios of its sides can be found.

This theorem presents no difficulty unless Th. 1 above was carelessly passed over, for the ratio of any two of the sides will be [that] which the sines of the angles opposite them have in inverted order, as was proven above.

Theorem 7

If the perimeter of a triangle is given with two of its angles, any one side can be found individually.

If the sum of the three sides of △ *ABG* is given together with its two angles *ABG* and *AGB*, then all its sides can be found individually.

Its three angles will be known;

*In the Latin text, for *non manifestabitur* read *nota manifestabitur*.
†For *BG* read *AG*.

noti,quare per argumentionem fæpe adductam propor=
tio a b ad a g nota erit, & ideo coniunctim aggregatũ
ex b a & a g ad lineã a g ,pportionẽ habebit notã. Itẽ
,pportio a g ad g b nota erit,unde & ,pportio b a,a g
ad g b cognita ,pueniet, & ideo cõiunctim tota pimeter
triãguli a b g ad lineã b g notã habebit ,pportionẽ. cũcʒ
pimetrũ ipfam dederit hypothefis,erit & linea b g cognita.hinc ǫcʒ reliqua duo
latera nota declarabunt.Poteris præterea idẽ cõcludere ducta ppendiculari a d :
nam per 3 o primi huius utriufcʒ linearuma b & a g ad perpendicularem a d pro
portio nota erit,quare earum inter fe proportio manifeftabitur. item a b ad b d
nota elicietur proportio,item ,pportio a g ad g d fimiliter nota erit,unde & utri=
ufcʒ duarum linearum a b & a g ad lineam b g proportio data proclamabitur:
hinc ut prius congeries trium laterum ad ipfam b g lineam ,proportionem habe
bit datam,cætera ut ante.

VIII.

Datis proportionibus trium laterum,perpendiculari̓cʒ nota, cun=
cta latera dimetiri.

Trianguli a b g bina latera proportiones habeant
cognitas,fitcʒ perpẽdicularis a d data.Dico,ǫ tria eius
latera innotefcent.Nam fi duo latera a b & a g fuerint
æqualia,erit b d æqualis ipfi d g,unde proportio a b ad
b d cognofcetur:& ideo quadrati a b ad quadratum b d
proportio fcita ueniet:quare etiam euerfim argumentã-
do quadrati a b ad quadratum a d nota dabitur proportio:cuncʒ quadrãtum a d
fit notum,propter coftam fuã ex hypothefi datam,erit quadratum a b notum, &
inde ipfa linea a b non ignorabitur,ex qua demum & proportionibus laterum p
hypothefim datis,reliqua latera innotefcent.Quòd ft alterum duorum laterum a
b & a g altero maius extiterit,fit a b breuius:eritcʒ ob hoc cafus b d breuior cafu
d g,abfcindatur d e æqualis ipfi b d .ex procefſu igitur   primi huius, quod fit
ex e g in g b eft æquale excefſui quadrati a g fupra quadratum a b , quorũ quidẽ
quadratoɤ ,pportio nota erit,unde & diuifim eius,ǭd fit ex e g in g b ad quadratũ a b ,pportio nota declarabit. Hæc aũt ,pportio per   6.elementoɤ cõponit ex
,pportiõe nota lineæ g b ad lineã a b.& ex ,pportione e g ad a b,cũcʒ tã ,pportio
compofita cʒ ipfa componens prima fint notæ,erit & reliqua componens nota :
unde & proportio b e ,& ideo medietatis eius b d ad lineam a b fcita confurget
quadratı̓cʒ a d ad quadratum b d innotefcet,& ideo euerfim quadratum a b ad
quadratum a d notam feret proportionem:quadrato igitur a d noto redundabit
quadratum lineæ a b cognitum,hinc ipfa a b lineæ cum reliquis trianguli lateri=
bus innotefcent.

IX.

Ex proportionibus trium laterum trianguli , tres angulos eius in=
ueftigare.

Refumpta priori figuratione concludemus propter hypothefim,ut in præmif
fa,proportionem a b ad b d notam,& ideo per   primi angulus b a d , & inde
angulus a b d cognofcetur,deinde propter angulum a b g iam notum cum proportione duorum laterum a b & a g data angulus a g b   huius arguente inno-
                                              G   tefcet.

Book II of *On Triangles*

therefore, by the reasoning often cited, the ratio of *AB* to *AG* will be known, and then by addition the sum of *BA* plus *AG* will have a known ratio to line *AG*. Similarly, the ratio of *AG* to *GB* will be known, and hence the ratio of *BA* plus *AG* to *GB* will be found. Thus by addition the total perimeter of △ *ABG* has a known ratio to line *BG*. And since the hypothesis gave the perimeter itself, line *BG* will be known. Hence the other two sides will also be declared known.

One could reach the same conclusion if perpendicular *AD* were drawn, for by Th. I.30 above the ratio of each of the lines *AB* and *AG* to the perpendicular *AD* will be known. Therefore the ratio of these [lines] to one another will be apparent. Similarly, the ratio of *AB* to *BD* may be found, and likewise the ratio of *AG* to *GD* will be known. Hence the ratio of each of the two lines *AB* and *AG* to line *BG* will be declared found. Whereupon, as previously, the sum of the three sides to line *BG* [alone] will have a given ratio, and so on as before.

Theorem 8

If the ratios of three sides are given and if the perpendicular is known, each side can be measured.

If in △ *ABG* the [various] pairs of sides have known ratios and the perpendicular *AD* is given, then the three sides may be found.

If the two sides *AB* and *AG* are equal, *BD* will be equal to *DG*; hence the ratio of *AB* to *BD* will be known. Then the ratio of the square of *AB* to the square of *BD* will be known. Therefore, when [this] is proven [for] the counterpart [ratio of *AB* to *AD*], the ratio of the square of *AB* to the square of *AD* will be found. Now since the square of *AD* is known, because its side [or square root] was given in the hypothesis, then the square of *AB* will be known; whereupon line *AB* it-

self will be known. Finally, from [*AB*], the other sides will be found because the ratios of the sides are given by the hypothesis.

But if one of the two sides *AB* and *AG* is greater than the other, let *AB* be the shorter. On account of this, segment *BD* will be shorter than segment *DG*. Let *DE*, equal to *BD*, be marked off [on *DG*]. Thus, from the procedure of Th. I.[44] above, the product of *EG* and *GB* is equal to the [difference between] the square of *AG* [and] the square of *AB*. Indeed, the ratio of these squares will be known, and therefore by division the ratio of the product of *EG* and *GB* to the square of *AB* becomes known. Moreover, according to [book not cited].6 of the *Elements*, this ratio is composed of the known ratio of line *GB* to line *AB* and of the ratio of *EG* to *AB*. And since, as the ratio is set up, its first factor [*GB/AB*] is known, so the other factor [*EG/AB*] will be known. Hence the ratio of *BE*, and therefore its half *BD*, to line *AB* is known; and since the ratio of the square of *AD* to the square of *BD* is known, then, by a [consideration of the] counterpart [ratio], the square of *AB* will have a known ratio to the square of *AD*. Therefore, since the square of *AD* is known, the square of line *AB* may be found; whence line *AB* itself may be found, together with the other sides of the triangle.

Theorem 9

If the ratios of the three sides of a triangle are given, the three angles may be found.

By reference to the previous figure, we infer, because of the hypothesis, that the ratio of *AB* to *BD* is known, as in the preceding. Therefore by Th. I.[54] ∠ *BAD* and hence ∠ *ABD* will be known. Next, because ∠ *ABG* is now known together with the given ratio of the two sides *AB* and *AG*, ∠ *AGB* will be known by the reasoning of [Th. 5] above.

teſcet.hinc & tertius angulus non poterit ignorari.Habes tamen in primo alium
modum,qui ſi planior uidetur,repetendus eſt.   Si libeat aliter hanc abſoluere ex
ponatur linea quantalibet notæ quantitatis reſpectu perpendicularis a d, ad quã
inueniantur duæ aliæ ſecundũ proportiones laterũ trianguli a b g datas :ex his
tribus intelligatur conſtitutus triangulus,& per primi huius inueniatur perpẽ
dicularis ſua,procedens â termino cõi duox laterum proportialiũ duobus late
ribus a b & a g .hæc em perpendicularis ſecundi trianguli habebit ſe ad perpen
dicularẽ a d ſicut latus quodlibet ſecundi trianguli notũ ad latus trianguli a b g
ſibi correlatiuum: cũ que tres huiuſmodi quantitatum ſint notæ,quartam cognitũ
iri neceſſe eſt.Hæc trahuntur ex ſimilitudine duorum triangulorum totalium at
que partialium,quam ex   ſexti elementorum facile eſt colligere.

X.

Data area trianguli cum proportionibus laterum,unumquodque eorũ
notificari. Vnde & angulos ſuos metiri licebit.

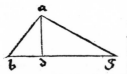

Repeto triangulum a b g cũ perpendiculari ſua a d,
quẽadmodũ apud   huius figurauimus,ubi cõcludebat
proportio a b ad perpendicularem a d nota:hinc & p
pter hypotheſim perpendicularis a d ad baſim b g &
ideo ad eius medietatem habebit notam proportionem:
cũque quod ſub ipſa perpendiculari & dimidia baſi continetur,ſit notũ, uidelicet
area ipſa trianguli,erit per primi huius tã perpendicularis a d que baſis b g no
ta. quam õrem & propter datas laterum proportiones reliqua latera & tandẽ an
guli ipſi non latebunt.Quõd ſi modus ille uel prolixius nimium uel difficilis uide
atur,alium aggrediaris:non dico tamen faciliorem,ſed fortaſſe tibi magis placi
turum.Ex tribus lineis quantiſcunque,notis per menſurã,ex qua area trianguli da
ta ſurrexit,habentibus tamen proportiones ueluti tria latera trianguli propoſiti
intellige conſtitutum triangulũ,cuius perpendicularem quamcunque uoles per
primi huius metiaris:quæ ducta in dimidiã baſim ſibi ſubſtratẽ ſuſcitabit aream
huiuſmodi trianguli ſecundi cognitã:cũque duo huiuſmodi triangulos conſtet eſ
ſe æquiangulos,erit area trianguli ſecundi ad aream trianguli primi,quæ iam no
tæ ſunt,ſicut quadratum lateris cuiuslibet ſecundi trianguli ad quadratum lateris
ſibi relatiui primi trianguli.unde quadratum illius lateris de primo triangulo,&
ideo latus ipſum notificabitur,hinc quoque reliqua non latebunt.

X I.

Data perpendiculari quacunque cum duobus angulis trianguli qui
buslibet omnia latera menſurare.

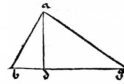

In triangulo a b g ſit perpendicularis a d cognita cũ
duobus angulis.Dico,que omnia latera innoteſcent. Ha
bebit enim triangulus a b d partialis rectangulus latus
a d cognitum cum uno angulo acuto,nam duobus angu
lis triãguli a b g cognitis,tertius latere nõ potuerit.qua
re per primi huius utraque linearum a b & b d menſurata ueniet. per eadẽ rur
ſus media utranque linearũ a g & g d metiemur:hinc tota b g,& ideo omnia late
ra trianguli propoſiti cognoſcentur,quod erat explanandum.

Data per

Consequently the third angle cannot be unknown. However, in the First Book you have another method which, if it seems clearer, should be recalled.

If one wishes to prove this differently, a line of any length that is known relative to the perpendicular *AD* is set down. Two other [lines] are found [relative] to this [line] according to the ratios given for the sides of △ *ABG*. A triangle is known to be formed by these three, and by Th. I.[46] above its perpendicular may be found, proceeding from the intersection point of the two sides which are proportional to the two sides *AB* and *AG*. For this perpendicular of the second triangle will be to the perpendicular *AD* as any known side of the second triangle is to the corresponding side of △ *ABG*. And since three of these quantities are known, the fourth is necessarily determined. These are derived from the similarity of the totalities and the portions of the two triangles, as is easy to infer from VI.[not given] of the *Elements*.

Theorem 10

If the area of a triangle is given together with the ratios of the sides, any one of these [sides] may be identified. Therefore its angles may be measured.

Recall △ *ABG* with its perpendicular *AD* as we constructed it in Th. [2] above where the ratio of *AB* to the perpendicular *AD* was proven known. From this and because of the hypothesis, the perpendicular *AD* will have a known ratio to the base *BG* and consequently to its half. And since the product of this perpendicular and half the base is known — namely, the area of the triangle itself — then by Th. I.[18] above, as the perpendicular *AD* will be known, so also will the base *BG* be known. For this reason and because the ratios of the sides were given, the other sides

and finally the angles themselves can be found.

But if this method seems either too extensive or too difficult, you can approach it in another way — still no easier but perhaps more agreeable to you. From three lines of any lengths that are known by the [same] measure through which the area of the triangle was given, yet having ratios just as the three sides of the given triangle, you know that a triangle is formed. Whichever of its perpendiculars you wish [to use], you may measure by Th. I.[46] above. This [perpendicular], multiplied by half the base on which it rests, reveals the area of this second triangle. And since these two triangles are established to be similar, the area of the second triangle to the area of the first triangle, which is already known, is as the square of any side of the second triangle to the square of the corresponding side of the first triangle. Hence the square of this side of the first triangle, and therefore the side itself, will be known; whereupon the others also can be found.

Theorem 11

If any perpendicular of a triangle is given together with any two angles, all the sides can be measured.

If in △ *ABG* perpendicular *AD* is known together with two angles, then all the sides can be found.

The right-triangle portion *ABD* will have side *AD* known along with one acute angle, for two angles of △ *ABG* are known and the third cannot be hidden. Therefore, by Th. I.[29] above, each of the lines *AB* and *BD* becomes measured. By the same means once again, each of the lines *AG* and *GD* may be measured; hence all of *BG* and therefore all the sides of the given triangle are known, which was to be explained.

XII.

Data perpendiculari atcp baſi,& proportione laterum cognitis, utruncp latus cognoſcere.

Hoc problema geometrico more abſoluere nō licuit hactenus,ſed per arte rei & cenſus id efficere conabimur. Habeat itacp triangulus a b g perpendicularem a d, & baſim b g cognitas,proportionemcp laterum a b & a g datam,quærimus utruncp eorum.Sit uerbĩ gratia,ppor- tio a b ad a g tancp 3 ad 5.ita,ut latus a b ſit breuius latere a g,quo demum euenit ut caſum b d breuiore caſu d g nemo inficiari poſſit,ſit ergo d e æqualis ipſi b d, deturcp perpendicularis a d 5,& baſis b g 20 pedes.pono lineam e g 2 res, ita, unde linea b e erit 20.demptis duabus rebus,& eius medietas b d 10 minus 1 re,re liqua uero d g,erit 10 & una res.duco b d in ſe,producitur 1 cenſus & 100, dem ptis 20 rebus,quibus addo quadratũ perpendicularis ſcilicet 25.colliguntur 1 cen ſus & 125.demptis 20 rebus,item b g in ſe,fiunt 1 cenſus,20 res & 100.quibus adij cio quadratum perpendicularis 25.colliguntur 1 cenſus 20 res & 125.ſic habebo duo quadrata linearum a b & a g, quorum proportio eſt ut 9 ad 25. duplicata ſcilicet proportio 3 ad 5,quæ erat pportio laterum.cum itacp proportio quadrati primi ad quadratum ſecundum ſit tancp 9 ad 25.ſi duxero 25 in quadratum pri mum,itemcp 9 in quadratum ſecundum,quæ producentur erunt æqualia,reſtau ranscp ut aſſolet defectibus,& ablatis æqualibus,utrobicp perducemur ad 16 cen ſus & 2000 æquales 680 rebus:quamobrem quod reſtat,præcepta artis edocebũt. Linea ergo g e quam poſui 2 res nota redundabit,hinc reſidua ex baſi b e & eius medietas b d,quæ cum perpendiculari a d,latus a b notum ſuſcitabũt,unde tan dē & latus a g notum pronunciabitur,quæ libuit efficere.

XIII.

Cognito utrocp caſuum, & proportione laterum data, quantitates laterum emoliri.

In triangulo a b g ducta perpendiculari a d,ſit uter cp caſuum b d & d g datus cum proportione laterum. Dico, cp utruncp latus cum perpendiculari ipſa innote ſcent.Sit caſus b d breuior,nam ſi eſſent æquales duo ca ſus,latera quocp haberentur æqualia,eorum tamen co gnitio non conſequitur caſus datos & pportionem laterum,quæ eſt æqualitatis. ſecetur ergo ex longiori caſu linea d e æqualis caſui breuiori:differentia quocp du orum laterum ſit h g.cum igitur pportio a g lateris ad a b ſit data,erit diuiſim p portio h g ad a h data,& ideo h g ad duplam ipſius a h ſcilicet congeriem dua rum linearum a b & a h data erit:quare etiam coniunctim pportio h g ad ſum mam duorum laterum a b & a g non erit ignota.quod autem ſub h g & duobus lateribus a b & a g cõiunctis continetur,æquum eſt ei,quod ſub e g & g b. illud autem notum eſt ppter duos caſus ex hypotheſi notos,unde & per proceſſum primi huius,quod ſub h g & g a a b continetur,notum erit:cuncp proportio line arum hoc continentium ſit nota,erit per   primi huius tam linea h g cp congeri es duorum laterum nota:hinc tandem dempta h g nota ex aggregato laterũ no to reſidui medietas pro latere breuiori reputabitur;unde & longius innoteſcet la tus,quæ

G 2

Theorem 12

If the perpendicular is given and the base and the ratio of the sides are known, each side can be found.

This problem cannot be proven by geometric means at this point, but we will endeavor to accomplish it by the art of algebra.* Thus, if △ *ABG* has perpendicular *AD* and base *BG* known, and if the ratio of sides *AB* and *AG* is given, we seek each of these [sides.]

For example, let the ratio of *AB* to *AG* be as 3 to 5, so that side *AB* is shorter than side *AG*. From this, it follows, next, that segment *BD* is shorter than segment *DG*, as no one can deny. Therefore let *DE* be equal to *BD*, and let the perpendicular *AD* be given as 5 [feet] and the base *BG* as 20 feet. Take line *EG* as $2x$; hence, line *BE* will be $20 - 2x$, and its half *BD* will be $10 - x$, while the rest [of the line] will be $10 + x$. Square *BD* and $x^2 + 100 - 20x$ results, to which add the square of the perpendicular, namely 25, and $x^2 + 125 - 20x$ is obtained. Similarly, *BG* squared makes $x^2 + 20x + 100$, to which add the square of the perpendicular, 25, and $x^2 + 20x + 125$ is obtained. Thus you will have the two squares of the lines *AB* and *AG*, whose ratio is as 9 to 25 — namely, the squared ratio of 3 to 5, which was the ratio of the sides. Therefore, since the ratio of the first square to the second square is as 9 to 25, then, if the first square is multiplied by 25 and the second square by 9, the products are equal. Restoring the deficits, as is customary, and subtract[ing] equals on both sides [of the equation], we obtain $16x^2 + 2000 = 680x$. Whereupon what remains [to be done] the rules of the art show. Therefore line *GE*, which was taken to be $2x$, will be known; hence the rest of the base, *BE*, and its half *BD* will be known. [*BD*] together with perpendicular *AD* reveals side *AB*. Whence, at last, side *AG* is pronounced known, as was promised.

Theorem 13

If each of the segments is known and the ratio of the sides is given, the lengths of the sides may be found.

If in △ *ABG*, when perpendicular *AD* is drawn, each of the segments *BD* and *DG* is given together with the ratio of the sides, then each side and the perpendicular can be found.

Let *BD* be the shorter segment, for if the two segments were equal, the sides also would have to be equal; however, knowledge of their equality does not follow from the segments [as] given and the ratio of the sides. Therefore line *DE*, equal to the shorter segment, is [marked off] on the longer segment. Also let the difference of the two sides be *HG*. Then since the ratio of side *AG* to [side] *AB* is given, then by division the ratio of *HG* to *AH* will be determined, and therefore *HG* to twice *AH* — namely, the sum of the two lines *AB* and *AH* — will be found. Therefore by addition the ratio of *HG* to the sum of the two lines *AB* and *AG* will be known. Moreover, the product of *HG* times the sum of the two sides *AB* and *AG* is equal to the product of *EG* and *GB*. Furthermore, the latter [product] is known because the two segments are known from the hypothesis. Hence by the procedure of Th. I.[16] above, the product of *HG* times [the sum of] *GA* plus *AB* will be known. And since the ratio of the lines forming this [product] is known, then by Th. I.[18] above, both line *HG* and the sum of the two sides will be known. Hence, finally, when known [value] *HG* is subtracted from the known sum of the sides, half of the remainder will be considered the shorter side. And therefore the longer side will become known. Q.E.D.

*Because of the phrase *per artem rei et census,* here translated "by algebra," Karpinski thinks that Regiomontanus was familiar with Robert of Chester's translation of al-Khowarizmi's algebra. See Louis C. Karpinski, *Robert of Chester's Latin Translation of the Algebra of al-Khowarizmi* (1915), p. 36.

tus,quæ fuere demonstranda.

XIIII.

Si uterq̃ duorum casuum inæqualium datus fuerit, aggregatumq̃
ex lateribus datum, utrunq̃ latus secernere.

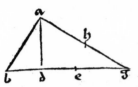

Resumo triangulum præcedentis, in quo datus sit uterq̃ casuum b d & d g, summa'q̃ duorum
laterum a b & a g sit nota. Dico, ꝙ utrunq̃ latus
agnoscetur. Erit enim quod sit ex e g in g b cognitum, & ideo quod sit ex h g in g a, a b coniun-
ctim notum erit. cūq̃ congeries ipsorum laterum sit data, erit per   primi huius
h g nota differentia scilicet laterum, quā si ex summa duorum laterum dempseris
reliqui medietas quantitatem lateris minoris proclamabit. hinc quoq̃ reliquum
latus non ignorabitur, quod erat absoluendum.

X V.

Basis trianguli data notum subtendens angulum cum aggregato
laterum cognito, utriq̃ laterum & utriq̃ angulorum sibi oppositorū
uiam mensurationis aperient.

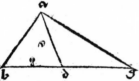

Triangulus a b g basim b g notam habeat cū
angulo b a g dato, sitq̃ congeries laterum a b &
a g cognita. Dico, ꝙ utrunq̃ latus eius cū utroq̃
angulorum eis oppositorum innotescent. Diuida
tur enim angulus b a g per mediū demissa linea
a d ad basim contingente in puncto d, erit igitur per tertiam sexti elementorum
ꝓportio b d ad d g sicut a b ad a g, & permutatim ꝓportio a b ad b d sicut a
g ad g d : quare per   quinti elementorū proportio aggregati ex lateribus a b
& a g ad basim b g, sicut lateris a b ad lineam b d : cunq̃ hæc proportio sit data
est enim uterq̃ terminus eius datus, erit proportio a b ad b d data, sed & angulus
b a d notus accipietur : cum sit medietas anguli b a g per hypothesim noti : quare
per   primi huius angulus a b d mensuratus habebitur : hinc quoq̃ reliquus de
duobus rectis angulus a d b latere non poterit, qui cum sit æqualis duobus angu
lis a g d & d a g, & angulus d a g sit notus, erit residuus angulus d a g mensura
tus. sic duo anguli a b g & a g b noti erunt. congeriem autem duorum laterum a
b & a g datam subiecit hypothesis : quare per   huius quarti utrunq̃ latus a
b & a g notum pronunciabitur, quæ fuere declaranda.   Illud autem aliter attin
gere poterimus hoc pacto. Inscribatur triangulo a b g circulus h n l, cuius cen=

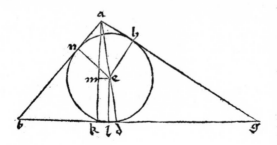

trum necessario erit in linea
a d, quĕadmodū ex   quar
ti elementorum trahitur, ꝙd
sit e, à quo ad tria puncta cō
tactuum h n & l, educantur
tres semidiametri e h, e n &
e l, deinde ex puncto a descendat perpendicularis a k
occurrens basi in puncto l :
oportet autem punctum k re
periri

Theorem 14

If each of two unequal segments is given, and if the sum of the sides is [also] given, each side may be determined.

Refer to the triangle of the preceding theorem. If in this [triangle] each of the segments *BD* and *DG* and the sum of the two sides *AB* and *AG* are known, then each side can be recognized.

The product of *EG* and *GB* will be known, and therefore the product of *HG* times [the sum of] *GA* plus *AB* will be known. And since the sum of these sides is given, then, by Th. I.[17] above, *HG* — namely, the difference of the sides — will be found. If [*HG*] is subtracted from the sum of the two sides, half of the remainder will be the length of the shorter side. Hence the other side may also be found. Q.E.D.

Theorem 15

If in a triangle the base subtending a known angle is given together with the known sum of the sides, each of the sides and each of the angles opposite them may be measured.

If △ *ABG* has a known base *BG* along with a given ∠ *BAG*, and if the sum of the sides *AB* and *AG* is known, then each side together with each of the angles opposite those [sides] may be found.

Let ∠ *BAG* be bisected by line *AD* [which is] dropped to the base [and] meets [the base] at point *D*. Therefore by VI.3 of the *Elements*

the ratio of *BD* to *DG* is as that of *AB* to *AG*, and by rearrangement the ratio of *AB* to *BD* is as that of *AG* to *GD*. Therefore, by V.[not given] of the *Elements*, the ratio of the sum of sides *AB* and *AG* to the base *BG* is as that of side *AB* to line *BD*. And since this ratio is given, for its every term is given, then the ratio of *AB* to *BD* will be determined. Furthermore, ∠ *BAD* is considered known since half of ∠ *BAG* is known by the hypothesis. Therefore, by Th. I.[56] above, ∠ *ABD* will be measured. Whence the other ∠ *ADB* may be found from two right angles. Since this [∠ *ADB*] is equal to the [sum of] angles *AGD* and *DAG*, and since ∠ *DAG* is known, the other ∠ *DAG** will be measured. Therefore the two angles *ABG* and *AGB* will be known. Moreover, the hypothesis gave the sum of the two sides *AB* and *AG*; therefore, by Th. [2] above of this quarto, each of the sides *AB* and *AG* is declared known. Q.E.D.

We can approach this in another way. Within △ *ABG* a circle *HNL* is inscribed, whose center will necessarily be on line *AD*, as is concluded from IV.[not given] of the *Elements*. Let this [center] be *E*, from which three radii, *EH*, *EN*, and *EL*, are drawn to the three points of tangency. Then from point *A* perpendicular *AK* descends, meeting the base at point *L*.† Moreover, point *K* must be found

*For *DAG* read *AGD*.
†For *L* read *K*.

periri in parte lateris a b, à linea diuidête angulū per æqualia puncta b & d, si latus ipsum breuius fuerit latere a g : erit enim angulus a b g maior angulo a g b, & ideo duo anguli a b g & b a d, qbus æquipollet angulus a d g, maiores erūt duobus angulis d a g & a g d, scilicet angulo a d b, adiectis utrobiꝗ æqualibus angulis b a d & d a g : angulus ergo a d g maior, & angulus a d b minor recto conuincetur. hinc etiam costat semidiametrū e l secuisse lineam b d, ducatur insuper linea e m æquedistans basis, & ideo perpendicularis ad lineam a d, praeterea basim b g æqualem esse duabus lineis b n & g h, nō negabis, si tertij elementorum satis didicisti, sublata ergo basi b g data ex congerie laterum data, relinquetur congeries duarum linearum a n & a h cognita, & ideo medietas eius scilicet linea a h mensurata: oportet enim duas lineas a n & a h circulū contingêtes esse æquales. triangulus ergo a e h rectangulus ex latere suo a h cognito cum angulo acuto e a h noto propter duplum eius notum, duo latera sua a e & e d cognita depromet. sic circuli triangulo proposito inscripti semidiameter nota colligetur, ex qua in medietatem perimetri trianguli notam resultat area trianguli nota: ex area autê nota & medietate basis, ꝓpter hypothesim cognita per

primi huius perpendicularis a k mesurata declarabitur, cui si lineā k m æqua lê ipsi e l semidiametro circuli dempseris, relinquetur linea a m cognita, ex qua demum & linea a e superius nota, angulum e a m metieris, quo tandem sublato ex medietate anguli b a g dati scilicet ex angulo b a d, relinquetur angulus b a k notus, qui deinde angulum a b g latere non sinet: sed & duo anguli b a g & a b g tertium socium suū angulum a g b notum suscitabunt, postremo igitur p̄ huius latera trianguli nota prosilient.

XVI.

Data basi alicuius trianguli cum perpendiculari cui subsistit, & aggregato laterum cognito, utrunꝗ eorum secernere.

Hæc partim conuertit præcedentem, & ideo figuram suā resumet, ubi ex perpendiculari nota cum medietate basis aream trianguli metiemur, cunꝗ perimeter trianguli sit nota, erit semidiameter e n circuli sibi inscripti nota: linea quoꝗ a n nota proclamabiꝯ, ut in præcedenti: quare & linea a e & angulus e a n notificabuntur, unde & duplus angulus n a h siue b a g non latebit, k m autem æqualis semidiametro e l siue e n cognitæ cum perpendiculari a k per hypothesim nota, differentiam suam scilicet a m lineam notificabunt: quæ rursus cum a e pridem cognita, angulum a e m notum reddent, æqualem uidelicet angulo a d b : ex duobus autem angulis n a e siue b a d & a d b cognitus, angulus quoꝗ a b d siue a b g notus declarabitur: erat autem b a g cognitus. quare residuus a g b non ignorabitur, & ideo per   huius utrunꝗ latus notum enunciabitur, quod placuit determinare. Non autem necesse est perpendicularem a k intra triangulum cadere, sed contingit eam cadere extra triangulum, nonnunꝗ etiam coincidere lateri minori, si fuerint inæqualia latera, maiori enim coincidere non potest: huius rei indicia erunt talia. Si acciderit lineam a n suo modo repertam cum semidiametro circuli inscripti triangulo coniunctim æquales esse perpendiculare a k datæ, necessario perpendicularis dicta coincidet lateri a b, id est, oportuit angulum a b g trianguli propositi esse rectum: si uero tale aggregatum minus fuerit, ipsa perpendiculari datæ signum est eam intra triangulum cecidisse, & si maius extra, super hoc autem demonstrationem conscribere non est consilium, cum faci

G 3   le quidem

in the direction of side *AB* from the angle bisector, between points *B* and *D*, if that side [*AB*] is shorter than side *AG*. For ∠ *ABG* will be greater than ∠ *AGB*, and thus the [sum of] angles *ABG* and *BAD*, to which [sum] ∠ *ADG* is equal, will be greater than the [sum of] angles *DAG* and *AGD* — namely, than ∠ *ADB* — when the equal angles *BAD* and *DAG* are added to both sides [of our inequality]. Therefore ∠ *ADG* is proven to be greater than a right angle and ∠ *ADB* is less [than a right angle]. Hence it is established that the radius *EL* has intersected line *BD*. Let line *EM* be drawn above and parallel to the base and, therefore, perpendicular to line *AD*.* Furthermore, it cannot be denied that the base *BG* is equal to the two lines *BN* and *GH* if you learned enough from III.[not given] of the *Elements*. Therefore, when the given base *BG* is subtracted from the given sum of the sides, the sum of the two lines *AN* and *AH* is left known. Thus its half — namely, line *AH* — is measured, for the two lines *AN* and *AH*, tangential to the circle, must be equal. Therefore [in] right △ *AEH*, since its side *AH* was found [and since] acute ∠ *EAH* is known because its double is known, the two sides *AE* and *ED*† [may be] found. Thus the radius of the circle inscribed within the given triangle is found. From the product of this [radius] times half the perimeter of the triangle, which is known, the area of the triangle results. Furthermore, from the known area and half the base, known because of the hypothesis, perpendicular *AK* may be declared measured by Th. I.[17] above. If you subtract line *KM*, equal to *EL* which is a radius of the circle, from this [perpendicular *AK*], line *AM* is left known. Finally, from [line *AM*] and line *AE*, already known, you measure ∠ *EAM*. At last, when this is subtracted from half of given ∠ *BAG* — namely, from ∠ *BAD* — ∠ *BAK* is left known, which then makes ∠ *ABG* known. Furthermore, the two angles *BAG* and *ABG* reveal their supplementary ∠ *AGB*. Then, finally, by Th. [7] above the sides of the triangle will become known.

Theorem 16

If the base of some triangle is given with the perpendicular that stands on [that base] and if the sum of the sides is known, each of them can be distinguished.

This [theorem] in part is the converse of the preceding [theorem], and therefore let its figure be recalled. As soon as we will measure the area of the triangle from the known perpendicular times half the base, then, since the perimeter of the triangle is known, the radius *EN* of the circle inscribed in it will be known. Also line *AN* will be declared known, as in the preceding [theorem]. Therefore both line *AE* and ∠ *EAN* will be identified; hence the double ∠ *NAH*, or *BAG*, will be found. Moreover, since *KM*, equal to radius *EL* or *EN*, is known together with perpendicular *AK*, known by hypothesis, then their difference — namely, line *AM* — is found. Again, this [line *AM*] together with *AE*, already known, makes ∠ *AEM* known, which is equal to ∠ *ADB*. Since the two angles *NAE*, or *BAD*, and *ADB* are known, then ∠ *ABD*, or *ABG*, is also declared known. Moreover *BAG* was [already] known. Therefore the other [angle] *AGB* will be found, and then by Th. [4] above each side will be known, as was promised.

It is not necessary for the perpendicular *AK* to fall within the triangle; it may happen to fall outside the triangle. Sometimes it can coincide with the shorter side if the sides are unequal, but it cannot coincide with the longer side. The criteria for this situation will be as follows: If it should happen that line *AN*, found in [the above] way, when added to the radius of the circle inscribed in the triangle, equals the given perpendicular *AK*, then the mentioned perpendicular necessarily coincides with side *AB*; that is, ∠ *ABG* of the given triangle must be a right angle. However, if this sum [of *AN* plus *EN*] is less, this signifies that the given perpendicular has fallen within the triangle, and if [the sum] is greater, [the perpendicular has fallen] outside. It is not [my] intention to write more about this proof since it is indeed easy

*For *AD* read *AK*.
†For *ED* read *EH*.

le quidem fit, parum autem utilitatis adducat: tuo igitur quod reliquum est ferua‑
tur ingenio. figuræ præterea aliter cadenti demonstrationē suam, nisi rudissimus
fueris, accomodare poteris.

XVII.

Datis duobus angulis & uno casu quocunꝗ, omnia latera cum per
pendiculari manifestare.

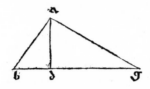

Sit triangulus a b g qualis proponitur, in quo p
pendicularis a d duos casus ex basi distinguat b d
& d g: quorum alter uerbi gratia b d sit cognitus
cū duobus angulis trianguli a b g. Dico, ꝗ omnia
latera sua noticiā non fugient. Erunt enim per hy‑
pothesim 32. primi elementorum suffragante tres
anguli trianguli a b g cogniti, quare triangulus partialis a b d rectangulus an
gulum a b d acutum habens notum cum latere b d, reliqua duo latera sua a b &
a d per primi huius notificabit: hinc in triangulo a g d partiali angulum a g
b acutum habente notum cum latere a d, utraꝗ linearum a g & g d innotescet
per eandem primi huius: collectis ergo duabus a d & d g, resultabit tota basis
cognita, & problematis integra consummabitur intentio.

XVIII.

Data proportione duorum laterū trianguli cum angulo alteri eo‑
rum opposito, reliquos duos angulos mensurare.

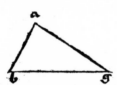

Talis esto triangulus a b g, cuius duo latera a b &
a g pportionem habeant notā, sitꝗ angulus a b g da
tus. Dico, ꝗ reliqui anguli non latebunt. Erit enim per
huius pportio a g lateris ad a b tanꝗ sinus anguli
a b g ad sinum anguli a g b, tribus autem harum no‑
tis existentibus quarta quātitas nota ueniet. inde ergo
angulus a g b notificabitur, & tandem tertius b a g angulus latere nō poterit.
Constat deniꝗ utrunꝗ laterum a b & a g ad ipsam basim b g notam habitum
ire proportionem, si supra memorata repetieris, quod quidem corollarij uice li‑
buit annectere.

XIX.

Datis duobus casibus cū differentia laterū utrūꝗ eorū percontari.

Est enim quod sub differentia casuum & ipsa basi continetur æquale ei, quod
sub differentia laterum atꝗ ipsorum congerie continetur: eo igitur cognito & dif
ferentia laterum data per primi huius congeries laterum mensurabitur: unde
etiam utrunꝗ eorum noticiæ subijcietur.

XX.

Si quis differentiam laterum dederit, differentiamꝗ casuum cū an
gulo quē basis subtendit, omnia latera cognita recipiet.

Esto triangulus a b g, in quo perpendicularis a d duos casus b d & d g secer
nat, quorum differentia e g sit data: differentia etiam laterū quæ sit h g nota sup‑
ponas

and adds too little usefulness. Therefore the rest is left to your ingenuity. Besides unless you are thoroughly unskillful, you can adjust your proof to a figure falling another way.

Theorem 17

If two angles and either segment are given, all the sides as well as the perpendicular may be found.

If *ABG* is a triangle of the type proposed, in which perpendicular *AD* separates the base into two segments *BD* and *DG*, of which one — for example, *BD* — is known along with two angles of △ *ABG*, then all its sides can be identified.

By the hypothesis with the support of I.32 of the *Elements*, the three angles of △ *ABG* are known. Therefore right-triangle portion *ABD*, having acute ∠ *ABD* known together with side *BD*, will disclose its other two sides *AB* and *AD* by Th. I.[29] above. Hence, in the triangular portion *AGD*, which has acute ∠ *AGB* known together with side *AD*, each of the lines *AG* and *GD* will be found by the same Th. I.[29] above. Therefore, when the two [lines] *AD* and *DG* are combined, the entire base results, and the whole problem is completed.

Theorem 18

If the ratio of two sides of a triangle is given together with the angle opposite one of them, the other two angles can be measured.

If *ABG* is such a triangle, whose two sides *AB* and *AG* have a known ratio, and if ∠

ABG is given, then the other angles may be found.

By Th. [1] above the ratio of side *AG* to side *AB* is as that of the sine of ∠ *ABG* to the sine of ∠ *AGB*. Furthermore, because three of these are known, the fourth quantity becomes known. Therefore ∠ *AGB* will be identified, and then the third ∠ *BAG* can be discovered. Finally, each of the sides *AB* and *AG* is established to have a known ratio to the base *BG*, if you will deduce it from the foregoing, which you may wish to append as a corollary.

Theorem 19

If two segments are given together with the difference of the sides, each of them may be investigated.

The product of the difference of the segments times the base itself is equal to the product of the difference of the sides times their sum. Therefore, because this is known and because the difference of the sides is given, by Th. I.[17] above the sum of the sides will be measured. Hence each of them will be determined.

Theorem 20

If the difference of the sides and the difference of the segments are given together with the angle which the base subtends, all the sides may be known.

If *ABG* is a triangle in which the perpendicular *AD* divides [the base into] two segments *BD* and *DG* whose difference *EG* is given, and if the difference of the sides, *HG*, is given

ponatur cum angulo b a g .Dico, ꝙ omnia latera &
omnes anguli cognoscendi uenient.Descendat nanꝗ
ex uertice trianguli à linea a k ,diuidens angulum b a
d per æqualia:erit itaꝗ ꝓportio a g ad g k ,sicut du
arum b a , a g coniunctorum ad basim b g ,quemad
modum superius in huius ratiocinati sumus: ꝓpor=
tio autē b a , a g ad basim b g est,ut e g ad h g per

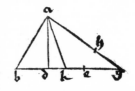

    primi huius,& secundā partem sexti elementorum,quæ cum sit nota ꝓpter ter=
minos suos ex hypothesi datos,erit & proꝑortio a g ad g k cognita:cunꝗ angu
lum b a g dederit hypothesis,erit eius medietas g a k cognita,quare per corolla=
rium huius angulum a g k siue a g b notum comparabimus, & ideo tertius
angulus b a g trianguli ꝓpositi non latebit.inde quoꝗ per huius proportio a
g lateris ad latus a b scietur, & diuisim proportio hæ differentiæ laterum notæ
ad latus breuius a b nota erit,unde & latus a b & tandem reliqua omnia ex supra
dictis cognoscemus .

X X I.

Datis duobus lateribus trianguli cuiuslibet cum proportione casu
um,quantitatem basis agnoscere.

    Sint duo latera a b & a g trianguli a b g
cognita,proportioꝗ casuum b d & d g sit da=
ta.Dico, ꝙ basis ipsa nota proueniet. Est enim
differentia quadratorum a b & a g nota ꝓpter
hypothesim, æqualis differentiæ quadratorum
b d & d g,quemadmodū in huius ostēdimus,
cunꝗ proportio casuum sit data,erit & ꝓportio quadratorum suorum data: & di
uisim differentia huiusmodi quadratoꝶ ad quadratum casus minoris b d notam
habebit proportionem,cūꝗ differētia ipsa sit nota,erit & quadratum casus mino=
ris cognitum,unde & casus ipse minor & deinceps reliquus innotescent, tota igit
basis nota elicietur.Non mireris autem, ꝙ hactenus ut plurimū perpendicularem
intra triangulū cadere supposuerim,quāuis nonnunꝗ extra triangulum cadere co
gatur,habet enim hoc omnis triangulus infallibiliter propriū , ꝙ ab aliquo pun=
ctorum eius angularium ad latus sibi oppositū ducibilis est perpendicularis una
intra angulum ipsum casura.Qđ si extra triangulum perpendicularis ceciderit,
paucis rebus mutatis & cognitu facilibus,quicquid facto opus est,cōsequeris :nol
lem equidem ingenium tuum pauculis quibusdam inueniendis non fatigari.

X X I I.

Datis duobus casibus cum proportione laterum , utrunꝗ eorum
dimetiri.

    Hæc uidetur conuertere præcedentem,deductionem autem eam prorsus ha=
bet quā præcedens,quāobrem tanꝗ satis lucubratā tibi relinquo.

X X I I I.

Data differentia duorum laterum , differentiaꝗ duorum casuū cū
ipsa perpendiculari cognita omnia latera propalare.

    Sit talis triangulus a b g,cuius duo latera a b & a g differentiā habeant notā
b g ,ductaꝗ ꝑpendiculari a d duorum casuum b d & d g,differentia sit e g : hæ
duæ diffe=

together with ∠ *BAG*, then all the sides and all the angles may be found.

Let *AK* descend from vertex *A* of the triangle, bisecting ∠ *BAD*.* Then the ratio of *AG* to *GK* will be as that of the sum of *BA* plus *AG* to the base *BG*, just as we have reasoned previously in Th. [7] above. Furthermore, the ratio of *BA* plus *AG* to the base *BG* is as that of *EG* to *HG* by Th. I.[45] above and the second part of VI of the *Elements*. Since this [ratio] is known because its terms were given by the hypothesis, the ratio of *AG* to *GK* will also be known. And since the hypothesis gave ∠ *BAG*, its half *GAK* will be known. Therefore, by the corollary of Th. [18] above, we will make ∠ *AGK*, or *AGB*, known. Thus the third angle, *BAG*,† of the given triangle will be revealed. Then by Th. [6] above the ratio of side *AG* to side *AB* will be known, and by division the ratio of the known difference of the sides to the shorter side *AB* will be known. Hence side *AB* and, finally, all the others will be found from what was said above.

Theorem 21

If any two sides of a triangle are given together with the ratio of the segments, the length of the base can be found.

If two sides *AB* and *AG* of △ *ABG* are known and the ratio of the segments *BD* and *DG* is given, then the base itself can be determined.

The difference of the squares of *AB* and *AG*, known because of the hypothesis, is equal to the difference of the squares of *BD* and *DG*, just as we showed in Th. [I.44] above. And since the ratio of the segments is given, the ratio of their squares will also be given. By division the difference of these squares will have a known ratio to the square of the shorter segment *BD*. And since the difference

itself is known, the square of the shorter segment will also be known. Hence the shorter segment itself and subsequently the other segment will be found. Therefore the total base will be known.

Do not wonder that up to this point the perpendicular of most [of the triangles] has been assumed to fall within the triangle although sometimes it must fall outside the triangle, for every triangle infallibly has this characteristic: that [at least] one perpendicular drawn from one of its vertices to the side opposite that [vertex] will fall within the [tri]angle. But if a perpendicular were to fall outside the triangle, when things are changed a little and [thus] more easily recognized, you will obtain whatever is to be done. For my part I am unwilling to tire your ingenuity with such insignificant findings.

Theorem 22

If two segments are given together with the ratio of the sides, each of them can be measured.

This is seen to be the converse of the preceding [theorem]. Moreover, it has a straightforward proof, as the preceding [did]. Whereupon I leave it to you for homework.

Theorem 23

If the difference of two sides is given and the difference of the two segments is known together with the perpendicular itself, all the sides can be found.

If *ABG* is such a triangle whose two sides *AB* and *AG* have a known difference *HG*, and if, when perpendicular *AD* is drawn, the difference of the two segments *BD* and *DG* is *EG*,

*For *BAD* read *BAG*.
†For *BAG* read *ABG*.

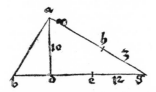

duæ differentiæ sint datæ,& ipsa perpendicula
ris a d data.Dico,ꝗ omnia latera trianguli no
ta concludentur.Per artem rei & census hoc ꝓ=
blema absoluemus.Detur ergo differentia late
rum ut 3 ,differentia casuum 12.& perpendicu
laris 10.pono pro basi unã rem,& pro aggrega
to laterum 4 res,nã proportio basis ad congeri
em laterũ est ut h g ad g e ,scilicet unius ad 4.erit ergo b d ¼ rei,minus 6,sed a b
erit 2 res demptis ½.duco a b in se,prꝙlucuntur 4 census & 2¼.demptis 6 re=
bus,itẽ b d in se facit ¼ census,& 36 minus 6 rebus:huic addo quadratum de 10
qui est 100.colliguntur ¼ census & 136 minus 6 rebus,æquales uidelicet 4 censi=
bus & 2¼ demptis 6 rebus.Restaurando itaꝗ defectus,& auferendo utrobiꝗ æ=
qualia,quemadmodum ars ipsa præcipit,habebimus census aliquot æquales nu=
mero,unde cognitio rei patebit,& inde tria latera trianguli more suo innotescẽt.

XXIIII.

Datis tribus lateribus trianguli rectilinei , diametrum circuli eum
circumscribentis inuenire.

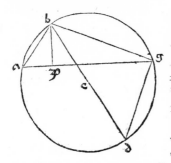

Hæc tametsi de angulis trianguli inueniendis
nihil proponat,utilis tamen sequentibus uidebit.
Sint tria latera a b, b g & g a;trianguli a b g
nota,quærimus diametrum circuli eum circũscri
bentis.Esto circulus huiusmodi a b g d,oportet
autem duos angulos quicunꝗ fuerint trianguli a
b g esse acutos,qui sint,uerbi gratia,a & g,quos
facile cognosces,si primo triangulorum libello sa
tis incubuisti;demittaturꝗ à puncto b perpendi
cularis b 3 ,& diameter circuli prædicti b e d,
cuius terminus d copuletur puncto g per lineã
d g .habes itaꝗ duos triangulos a b 3 & b d g æquiangulos,uterꝗ enim angulo
rum b a g & b d g in circumferentia consistens suscipit arcum b g , sed uterꝗ
angulorum a 3 b & b g d rectus est: a 3 b quidem ex dispositione figuræ,b g d
autem ex dispositione & 30.tertij elementorum,quare & tertius tertio æqualis cõ
uincetur;unde & per quartã sexti proportio 3 b ad b g est ut a b ad b d ,tres au
tem harum notæ sunt,duæ scilicet a b ad b g per hypothesim : perpendicularis
uero b 3 ex primi huius inuenitur,ergo & quarta,quæ est diameter circuli, no=
ta ueniet,quod expectabas ostendendum.Elegimus autem duos angulos acutos,
ut perpendicularis intra triangulum coartaretur facilitatis gratia,nam si caderet
extra,parumper uariandus esset processus.Quòd si acciderit quadratum alicuius
trium laterum quadratis duorum reliquorum laterum simul iunctis æquiualere:
uerbi gratia, quadratum a g æquari duobus quadratis linearum a b & b g,
erit a g diameter circuli circumscribentis triangulũ,neꝗ ampliori quæsito opus
est:fiet enim per ultimã primi elementorum angulus a b g rectus,& ideo per cõ
uersionem 30.tertij a b g semicirculus habebitur,& a g diametẽr circuli, quod
erat exequendum.

XXV.

Si basim trianguli notam acceperimus cũ perpendiculari siue area
trianguli,eum quoꝗ quem basis subtendit angulum datum habueri=
mus,utri

and if these two differences are given and perpendicular *AD* itself is given, then all the sides of the triangle can be determined. This problem we will prove by algebra.

Therefore let the difference of the sides be given as 3, the difference of the segments as 12, and the perpendicular as 10. Take the base as x, and take the sum of the sides as $4x$, for the ratio of the base to the sum of the sides is as *HG* to *GE*, or as unity to 4. Therefore *BD* will be $\frac{1}{2}x - 6$, while *AB* will be $2x - \frac{3}{2}$. Square *AB*, producing $4x^2 + 2\frac{1}{4} - 6x$. Similarly, square *BD* to make $\frac{1}{4}x^2 + 36 - 6x$. To this add the square of 10, which is 100, giving $\frac{1}{4}x^2 + 136 - 6x$, which is equal to $4x^2 + 2\frac{1}{4} - 6x$. Therefore by restoring the deficits and subtracting the equalities on both sides, as the art dictates, we have so many x^2 equal to a number. Hence x will be found, and therefore the three sides of the triangle will be known, [each] by its function.

Theorem 24

If three sides of a rectilinear triangle are given, the diameter of the circle circumscribed around it can be found.

Although this proposes nothing for finding the angles of a triangle, still its usefulness will be seen in the subsequent [theorems]. If the three sides *AB*, *BG*, and *GA* of △ *ABG* are known, we shall seek the diameter of the circle circumscribed around it.

Let this circle be *ABGD*. Moreover, any two angles — *A* and *G*, for example — that belong to △ *ABG* must be acute. These are easily recognized if you have pondered over the First Book of the *Triangles*. Let perpendicular *BZ* be dropped from point *B*, and let the diameter of the aforementioned circle be *BED*, whose end [point] *D* is joined to point

G by line *DG*. Thus you have two similar triangles *ABZ* and *BDG*, because each of the angles *BAG* and *BDG*, resting on the circumference, determines arc *BG*; furthermore, each of the angles *AZB* and *BGD* is a right angle, *AZB* from the construction of the figure and *BGD* from the construction and III.30 of the *Elements*. Therefore the third [angle of △ *ABZ*] is proven equal to the third [angle of △ *BDG*]. From this and by [Euclid] VI.4 the ratio of *ZB* to *BG* is as that of *AB* to *BD*. Furthermore, three of these are known — namely, the two [lines], *AB* to *BG*, [are known] by hypothesis while the perpendicular *BZ* is found from Th. I.[29] above. Thus, the fourth, which is the diameter of the circle, will become known. Q.E.D.

We chose the two angles to be acute so that the sample perpendicular would fall easily within the triangle, for if it had fallen outside, the procedure would have been changed [only] a little.

But if it were to happen that the square of one of the three sides was equal to the squares of the two other sides added together — for example, if the square of *AG* was equal to [the sum of] the squares of lines *AB* and *BG* — then *AG* would be the diameter of the circle circumscribed around the triangle, and with no further search the work is [completed]. This is so because, by the last theorem of I of the *Elements*, ∠ *ABG* is a right angle, and therefore, by the converse of [Euclid] III.30, *ABG* will be a semicircle and *AG* will be the diameter of the circle. Q.E.D.

Theorem 25

If we have the base of a triangle known along with the perpendicular — which is the area of the triangle — and if we also have the given angle which the base subtends, mus, utriuſcp lateris noticiam extemplo reddemus.

Sit triangulus a b g, baſim b g no=
tam habens cum perpendiculari a d, ſi=
ue cum area ſua, nã alterum ex altero pen
det:ſit quocp angulus b a g datus.Dico
cp utruncp laterum eius noticiã attinge=
mus. Detur enim angulus b a g obtu=
ſus,extendaturcp g a donec perpendicu̅
laris b k ex puncto deſcenſura reſidere
poſſit in eadem, erit itacp propter angu=
lum b a k notum proportio a b ad b k
nota per   huius:proportio autẽ a b
ad b k eſt tancp eius quod a b & a g ad
id,quod ſub b k & a g continetur,qua=

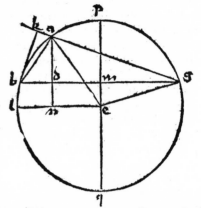

re & illa proportio nota erit:quod autem fit ex b k in a g notum eſt,quia duplũ
areæ trianguli,igitur & quod ſub a b & a g continetur,rectangulum erit notũ:
exarata demum circulo triangulũ a b g circumſcribente,diameter eius notifica=
bitur ex præcedenti propter perpendicularem a d notã cum eo, quod ſub a b &
a g continetur noto:huiuſmodi diameter ſit p q,ſecans cordã b g per medium,
& ideo orthogonaliter in puncto m,eductiscp ex centro e tribus ſemidiametris
e g, e a & e l,quæ æquediſtet cordæ b g, cui incidat perpendicularis a d ſatis
prolongata in puncto n .propter ſemidiametrum igitur e g & dimidiã baſi m g
notas cum angulo e m g recto,notificabitur linea e m,cui æqualis eſt ipſa d n:
hinc tota a n cognita reſultabit,quæ deinceps auxilio ſemidiametri e a notæ,
angulo apud n recto exiſtente,ſuſcitabit lineã n e cognitã,cui æqualis habetur
linea d m,qua dempta ex dimidia baſi b m,relinquetur caſus minor:ea uero ad
iecta,reſultabit caſus maior cognitus:hinc ex caſibus notis & perpendiculari a d
latera duo cognita fient.Hæc autem tenent,quando duorum latẽ a b & a g al
terum altero maius extiterit,cuius rei indicium erit,ſi linea d n ex perpẽdiculari
uidelicet a d & linea m e reſultãs,minor fuerit ſemidiametro:ſi enim eſſet æqua
lis ſemidiametro,duo latera a b & a g neceſſario fuiſſent æqualia,unde etiã per=
pendicularis a d diuiſiſſet baſim per æqualia:ſic ex perpendiculari nota cum di=
midia baſi reſultaret utruncp latus cognitum.In his rebus demonſtrandis non ſi=
ſto gradum,cum facilia admodum oſtenſu reputentur.   Si autem angulus b a g
fuerit rectus,erit b g neceſſario diameter circulo trian
gulum b a g circumſcribentis,cuius medietati æqua
lis ſi fuerit perpendicularis a d data,erunt duo latera
trianguli æqualia,utruncp uidelicet corda quadrantis,
unde facillime cognoſcetur.Si uero perpẽdicularis fue
rit minor dimidiæ baſi,erũt duo latera inæqualia, edu
cta igitur ſemidiametro e a ,quæ eſt æqualis dimidiæ
baſi,nota erit d e, hinc ut prius utercp caſuum cum
utrocp laterũ innoteſcent.Quod ſi angulus b a g acu
tus offeratur,erit portio b a g maior ſemicirculo, &

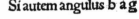

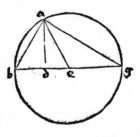

ideo ſemidiameter e l ſecabit perpendicularem:quare cæteris omnibus ut antea
procedentibus,niſi cp perpendicularis b k intra triangulũ cadit,linea d n æqua=
lem ipſi e n ex perpendiculari a d nota minuemus,inuentamcp tandem lineam
                                                            H   d m ex

each side will be identified.

If *ABG* is a triangle which has base *BG* known together with perpendicular *AD* — that is, together with its area for the one depends upon the other — and if ∠ *BAG* is also given, then each of the sides may be found.

Let ∠ *BAG* be obtuse, and let *GA* be extended until a perpendicular *BK*, which will be drawn from the vertex [*B*], can rest on [*GA* extended]. Then, because ∠ *BAK* is known, the ratio of *AB* to *BK* will be known by Th. [1] above. Furthermore, the ratio of *AB* to *BK* is as that of the product of *AB* times *AG** to the product of *BK* times *AG*. Therefore this latter ratio will be known. Furthermore, the product of *BK* times *AG* is known because it is twice the area of the triangle; hence the rectangular product of *AB* times *AG* will be known.

Now, once the circle circumscribing △ *ABG* has been drawn, its diameter will be disclosed by the preceding [theorem] because perpendicular *AD* is known [and because] the product of *AB* and *AG* is known. Let this diameter be *PQ*, bisecting chord *BG* perpendicularly at point *M*. From the center *E* three radii, *EG*, *EA*, and *EL*, are drawn, the last [being] parallel to chord *BG* and [being] intersected [by] perpendicular *AD*, sufficiently extended, at point *N*. Then because radius *EG* and half the base, *MG*, are known together with right ∠ *EMG*, line *EM* will be found; *DN* itself is equal to [*EM*]. Hence the total [line] *AN* will be known, which, in turn, reveals line *NE* because radius *EA* is known and the angle at *N* is a right angle. Line *DM* is equal to [line *NE*]. When [*DM*] is subtracted from half the base, *BM*, the shorter segment, is left, [and when *DM* is] added [to half the base], the longer segment results. Hence from the known segments and perpendicular *AD*, the two sides may be found.

These statements hold [true] when one of the two sides *AB* and *AG* is longer than the other. The criterion for this situation will be whether line *DN*,† consisting of the perpendicular *AD* plus line *ME*, is shorter than the radius. If [*AN*] is equal to the radius, the two sides *AB* and *AG* are necessarily equal and hence perpendicular *AD* has bisected the base. In that case, from the known perpendicular together with half the base, each side becomes known. In showing these [following] conditions, I am not presenting [every] step since [all of them] can be determined with complete ease. If ∠ *BAG* were a right angle, *BG* would necessarily be the diameter of the circle circumscribing △ *BAG*. And if the given perpendicular *AD* were equal to half this [diameter, or base *BG*], the two sides of the [right] triangle will be equal, and clearly each will be a chord of the quadrant and hence will be easily known. If, however, the perpendicular is less than half the base, the two sides will be unequal. Thus, when radius *EA*, which is equal to half the base, is drawn, *DE* will be known. Hence as above, each of the segments with each of the sides may be found.

But if ∠ *BAG* were given as acute, then portion *BAG* [of the circle] would be larger than a semicircle, and then the radius *EL* would intersect the perpendicular. Therefore because all the other [parts of the proof] proceed as above, unless perpendicular *BK* falls within the triangle, we subtract line *DN*, equal to *EN*,‡ from known perpendicular *AD*. And we subtract line *DM*, which is ultimately found,

*For *eius quod* read *eius sub quod*.
†For *DN* read *AN*.
‡For *EN* read *EM*.

d m, ex dimidia bafi dememus, ut relinquatur cafus minor, perpendiculari faltem
a d intra angulum cadente, quod accidit, dum linea d m minor dimidia bafi re=
fultabit: aut econtra dimidiā bafim ex linea d m minuemus, fi perpendicularis ex
tra triangulum ceciderit, id eft, fi d m dimidiā bafim fuperauerit, quæ duæ fi fue=
rint æquales, conftabit perpendicularem a d coincidiffe lateri breuiori: quod au=
tem in hac demonftratione reftat, fi pauiculum habes ingenium, eniti poteris.

X X V I.

Data area trianguli cum eo, quod fub duobus lateribus continetur
rectangulo, angulus quem bafis refpicit, aut cognitus emerget, aut cū
angulo cognito duobus rectis æquipollebit.

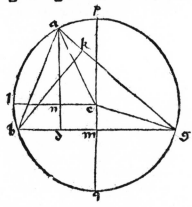

Refumptis figurationibus præcedētis,
fi perpendicularis b k uerfus lineā a g p
cedens, extra triangulum ceciderit, erit per
ea, quæ in præcedenti commemorauimus,
proportio b k ad b a nota, & ideo per
primi huius angulum b a k notum acci=
piemus, fic angulus b a g cum angulo b
a k noto duobus rectis æquiualebūt. Siue
ro perpendicularis b k intra triangulum
ceciderit, quemadmodū in tertia figuratio
ne præcedentis cernitur, erit ut prius a b
ad b k notā habens proportionē, & ideo
angulus b a k fiue b a g notus conclude
tur. At fi perpendicularis b k coinciderit lateri a b, neceffe eft angulum b a g
fuiffe rectum, & ideo cognitum, quod quidem accidit, quando area trianguli pro=
pofiti æquatur ei, quod fub duobus lateribus eius continetur rectangulo.

X X V I I.

Data differentia duarum linearum, quæ rectangulum fpaciū con=
tinent notum, utranq earum dimetiri.

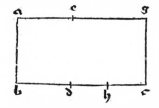

Sint duæ lineæ a b & b c inæquales, rectangu
lum fpacium a c continentes notum, fitq diffe=
rentia earum d c cognita. Dico, ꝗ utraq earum
nota reddet. Erit enim quod fit ex d b in b c pa=
rallelogramū rectangulū notum, ꝗ b d fit æqua=
lis ipfi a d : diuifaq d c differentia per medium
in h, erit quadratum lineæ d h cognitum, quod
quidem adiectum rectangulo a c noto, per fextā fecundi conficiet quadratum b
h cognitum: hinc ergo cofta fua fcilicet linea b h notificabitur, ex qua fi reiecerī=
mus lineā d h notā, manebit b d, & ideo ipfa a b nota: fed & eidē b h notæ ad=
demus medietatem differentiæ, fcilicet lineam h c notam, ut refultet tota b c co
gnita, quod erat abfoluendum.

X X V I I I.

Data area trianguli, & angulo quem bafis refpicit cognito cum dif
ferentia laterum, utrunq eorum innotefcet.

Per mo=

from half the base so that the shorter segment is left. In any event, this happens if perpendicular *AD* falls inside the [tri]angle [or] as long as line *DM* will be less than half the base. On the other hand, if the perpendicular falls outside the triangle, that is, if *DM* is greater than half the base, then we subtract half the base from line *DM*. If the two [*DM* and half the base] are equal, it is established that the perpendicular *AD* coincides with the shorter side. What remains in this proof, if you have a little ingenuity, you can find.

Theorem 26

If the area of a triangle is given together with the rectangular product of the two sides, then either the angle opposite the base becomes known or [that angle] together with [its] known [exterior] angle equals two right angles.

When the preceding constructions are recalled, if perpendicular *BK*, proceeding toward side *AG*, falls outside the triangle, then, by that [theorem] which was mentioned in the preceding [theorem], the ratio of *BK* to *BA* will be known, and therefore, by Th. I.[28] above, ∠ *BAK* is found. Thus ∠ *BAG* together with known ∠ *BAK* will be equal to two right angles.

However, if perpendicular *BK* falls within the triangle, just as was noted in the third construction of the preceding theorem, then, as before, the ratio of *AB* to *BK* is a known ratio. Hence ∠ *BAK*, or *BAG*, is found.

And if perpendicular *BK* coincides with side *AB*, ∠ *BAG* is necessarily a right angle and therefore known. This will happen when the area of the given triangle is equal to the rectangular product of the two sides.

Theorem 27

If the difference of two sides, which contain a known rectangular area, is given, then each of them can be measured.

If the two lines *AB* and *BC* are unequal and contain a known rectangular area *AC*, and if their difference *DC* is known, then each of them can be found.

The rectangular parallelogram which is the product of *DB* and *BC* will be known because *BD* is equal to *AD*.* After the difference *DC* is divided in half at *H*, the square of line *DH* will be known. This, added to the known rectangle *AC*, makes square *BH* known by [Euclid] II.6. Therefore its side — namely, line *BH* — may be determined. From this, if we subtract the known line *DH*, *BD* is found, and therefore *AB* itself is known. Furthermore, we add half the difference — namely, known line *HC* — to the same known *BH* to make the total [line] *BC* known. Q.E.D.

Theorem 28

If the area of a triangle is given and the angle opposite the base is known together with the difference of the sides, each of [the sides] can be found.

*For *AD* read *AB*.

Per modum enim circa huius explanatum concludemus, quod sub duobus lateribus cōtinetur cognitum: cuncꝗ differentiā eorum notā attulerit hypothesis, erit per præcedentem utruncꝗ eorum cognitū, cuius gratia fatigati sumus.

XXIX.

Si ab angulo trianguli noto descendat linea quædam cognita, basim datam diuidens per æqualia, utruncꝗ latus, reliqui etiam anguli non erunt ignoti.

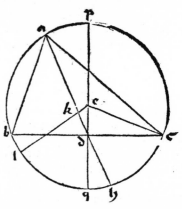

Sit triangulus a b c, angulum b a c notum habens, à cuius uertice a descendat linea a d nota respectu basis b c, quā per medium scindit in puncto d. Dico, ꝗ utruncꝗ laterum a b & a c notum ueniet cum reliquis duobus angulis. Circumscribo enim huic triangulo circulum a b h c super centro e, cuius diameter p q p punctum d transeat, orthogonaliter secans ipsam b c datam lineā; continueturcꝗ linea a d, donec occurret circumferentiæ circuli in h, educantur deniꝗ duæ semidiametri e c quidem cōterminalis duobus lateribus trianguli propositi: e l autem secans cordā a h si possibile sit per medium & orthogonaliter: erit itaꝗ angulus c e d æqualis angulo b a c dato, & ideo per primi huius propter angulum apud d rectum, proportio d c notæ ad lineam d e nota erit, quare linea d e nota habebitur. est autem quadratum lineæ b d notæ ꝓpter hypothesim æquale ei, quod fit ex a d in d h per tertij elementorū, cuncꝗ a d sit nota, erit & d h inde quoꝗ nota a h cum eius medietate a k non ignorabitur: hinc residua k d scita dabitur. ex duabus itaꝗ lineis d e & d k cū angulo k recto per primi huius cognoscetur angulus d e k, & ideo arcus l q notus accipietur, propter triangulum autem d e c notorum angulorum linea d c, & ideo etiam dupla eius b c respectu semidiametri circuli e c notam habebit proportionem, unde & corda a h, quæ nota prius erat respectu lineæ b c, iam respectu diametri circuli huius cognita dabitur: unde & per tabulam sinuū aut cordarum arcus a b h innotescet. oportuit autem arcum b q c esse notum propter angulum b a c datū, dempto igitur arcu l q prius noto ex arcu b q scilicet medietate arcus b q c, relinquetur arcus b l notus, qui deinde reiectus ex arcu a l scilicet medietate arcus a l h, manebit arcus a b notus, cui si addideris totum arcum b q c notum, resultabit arcus a b c cognitus: horum duorum arcuum cordas ex tabula colligemus, sic duo latera trianguli propositi respectu diametri circuli nota uenient: erat autem & basis b c respectu eiusdem nota, unde & ipsa latera respectu basis mensuratas habebunt longitudines, angulos autem trianguli ꝓpositi reliquos latere non sinunt duo arcus a b & a c, in quos ipsi cadunt supra circumferentiam consistentes. Contingit autem lineam a h esse diametrum circuli, quod quando fiet, iam cōmemorata satis edocebunt, conclusimus enim tam ipsam a h, ꝗ diametrum circuli respectu basis b c notam habere quantitatem. Poterit etiam quispiam dare angulum b a c rectum, unde basis b c fieret diameter circuli, & a d semidiameter: tunc autem problema erit uarium, nisi alia quædā

H 2   conditio

By the method explained in Th. [25] above, we can find the product of the two sides. And since the hypothesis gives their difference, then each of them may be found by the preceding theorem. The ease of this bores us.

Theorem 29

If, from a known angle of a triangle, a known line is drawn that bisects the given base, each side and the remaining angles can be found.

If △ *ABC* has a known ∠ *BAC* from whose vertex *A* the line *AD*, known in the same terms as base *BC*, is drawn [such that] it bisects [the base] at point *D*, then each of the sides *AB* and *AC* becomes known along with the other two angles.

Around this triangle circumscribe circle *ABHC* with *E* as center. Let a diameter *PQ* of this [circle] pass through point *D*, intersecting the given line *BC* perpendicularly. Let line *AD* be extended until it meets the circumference of the circle at *H*. Finally, let two radii be drawn, with *EC* [to the point of intersection of] two sides of the given triangle and *EL* intersecting chord *AH*, if possible, equally and perpendicularly. Then ∠ *CED* is equal to given ∠ *BAC*, and thus, because the angle at *D* is a right angle, by Th. I.[30] above the ratio of known line *DC* to line *DE* will be known. Hence line *DE* will be found. Furthermore, the square of line *BD*, known because of the hypothesis, is equal to the product of *AD* and *DH* by III.[not given] of the *Elements*. And since *AD* is known, then both *DH* and *AH* will be known, together with half [of *AH*], *AK*. Hence, the rest [of the line], *KD*, will be found. Thus, from the two lines *DE* and *DK* with ∠ *K* a

right angle, ∠ *DEK* is found by Th. I.[27] above, and therefore arc *LQ* is known. Because △ *DEC* has known angles, line *DC*, and therefore its double *BC*, will have a known ratio to radius *EC* of the circle. Hence chord *AH*, which was known previously in terms of line *BC*, will now be known in terms of the diameter of this circle. Therefore by the table of sines, or chords, arc *ABH* will be found. Furthermore, arc *BQC* is necessarily known because of the given ∠ *BAC*. Thus when arc *LQ*, known previously, is subtracted from arc *BQ* — namely, half of arc *BQC* — arc *BL* is left known. [Arc *BL*] subtracted from arc *AL* — namely, half of arc *ALH* — will leave arc *AB* known. If you add the whole known arc *BQC* to [arc *AB*], arc *ABC* will result. We find the chords for these two arcs from the table. Thus the two sides of the given triangle become known in terms of the diameter of the circle. Moreover, base *BC* was known in terms of the same [diameter]. Hence the sides themselves will have lengths measured in terms of the base. Furthermore, the two arcs *AB* and *AC*, which are determined on the circumference of the circle by the [angles *ACB* and *ABC*] themselves, will reveal those other angles [*ACB* and *ABC*] of the given triangle.

When it happens that line *AH* is the diameter of the circle, as the previously mentioned [theorems] will show sufficiently, we may conclude that both *AH* and the diameter of the circle have a known length in terms of base *BC*. Still some [other proposal] could give ∠ *BAC* as a right angle; then base *BC* would be the diameter of the circle and *AD* the radius. Indeed, in [each] problem [the situation] will vary unless in another [problem] the [same] particular

conditio accefferit. In hac autem figuratione angulum b a c acutum fubiecimus
qui fi obtufus efferretur, quáuis figuratio paulifper uariaretur, proceffus tamen
idem fermè ad intentum nos perducet.

XXX.

Si quis triangulus duo latera habeat inæqualia, à quorum commu
ni termino defcendat linea angulum per æqualia diuidens, bafim autem per inæqualia, fitʠ ipfa linea diuidens nota cum portionibus bafis diuifæ utrunʠ latus cognitum fri.

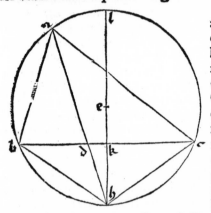

Sit talis triangulus a b g , cuius latus
a b breuius latere a c: à cuius angulo a
defcendat linea a d nota, angulum quidē
b a c diuidens per medium, bafim autem
in duas partiales a d & d c notas. Dico,
ʠ utrunʠ laterum a b & a c innotefcet.
Circumfcribatur enim huic triangulo cir
culus a b h c centrum e habens: proten
faʠ a h ufʠ ad occurfum circüferentiæ
in h, ducant duæ cordæ b h & c h, quas
côstat esse æquales ʠpter angulum b a c
per æqualia diuifum. ducatur rurfus diameter circuli l h, quæ cum diuidat arcum
b c per medium, diuidet etiam per tertij

cordam eius b c per æqualia, unde & per tertij orthogonaliter eam fecabit. cū
autem utraʠ linearum b d & d c fit nota, erit per tertij fexti & primi huius
linea d h nota: eſt autem & d k cognita uidelicet differentia dimidiæ bafis & mi
noris fectionis: angulo igitur k recto exiſtente, linea k h per primi huius nota
pronunciabitur, ex qua demum & dimidia bafi cognita, nifi primi huius mentia
tur, utraʠ cordarum b h & h c æqualium nota refultabit. quadrangulum itaʠ a
b h c circulo infcriptum, duas diametros a h & b c notas habebit, quod autem
fub eis continetur, æquū eſt duobus rectangulis, quorum alterum fub b h & a c,
alterum fub h c & a b continetur: hoc enim alibi demonſtratum eſt. Hæc autem
duo rectangula parallelograma æquantur ei, quod fub b h & congerie duorū laterum continetur, ʠpter æqualitatem linearum b h & h c. quod igitur fub b h
& aggregato laterum a b & a c continetur, erit cognitum: unde & propter line
am b h cognitā primi huius ratiocinante, côgeries duorum laterum nota pro-
ueniēt. eſt autē ʠportio b d ad d c ficut a b ad a c per tertiā fexti, & côiunctim
b c ad c d ficut congeries duorū laterū ad ipfum latus a c: cunʠ tres huiufmo
di quantitatū fint notæ, erit & quarta fcilicet linea a c inuenta: unde & reliquum
latus a b non poterit latere. hæc pro lateribus cognofcendis, ad angulos autem
inueniendos, iam paratum habes iter, fi huius intento tuo accomodaueris.

Poterimus 3 o huius aliter abfoluere hoc pacto. Sit triangulus a b c, aream ha
bens notam cum bafi b c & angulo b a c. Dico, ʠ utrunʠ laterū a b & a c no
tum prodibit. Intelligo enim huic triangulo circumfcriptū circulum a b d c, in
quo produco cordam a l æquediſtantem ipfi b c, & diametrum d k utriʠ dicta
rum cordarum perpendiculariter incidentem: huic quidem in puncto 3 , illi autē
in puncto h: fitʠ a e perpendicularis ac b c, fi oportuerit prolongatā. quoniā
igitur aream trianguli cum bafi b c notam fubiecimus, erit perpendicularis a e
               scita,

condition is approximated. In this construction we made ∠ *BAC* acute; if it had been given obtuse, although the construction would be slightly changed, the procedure would still take us on nearly the same course.

Theorem 30

If any triangle has two unequal sides from whose point of intersection an angle bisector is drawn which divides the base unequally, and if this bisector is known together with the segments of the divided base, each side may be found.

If *ABG** is such a triangle whose side *AB* is shorter than side *AC* and from whose ∠ *A* a known line *AD* is drawn, dividing ∠ *BAC* in half and dividing the base into two known portions *AD* and *DC*, then each of the sides *AB* and *AC* may be found.

Let a circle *ABHC*, having *E* as center, be circumscribed around this triangle, and when [line] *AH* is extended all the way to meet the circumference at *H*, let the two chords *BH* and *CH* be drawn. It is established that they are equal because ∠ *BAC* is divided equally. Now, let the diameter of the circle, *LH*, be drawn; this [diameter], since it bisects arc *BC*, will bisect its chord *BC* by [Euclid] III.[not given], and therefore by [Euclid] III.[not given] it will intersect [chord *BC*] perpendicularly. Moreover, since each of the lines *BD* and *DC* is known, then by [Euclid] III.[not given], [Euclid] VI.[not given], and Th. I.[52] above, line *DH* will be known. Furthermore, *DK* — namely, the difference between half the base and the shorter segment — is known. Therefore, because ∠ *K* is a right angle, line *KH* will be declared known by Th. I.[26] above. From this and because half the base is known, unless Th. I.[26]

above lies, each of the equal chords *BH* and *HC* may be found. Therefore quadrangle *ABHC*, inscribed within the circle, will have two known diagonals *AH* and *BC*. Furthermore, their product is equal to two rectangular products, one of which is the product of *BH* and *AC* and the other is the product of *HC* and *AB*. This has been proven elsewhere. Moreover, these two rectangular parallelograms are equal to the product of *BH* times the sum of the two sides because of the equality of lines *BH* and *HC*. Therefore the product of *BH* times the sum of sides *AB* and *AC* will be known. For this reason and because line *BH* is known, then, by the argument of Th. I.[17] above, the sum of the two sides may be found. Furthermore, the ratio of *BD* to *DC* is as that of *AB* to *AC* by [Euclid] VI.3, and by addition *BC* to *CD* is as the sum of the two sides to side *AC* itself. And since three of these quantities are known, the fourth — namely, line *AC* — will be found. Hence the other side *AB* cannot be hidden. These sides being known, you have prepared the way toward finding the angles if you adapt Th. [9] above to your attempt.

We can prove Th. 30 differently in this way. If △ *ABC* has [its] area known together with base *BC* and ∠ *BAC*, then each of the sides *AB* and *AC* may be found. Circle *ABDC* is understood to be circumscribed around this triangle, and in this [circle] chord *AL* is drawn parallel to *BC*. Diameter *DK* [is understood] to meet each of the mentioned chords perpendicularly, the one at point *Z* and the other at point *H*. Let *AE* be the perpendicular to *BC*, extended if necessary. Therefore, because the area of the triangle was known together with base *BC*, perpendicular *AE* will be

*For *ABG* read *ABC*.

scita,ipsa enim in basim dimidiam ducta,
conficit aream trianguli datā:hinc æqua‑
lis 3 h ,ob eam rem non erit ignota . pro‑
pter angulum autem b a c & ideo arcum
b c cognitum,erit corda b c cognita per
tabulam sinuũ aut cordarum respectu dia
metri circuli : unde & sagitta 3 d eadem
relatione nota fiet:cum autem perpendi‑
cularis a e cognita habeatur respectu cor‑
dæ b c ,erit & ipsa respectu diametri cir‑
culi nota,& ei æqualis 3 h : tota igitur sa
gitta d h respectu diametri circuli nota
consurget,& ideo arcus a d cognitus da‑
bitur,à quo si dempsero arcum b d ,medi‑

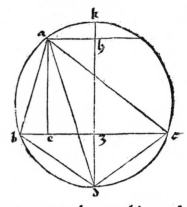

etatem scilicet arcus b c pridem noti,relinquetur arcus a b notus: hinc corda a
b ·cognoscetur respectu diametri circuli,& consequenter respectu lineæ b g . ar‑
cus deniq a b iam notus arcui b c adiectus,totum arcum a d c cognitum su‑
scitabit,& ideo corda eius respectu diametri circuli,tandemcq respectu lineæ b e
manifestabitur.Angulos autem a b c & a c b propter arcus a b & a c iam no
tos nemo ignorabit.

XXXI.

Duobus triangulis supra basim unam notam constitutis, binacq la‑
tera æqualia ac cognita habentibus,quantum uertices eorum distent
inquirere.

Supra basim b c notam constituantur duo trian‑
guli a b c & d b c,quorum utriusq latera sint cogni
ta:ducatur linea a d uertices eorum coniungens,quã
præsens quærit problema.Ex uerticibus a & d duæ de
scendant perpendiculares ad basim,si opus fuerit, sa‑
tis extensam,quæ sint a e & d g ,has perpendicula‑
res quo pacto inuenias,superius cōmemoratum est: si
igitur fuerint æquales ,distantia duorum uerticũ erit

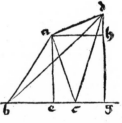

per   primi elementorũ æqualis lineæ e g ,quæ est ag
gregata ex duobus casibus e c & c g in hac figuratione,quos superius mensura‑
re docuimus:unde & distantia uerticum nota erit.Si uero perpendiculares nõ fue
rint æquales,à summitate minoris earum quæ sit,uerbi gratia, a e ,ducatur a h
æquedistans basi,& occurrens reliquæ perpendiculari in puncto h ,quæ nota erit
quoniã æquali lineæ e g aggregato duorum casuũ:lineam quocq d h non igno‑
rabit Geometra,cum h g sit æqualis a e perpendiculari notæ:hinc & ppter an‑
gulum a h d rectum   primi huius ratiocinãte,linea a d latere nõ poterit , quæ
hactenus quærebatur.Ad lineã autem a h cognoscendã nõ semper oportebit col
ligere duos casus,quemadmodũ in præsentiarum iussimus:sed nonnuncq alterũ
ex altero demi,ut relinquatur linea e g siue a h cognita,& sicut hic cõiunximus
casum minorem unius trianguli minori casui alterius,ita interdum casum mino‑
rem unius maiori casui alterius coniungere opus erit:quod quando fiet,tuo inge‑
nio relinquitur discernendum .

H   3   Si duo

known, for when the [perpendicular] is multiplied by half the base, it gives the area of a triangle. Hence *ZH*, equal [to the perpendicular], will be known. Furthermore, because ∠ *BAC*, and therefore arc *BC*, is known, chord *BC* will be known by the table of sines, or chords, in terms of the diameter of the circle. Hence shaft *ZD* is made known in the same terms. Furthermore, since perpendicular *AE* is known in terms of the chord *BC*, it will also be known in terms of the diameter of the circle and [in terms of] its equal *ZH*. Therefore, the total shaft *DH* becomes known in terms of the diameter of the circle. Thus arc *AD* will be found, and if arc *BD* — namely, half of arc *BC*, previously known — is subtracted from [*AD*], arc *AB* is left known. Hence chord *AB* is known in terms of the diameter of the circle and consequently in terms of line *BG*.* Next, arc *AB*, now known, added to arc *BC*, gives the total arc *ADC*, and therefore its chord will be found in terms of the diameter of the circle and, finally, in terms of line *BC*.† Furthermore, no one will be ignorant of angles *ABC* and *ACB* because arcs *AB* and *AC* are already known.

Theorem 31

If two triangles are established upon one known base, and if they each have two respectively unequal,‡ known sides, the distance between their vertices may be found.

Let the two triangles *ABC* and *DBC*, each of whose sides are known, be constructed upon known base *BC*. Let line *AD* be drawn connecting their vertices. The present problem [is to] find this [line].

From vertices *A* and *D* let two perpendiculars fall to the base, extended if the work demands it; these [perpendiculars] are *AE* and *DG*. You will find these perpendiculars in that way explained above. Then if they are equal, by I.[not given] of the *Elements* the distance between the two vertices will be equal to line *EG*, which is the sum of the two segments *EC* and *CG* in this construction. It was shown above [how] to measure these [segments]. Hence the distance between the vertices will be known.

However, if the perpendiculars are not equal, then from the top of the shorter [perpendicular] — for example, *AE* — let *AH* be drawn parallel to the base and meeting the other perpendicular at point *H*. [*AH*] will be known because it is equal to line *EG*, the sum of the two segments. Geometry will also determine line *DH* since *HG* is equal to known perpendicular *AE*. Hence, because ∠ *AHD* is a right angle, then, by the argument of Th. I.[26] above, line *AD* can be found, which was desired at this point.

To obtain line *AH*, it is not always necessary to add the two segments, as suggested in this presentation. But sometimes when the one is subtracted from the other, line *EG*, or *AH*, is left known. And as herein we added the shorter segment of the one triangle to the shorter segment of the other, so sometimes it is necessary to add the shorter segment of the one to the longer segment of the other. When this should be done is left to your ingenuity to ascertain.

*For *BG* read *BC*.
†The print of the original MS is unclear here, but it should read, *tandemque respectu lineae bc.*
‡In the Latin text, for *aequalia* read *inaequalia*.

XXXII.

Si duo triànguli fupra bafim unam notam conftituti fuerint, quo=
rum unus æquicruris, alter autem uarius, & anguli, quos bafis fubten=
dit, dati fuerint cum diftantia duorum uerticum fuorum, bina latera
eorum cognitum iri.

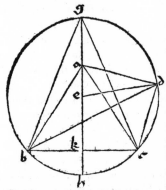

Supra bafim datam b c conftituantur duo tri
anguli, a b c quidem duo latera a b & a c ha=
bens æqualia, d b c autem duo latera habens in=
æqualia, & fit utercʒ angulorū b a c & b d c da=
tus, ductáʒ linea a d diftantia fcilicet duorū ca=
pitum fit data. Dico, ꝗ utriufcʒ eorum duo latera
erunt data. Sit enim pro libito angulus b a c ma
ior angulo b d c, demittatúrcʒ ex puncto a per
pendicularis ad bafim cōmunem, quæ neceffario
diuidet bafim per æqualia, quod fiat in puncto k:
hæc perpendicularis intelligatur utrincʒ indefini
tæ quantitatis: circuli itacʒ circumfcripti triangu
lo d b c circumferentia fecet eam fuperius in g,

inferius autem in puncto h: conftabit itacʒ lineam g h effe diametrum huius cir
culi per corollarium primæ tertij, in qua fit centrum circuli e: ductácʒ femidiame
tro e d & duabus cordis g b & g c, erit angulus b g c æqualis angulo b d c: cū
igitur utercʒ triangulorum a b c & g b c æquicrurium habeat angulum notum
cum bafi data, erunt per   primi huius bina latera eorum nota cum perpendicula
ribus fuis: unde etiam tota diameter g h nota fiet, quadratum enim dimidiæ ba=
fis k c notæ æquatur ei, quod fit ex g k iam nota in k h, hinc k h & confequen
ter tota g h diameter nota ueniet: dempta autem differentia perpendiculariū g a
propter ipfas perpendiculares cognita ex femidiametro g e nota refiduabitur li
nea a e fcita: fic triangulus a e d tria latera nota habens, angulum fuum e a d
cognitum efficiet, ex quo demptus angulus e a c notus, quoniā medietas b a c
dati, relinquet angulum c a d menfuratum: is demum cum duabus lineis c a &
a d notis, tertiam quocʒ c d manifeftabit. Item dimidius angulus b a c, qui eft
b a k adiectus angulo e a d prius cognito, conflabit totum angulum b a d notum, qui cum duobus lateribus a b & a d iamdudum cognitus, lineam b d fu=
fcitabit notam: fic ergo duo latera trianguli d a c dimenfi fumus, quod libuit at=
tingere. Paulo aliter ratiocinabimur circumferentia circuli b c d g fecante per=
pendicularem infra punctum a, quod reuera contingit, dum angulus b d c
maior angulo b a c oblatus fuerit. Quòd fi duo dicti anguli æquales fubijciātur,
circumferentia circuli b c d g per punctum a tranfeat neceffe eft: cognita igit
diametro circuli ut antea, itémcʒ duabus lineis a b & a d, quæ erūt cordæ dicto
circulo infcriptæ, non erit difficile menfurare utrancʒ cordarū b d & c d, fi circa
cordas circuli inueniendas parumper exercearis. quod igitur in hac re fupereft,
tuæ relinquimus induftriæ.

XXXIII.

Si cuiuslibet trianguli angulum per æqualia diuidat linea ad bafim
notā defcendens, fuerintcʒ fectiones bafis inæquales notæ: itémcʒ an=
gulus acutus, quē linea diuidens, cum ipfa bafi continet notus datus,
utruncʒ latus trianguli cognitum reddetur.   In trian=

Theorem 32

If two triangles are established upon one known base [and] one of them is isosceles and the other scalene, and if the angles which the base subtends are given together with the distance between their two vertices, the two sides of each of them can be found.

If two triangles, *ABC* having two sides *AB* and *AC* equal and *DBC* having two sides unequal, are established upon given base *BC*, and if each of the angles *BAC* and *BDC* is given, and if, when drawn, line *AD* — namely, the distance between the two vertices — is given, then the two sides of each of them will be found.

Let ∠ *BAC* arbitrarily be larger than ∠ *BDC*, and from point *A* let the perpendicular to the common base be dropped. [The perpendicular] necessarily bisects the base at point *K*. This perpendicular is understood to be of a length not limited by the two sides. Then the circumference of a circle circumscribed around △ *DBC* transects [the perpendicular] above at *G* and below at point *H*. Thus it will be established by the corollary of [Euclid] III.1 that line *GH* is the diameter of this circle, [and] upon this [line] is the center, *E*, of the circle. When a radius *ED* and two chords *GB* and *GC* are drawn, ∠ *BGC* will be equal to ∠ *BDC*. Therefore, since each of the isosceles triangles *ABC* and *GBC* has a known angle with a given base, by Th. I.[38] above both the sides of each of them will be known together with the perpendicular [of each]. Hence the total diameter *GH* will be known, for the square of half the known base, *KC*, is equal to the product of *GK*, now known, times *KH*. Hence *KH*, and consequently the total diameter *GH*, becomes known. Moreover, when the

difference of the perpendiculars, *GA*, known because of the perpendiculars themselves, is subtracted from the known radius *GE*, line *AE* will be left known. Then △ *AED*, having three known sides, has its angle *EAD* known. And when ∠ *EAC*, known because it is half of the given [∠] *BAC*, is subtracted from [∠ *EAD*], ∠ *CAD* is left measured. This [angle] together with the two known lines *CA* and *AD* reveals the third [line] *CD* also. Similarly, when half of ∠ *BAC*, which is *BAK*, is added to the already known ∠ *EAD*, the entire ∠ *BAD* will be found. Because this [angle] is now known together with the two sides *AB* and *AD*, line *BD* will be found. Thus the two sides of △ *DAC* are measured. Q.E.D.

We will argue slightly differently if the circumference of circle *BCDG* cuts the perpendicular below point *A*; this would happen, in truth, as long as ∠ *BDC* were given greater than ∠ *BAC*. But if the two given angles are equal, then the circumference of circle *BCDG* necessarily passes through point *A*. Therefore, since, as before, the diameter of the circle is known as well as the two lines *AB* and *AD*, which will be chords inscribed within the mentioned circle, it will not be difficult to measure each of the chords *BD* and *CD*, if you work for a while at finding the chords of the circle. Therefore what remains in this situation we leave to your industry.

Theorem 33

If in any triangle the angle bisector descends to a known base, and if the sections of the base are unequal [and] known, and similarly if the acute angle, which the dividing line forms with the base, is given, then each side of the triangle can be found.

In triangulo a b g ducatur linea a d, diuidens angulum quidem b a c per æqualia,basim autem b c notam per inæqualia in puncto d : sitcq utracq sectionum b d minor & d c maior data cum angulo a d b acuto.Dico,cp utruncq latus cognoscetur.Oportet autem angulum a d b,quemadmodum comemorauímus,esse acutum propter b d minorem sectionê cui insidet, erit enim tertia sexti ratiocinante a b minor a c,& ideo angulus a b c maior angulo a c b conuincetur.angulus a d c ualet duos b a d & a b d . item angulus a d b æquiualet duobus a c d & d a c : cuncq angulus b a d sit æqualis angulo c a d, resultabit angulus a d c maior angulo a d b,& ideo hic quidem obtusus, ille uero acutus enunciabitur.Circumscribatur ergo triangulo a b c circulus a b h c, reliquacq disponantur quemadmodû in huius:ex angulo itacq a d b siue h d k noto cû angulo k recto,& linea d k differentia dimidiæ basis datæ & minoris sectionis, utracq linearum d h & h k nota proueniet cum angulo d h k : deinceps utracq linearum b h & h c propter dimidiam basim notam cum linea h k cognita dabitur:cuncq quod sub duabus b d & d c continetur sit æquale ei, quod sub a d & d h continetur:tres autem harum sunt notæ,erit per & primi huius linea a d nota,sic duas diametros a h & b c quadrangulo a b h c circulo inscripti notas habemus:unde & reliqua sicuti in huius absoluere licebit.

⚶ LIBER TERTIVS
TRIANGVLORVM.

I.

Si sphæra plano secetur, communis sectio superficiei sphæricæ & plani secantis erit circumferentia circuli.Vnde constabit pedem perpendicularis à centro sphæræ ad superficiem secantem descendentis circuli huiusmodi centrum esse.

Cômunis sectio superficiei sphæricæ & plani eam secantis sit linea a b g,quam dico esse circumferentiam circuli.Planum enim secans aut per centrum sphæræ incedit,aut non. Si p̄ centrum eius,quoniã omnes rectæ lineæ à centro sphæræ ad ipsam sectionem cômunem in plano huiusmodi eductæ,æquales sunt,diffinitione sphæræ id confirmante,manifestû cp pla num intra lineam a b g conclusum est circulus,ipsacq linea a b g circumferentia eius . Si uero planum prætereat centrû sphæræ, demittatur à centro sphæræ,quod sit 3,ad ipsum planum perpêdicularis recta linea 3 h, à cuius pede scilicet puncto h,lineæ rectæ quotlibet educantur ad sectionem præ
dictam

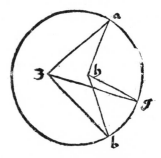

If in △ ABG* line AD is drawn, dividing ∠ BAC equally and dividing known base BC unequally at point D, and if each of the segments, BD the shorter and DC the longer, is given together with acute ∠ ADB, then each side may be found.

∠ ADB, as we mentioned, must be acute because it rests on the shorter segment BD. Then by the argument of [Euclid] VI.3, AB is shorter than AC and therefore ∠ ABC is proven to be larger than ∠ ACB. Angle ADC is equal to the two angles BAD and ABD. Similarly, ∠ ADB equals the two angles ACD and DAC. And since ∠ BAD is equal to ∠ CAD, ∠ ADC will be larger than ∠ ADB, and therefore the former will be obtuse while the latter will be acute. Then let circle ABHC be circumscribed around △ ABC, and the rest

[of the theorem] will be handled just as in Th. [30] above. Therefore, because ∠ ADB, or HDK, is known together with right ∠ K and because line DK is the difference between half the given base and the shorter segment, each of the lines DH and HK will be found, together with ∠ DHK. Finally, each of the lines BH and HC is known because half the base [and] line HK [are] known. And since the product of BD and DC is equal to the product of AD and DH and since three of these [quantities] are known, then by Ths. I.[17] and I.[19] above line AD will be known. Thus we have two known diagonals, AH and BC, of the quadrangle ABHC inscribed in the circle. Hence the rest can be proven just as in [Th. 30] above.

*For ABG read ABC.

The End [of Book Two]

BOOK THREE OF THE TRIANGLES

Theorem 1

If a sphere is intersected by a plane, the line of intersection of the spherical surface and the intersecting plane will be the circumference of a circle. Hence it will be established that the bottom of the perpendicular which descends from the center of the sphere to the cut surface is the center of this circle.

If the line of intersection of the spherical surface and the plane intersecting it is line ABG, then [that line] is the circumference of a circle.

The intersecting plane may pass through the center of the sphere or it may not. If [the plane passes] through its center, then, because all straight lines drawn in this plane from the center of the sphere to the line of intersection are equal by the confirmation of the definition of a sphere, it is apparent that the plane enclosed within line ABG is a circle, and line ABG is its circumference.

However, if the plane by-passes the center of the sphere, let rectilinear perpendicular ZH be drawn from the center of the sphere, which is Z, to that plane. From the bottom of this [perpendicular] — namely, from point H — let any number of straight lines be drawn to the mentioned line of intersection,

dictam,sintq̃ tres huiusmodi h a, h b & h g,protractis semidiametris sphæræ
3 a, 3 b & 3 g. Trium itacq̃ triangulorum a h 3, b h 3 & g h 3 unusquisq̃ angulum iuxta 3 rectum habet,propter lineam 3 h perpédiculariter plano incidẽ
tem,latera autem rectos angulos subtendentia,sunt semidiametri sphæræ æquales,dempto igitur quadrato perpendicularis singulatim â quadratis semidiame-
trorum remanebunt per penultimã primi & cõmunem scientiã quadrata trium li
nearum h a, h b & h g æqualia:unde & lineas ipsas æquales esse oportet.Non
aliter,pbabis alias lineas quotlibet â puncto h ad sectionem cõmunem eductas
sibi & tribus lineis iam memoratis æquales esse,diffinitio igitur circuli theorema
tis concludet ueritatem.　　▼ Ex his aũt & diffinitione centri trahitur pedem demissæ perpendicularis esse centrum circuli iam dicti,quod pollicebaˉt corollariũ.

II.

Omnis linea recta à centro sphæræ ad centrum circuli minoris in
ea producta,perpendicularis est ad superficiem ipsius circuli.

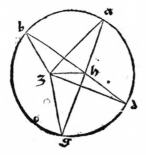

Hæc conuertit corollarium præcedentis. Sit cir
culus minor a b g d in sphæra signatus,cuius cen
trum h,quod cum cum centro sphæræ 3 copulabo
per lineam 3 h.Dico,qp linea 3 h perpendicularis
est ad superficiem huius circuli.Productis enim duabus semidiametris a g & b d circuli minoris, ter
minos earum centro sphæræ copulabo per semidia
metros sphæræ a 3, b 3, g 3 & d 3.per diffinitionẽ
igitur circuli & sphæræ linea 3 h cõmuni existente
ex 8 primi omnes anguli,quos facit linea 3 h cum
lineis sibi in superficie circuli minoris contermina-
libus sunt recti.quare linea 3 h perpendiculariter incidit duabus diametris a g &
b,& ideo per 4.undecimi perpendicularis est ad superficiem circuli a b g d, quod
libuit deducere.Qdˀ autem linea à centro sphæræ superficiei circuli minoris perpendiculariter incidens,ad centrum ipsius circuli minoris terminetur,ex processu
primæ huius satis didicimus.

III.

Linea recta,quæ à centro circuli in sphæra positi, orthogonaliter
egreditur,centrum sphæræ necessario continebit.

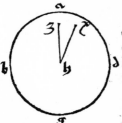

Sit circulus in sphæra a b d g,cuius centrum h,â quo
egrediatur orthogonalis linea h k utruncq̃ indefinita.Di
co,qp in ea centrum sphæræ reperietur.Si enim circulus il
le maior extiterit,cum centrum eius sit centrum etiã sphæ
ræ,à quo orthogonalis ipsa nascitur,planum est quod ,p-
posuimus.Si autem fuerit circulus minor,& centrũ sphæræ extra orthogonalem,sententia quidem aduersarj̃ habeatur,sit ipsum x,producta igitur linea x h, per præ-
missam erit perpendicularis ad superficiem circuli a b g d .ab uno itacq̃ puncto
h superficiei a b g d duæ orthogonales egrediuntur,quod est impossibile.& con
trã　ũndecimi,quo interempto,relinquitur ueritas conclusionis nostræ.

Omnis

and let three of these [lines] be *HA, HB,* and *HG,* [meeting three] drawn radii of the sphere, *ZA, ZB,* and *ZG.* Then each of the three triangles, *AHZ, BHZ,* and *GHZ,* has a right angle at *Z* because line *ZH* meets the plane perpendicularly. Moreover, the sides subtending the right angles are equal radii of the sphere. Therefore, when the square of the perpendicular is subtracted from the squares of the radii, one at a time, then, by the penultimate theorem of [Euclid] I and the axioms, the squares of the three lines *HA, HB,* and *HG* will be left equal. Hence the lines themselves must be equal. You will prove this no differently, no matter how many other lines are drawn from point *H* to the line of intersection, [for] they also will be equal to the three lines already mentioned. Therefore the definition of a circle concludes the truth of this theorem.

From these [statements] and the definition of a center, it is understood that the bottom of the perpendicular that was dropped is the center of the circle just mentioned, which was offered as a corollary.

Theorem 2

Every straight line drawn from the center of a sphere to the center of a small circle within that [sphere] is perpendicular to the surface of the circle itself.

This is the converse of the corollary of the preceding [theorem]. If the small circle *ABGD,* whose center is *H,* is inscribed in the sphere and [*H*] is joined to the center, *Z,* of the sphere by line *ZH,* then line *ZH* is perpendicular to the surface of that circle.

When two radii *AG* and *BD* of the small circle are drawn, their ends are joined to the center of the sphere by radii *AZ, BZ, GZ,* and

DZ of the sphere. Then by the definition of a circle and of a sphere, because line *ZH* is shared, from [Euclid] I.8 all the angles that line *ZH* makes with the lines meeting it on the surface of the small circle are right angles. Therefore line *ZH* meets the two diameters *AG* and *B** perpendicularly, and then by [Euclid] XI.4 [*ZH*] is perpendicular to the surface of circle *ABGD.* Q.E.D. We have learned sufficiently, from Th. 1 above, that the line from the center of a sphere, meeting the surface of a small circle perpendicularly, terminates at the center of that small circle.

Theorem 3

A straight line which passes perpendicularly from the center of a given circle in a sphere will necessarily include the center of the sphere.

Let the circle in the sphere be *ABDG,* from whose center *H* a perpendicular line *HK* emerges, unbounded on either side; then the center of the sphere will be found on that [line].

If the circle is a great circle, then, since its center, from which the perpendicular arose, is also the center of the sphere, what we proposed is obvious. If, however, the circle is a small circle, let the center of the sphere, *X,* not be on the perpendicular, indeed an assertion of the contrary. Then when line *XH* is drawn, by the preceding theorem it will be perpendicular to the surface of circle *ABGD.* Thus two perpendiculars would emerge from the one point *H* on the surface *ABGD,* which is impossible and contrary to [Euclid] XI.[not given]. With this [possibility] eliminated, the truth of our conclusion remains.

*For *B* read *BD.*

IIII.

Omnis linea recta à polo circuli ad eius centrum demissa,perpendi
cularis ad superficiem circuli conuincitur,productacȝ ultra circuli cen
trum, donec superficiei sphæricæ obuiabit, reliquum circuli polum
offendet.

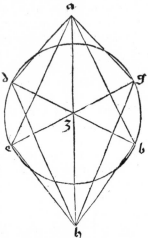

Sit circulus d e b g , cuius quidem polus sit à
punctus,centrum uero 3 ,demittaturcȝ à polo a ad
centrum 3 linea a 3 quã dico esse perpendicularem
ad superficiem circuli.Duarum enim diametrorum
d b & e g terminos cum puncto a copulabo per li
neas a d, a e, a b & a g ,quo sit,ut trianguli a b 3
& a d 3 bina latera sint æqualia, a 3 enim commu=
ne est ambobus, 3 b autem & 3 d sunt semidiame=
tri eiusdem circuli,duas demum bases a d & a b æ=
quales affert poli diffinitio.quare angulus a 3 d æ=
qualis est angulo a 3 b ,& ideo a 3 perpendicularis
est ad lineam d b .Similiter probabimus eandē a 3
perpendicularem esse ad lineam 3 g .cum itacȝ triũ
linearum conterminaliũ una duabus reliquis ortho
gonaliter insistat,erit ipsa perpendicularis ad super
ficiem reliquarum duarum argumento 4.undecimi.hæc autem superficies est ipse
circulus d e b g ,primam igitur theorematis partem ostendisse uidemur. Secun=
dæ uero parti assenties,si linea a 3 uscȝ ad punctũ superficiei sphæricæ ħ produ=
cta,ipsum h punctum omnium diametrorũ terminis coniunges.omnes enim fa=
cti trianguli,quibus uertex cõmunis est,centrum circuli 3 bina latera habet æqua
lia, h 3 enim commune,reliqua uero sunt semidiametri eiusdem circuli, sed & an=
guli eorum apud punctum 3 æquales sunt,quia recti:quare bases dictorum trian=
gulorum uniuersæ æquabuntur,per diffinitionem igitur h punctus circuli dicti
polus habebitur,quod erat lucubrandum.

V.

Omnis recta linea à centro circuli in sphæra orthogonaliter exiēs,
per polos eius utrincȝ continuata transibit.

In figuratione præhabita intelligatur lineam 3 a orthogonaliter à centro cir
culi 3 exiuisse,reliquis ut antehac conseruatis . Dico, duo puncta a & h esse po=
los circuli d e b g .Quia enim linea a 3 orthogonalis est ad superficiē circuli,per
centrum eius transiens,erit ipsa per conuersionē diffinitionis orthogonalis ad om
nem circuli semidiametrũ.posito igitur quadrato 3 a communi,cum omnia qua
drata semidiametrorũ sint æqualia,erunt per penultimã primi & cõmunem ani=
mi conceptionem quadrata omnium linearũ ab a puncto ad circumferentiã cir=
culi demissarum æqualia,& ideo ipsæ lineæ æquales. per diffinitionem ergo poli
constabit ueritas propositionis.Idem similiter concludere de puncto h licebit.

VI.

In linea recta centrũ sphæræ cũ centro circuli minoris in ea signati
cõtinuãte,si quãtũ oportet utrincȝ ꝓlõget,ipsius circuli polos iueniri

I   Circuli

Theorem 4

Every straight line drawn from a pole of a circle to its center is perpendicular to the surface of the circle, and when it is extended beyond the center of the circle until it meets the surface of the sphere, it hits the other pole of the circle.

If *DEBG* is a circle whose pole is point *A* and whose center is *Z*, and if line *AZ* is dropped from pole *A* to the center *Z*, then [this line] is perpendicular to the surface of the circle.

The ends of two diameters *DB* and *EG* are joined with point *A* by lines *AD*, *AE*, *AB*, and *AG*. Thus two sides respectively of each of triangles *ABZ* and *ADZ* are equal, for *AZ* is common to both [triangles] and *ZB* and *ZD* are radii of the same circle. The definition of a pole shows the two bases *AD* and *AB* to be equal. Therefore ∠ *AZD* is equal to ∠ *AZB*, and therefore *AZ* is perpendicular to line *DB*. We may prove similarly that the same [line] *AZ* is perpendicular to line *ZG*. Then since one of the three adjacent lines meets the others perpendicularly, that [line] will be perpendicular to the surface [formed] from the other two by the reasoning of [Euclid] XI.4. Moreover, this surface is the circle *DEBG* itself. Thus we are seen to have proven the first part of the theorem.

You will agree to the second part if, when line *AZ* is drawn all the way to point *H* on the surface of the sphere, you join point *H* to the ends of all the diameters. All the triangles [thus] formed, in which the center, *Z*, of the circle is the common vertex, have two respectively equal sides each, for *HZ* is common [to all] while the other [sides] are radii of the same circle. Furthermore, their angles at point *Z* are equal because they are right angles. Therefore the bases of the mentioned triangles are all equal. Then by definition point *H* will be the pole of the mentioned circle. Q.E.D.

Theorem 5

Every straight line perpendicularly leaving the center of a circle within a sphere will, when extended on each end, pass through the poles of that [circle].

If in the above-supplied construction it is understood that line *ZA* leaves the center of the circle, *Z*, perpendicularly when the rest is kept as before, then the two points *A* and *H* are poles of circle *DEBG*.

Because line *AZ* is perpendicular to the surface of the circle, passing through its center, then, by the converse of the definition, it will be perpendicular to every radius of the circle. Therefore, because the square of *ZA* is common, and since all the squares of the radii are equal, then, by the penultimate theorem of [Euclid] I and the axioms, the squares of all the lines drawn from point *A* to the circumference of the circle will be equal. Therefore the lines themselves will be equal. Thus the truth of the proposition is established by the definition of a pole. The same can be concluded in a similar way concerning point *H*.

Theorem 6

On the straight line connecting the center of the sphere with the center of a small circle inscribed within it, if [that line] is extended as far as is necessary on both sides, the poles of the circle itself are found.

Circuli minoris a b g d in sphæra constituti cētrum sit h ,quod cum centro sphæræ 3 continuetur per líneam h 3 .Dico,ꝙ in línea h 3 quantū sat est proló= gata,duos poꝛos circulí a b g d reperiemus.Erit enim per 2 huius línea h 3 per pendicularis ad superficiē circuli dicti,quare per tertiā huius utrinꝗ continuata, donec superficieï sphæricæ occurret ad duos circuli memorati polos terminabi= tur,Quatuor itaꝗ huiusmodi puncta,centrum uidelicet circulí minoris in sphæ= ra intellecti,centrum sphæræ & duos circuli minoris polos in una semper recta li= nea reperiri necesse est,quod erat explanandum.

VII. ☿

Quæcunꝗ línea recta per polos circuli in sphæra signati transiue= rit,per centrum quoꝗ ipsius transire,ipsiꝗ circulo orthogonaliter in cidere cogitur.

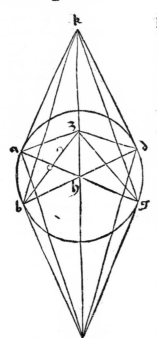

Figuratio 7.ʊ 8.

Línea k l transiens per duos polos k & l circu= li a b g d ,secet eum in puncto 3. Dico,ꝙ punctus 3 sit centrum circuli a b g d ,& ꝙ línea k l ortho= gonaliter incidat circulo a b g d . A puncto enim 3 egrediantur plures ꝗ duæ rectæ líneæ ad circum= ferentiā circuli dicti ,quæ sint 3 a , 3 b , 3 g & 3 d , quarum terminos utriꝗ polorum connectemus per líneas polares duplices concludendo quatuor trian gulos uerticem cōmunem habentes polum k ,bases autē quatuor polares líneas à polo l deriuatas, quo sit,ut anguli eorum apud polum k 8.primi arguen te,æquales habeātur.angulus uidelicet a k l æqua lis angulo b k l,& ita de cæteris.Trasferamus nos demum ad quatuor triangulos,quibus latus k 3 cō mune est,reliqua uero latera æqualia habentes qua tuor scilicet polares líneas ab ipso polo k descēden tes,qui cum angulos suos apud k æquales habeāt, quemadmodum ostendimus per 4 .primi, & ba= ses habebunt æquales,líneas uidelicet 3 a , 3 b , 3 g & 3 d .quare per 9.tertij punctus 3 erit centrum cir culí a b g d .primam igitur theorematis partem rō borauimus,unde & per quartam huius secundæ par ti assentire compellimur, línea k l per polum cen= trumꝗ circuli dicti transeunte,quæuolebamus declarare.

VIII.

Línea recta quæ per duo quatuor punctorum dictorum incedit,re liqua duo præterire non poterit.

Quatuor illa puncta notamus centrum sphæræ,centrum circulí minoris in ea & duos polos eius.Sit itaꝗ circulus minor in sphæra quatuor characteribus a b g d repræsentatus,cuius centrum h,duoꝗ poli k & l,centrum autem sphæræ sit 3 .Dico,ꝙ línea recta duo eorum quæcunꝗ continens,& reliqua duo,si satis por= recta fuerit ,complectetur. Nam,ut à capite initium sumamus,polum k & cen= trum sphæræ 3 in línea k 3 statuamus,extensaꝗ k 3 occurrat circulo a b g d in
puncto

If the center of a small circle *ABGD* established within a sphere is *H*, which is joined with the center of the sphere, *Z*, by line *HZ*, then on line *HZ*, extended far enough, we will find the two poles of circle *ABGD*.

By Th. 2 above line *HZ* will be perpendicular to the surface of the mentioned circle. Therefore by Th. 3 above, when it is extended on both sides until it meets the surface of the sphere, it will be terminated at the two poles of the mentioned circle. Thus these four points — namely, the center of the small circle understood to be within the sphere, the center of the sphere, and the two poles of the small circle — are necessarily always found on one straight line. Q.E.D.

Theorem 7

If any straight line passes through the poles of a circle inscribed within a sphere, it must also pass through the center of that [circle] and meet the circle perpendicularly.

If line *KL*, passing through the two poles *K* and *L* of circle *ABGD*, intersects [the circle] at point *Z*, then point *Z* is the center of circle *ABGD* and line *KL* meets circle *ABGD* perpendicularly.

Let more than two straight lines pass from point *Z* to the circumference of the mentioned circle. These [lines] are *ZA*, *ZB*, *ZG*, and *ZD*, whose ends we connect to each of the poles by two [sets of] polar lines to form four triangles which have pole *K* as a common vertex and [which have] the four polar lines drawn from pole *L* as [their] bases. It is for

this [reason] that their angles at pole *K* are equal by the proof of [Euclid] I.8; that is, ∠ *AKL* is equal to ∠ *BKL*, and so on. Now let us turn to the four triangles in which side *KZ* is shared: indeed [these triangles] have their other sides equal -- namely, the four polar lines descending from pole *K*. And since [the triangles] have their angles at *K* equal, then, as we showed by [Euclid] I.4, the bases are equal — namely, lines *ZA*, *ZB*, *ZG*, and *ZD*. Therefore by [Euclid] III.9 point *Z* will be the center of circle *ABGD*. Thus the first part of the theorem is proven.

From this and [from] Th. 4 above we are compelled to agree to the second part because line *KL*, passing through the pole, is the center of the mentioned circle. Q.E.D.

Theorem 8

A straight line which passes through two of the four mentioned points cannot by-pass the other two.

We recognize those four points [to be] the center of a sphere, the center of a small circle within [the sphere], and the two poles [of that circle]. Thus, if the small circle within the sphere is represented by the four letters *ABGD*, with its center as *H*, the two poles by *K* and *L*, and the center of the sphere by *Z*, then a straight line joining any two of these will include the other two, if [the line] is extended sufficiently.

Now, so that we may start at the top, let us place pole *K* and the center of the sphere, *Z*, on line *KZ*. Let *KZ* extended meet the circle *ABGD* at

puncto h ,cui nondum nomen centri imponimus.producantur deniꝗ lineæ pola
res k a, k b, k g & k d .ité femidiametri fphæræ 3 a, 3 b, 3 g & 3 d,fed & lineæ
a h, b h, g h & d h ,fatis tamen erat trinas huiufmodi non quatuonas produce=
re.quatuor igitur trianguli in uertice k cōmunicantes,quorum bafes funt quatu
or femidiametri fphæræ,cum fint æquilateri per diffinitionem fphæræ & poli , li=
nea k 3 cōmuni exiftente,per 8.primi erunt æquianguli , poftea uero ꝗ̃ angulos
eorum apud k æquales effe didicimus ad quatuor triangulos,quibus & dicti an=
guli cōmunes funt,bafes autem quatuor in puncto h confluentes , tranfeundum
eft,qui cum bina latera,quatuor dictos æquales angulos ambientia,habeāt æqua
lia per diffinitionem poli,linea k h cōmuni exiftente,per quartā primi bafes ha
bebunt æquales.à puncto igitur h plures ꝗ̃ duæ lineæ æquales exeunt ad circum=
ferentiā circuli a b g d :quare per 9.tertij h centrū erit circuli prædicti, & ipfum
eft in linea k h indeterminata,in qua cum habeatur etiam centrum fphæræ,con=
cludimus per præmiffam reliquū quatuor punctorum uidelicet polum l in ea re=
periri.Ponantur demum duo puncta, k polus & h centrum circuli minoris in li
nea k h, erit itaꝗ per 4.huius linea k h perpendicularis ad fuperficiem circuli a
b g d,quare per tertiam & quintam huius reliqua duo puncta,centrum uidelicet
fphæræ & polus l in linea k h indefinita reperientur.Qđ fi duos polos k & l in
linea k l ftatuerimus,conclufionem noftram probaturi quatuor puncta a b g d
ipfi polo l per lineas fuas,quemadmodum figura docet,connectemus.quæ cū fint
æquales diffinitione poli id exigente,fimiliter & quatuor lineæ polares à polo k
deriuatæ,fibi inuicem æquentur,linea k l cōmuni exiftente quatuor triangulis a
k l,b k l, g k l& k l ,erunt quatuor eorum anguli apud polum k æquales .
Rurfus propter angulos huiufmodi æquales,lineasꝗ polares fuperiores æquales,
linea k h communi affumpta,ex quarta primi quatuor bafes a h, b h, g h & d
h æquales conuincemus.per 9.igitur tertij h centrum erit circuli a b g d :unde
& per fupradicta centrum fphæræ in ipfa linea k l neceffario reperietur . Poftre=
mo in linea 3 h, quemadmodum ex præmiffa trahitur , neceffario reperiuntur duo
poli circuli a b g ,fed in linea 3 l continebitur,& centrum circuli h & polus eius
k ,quod non aliter ꝗ̃ de linea k 3 confirmandum erit.fimiliter quod circa lineā k
h diximus,lineæ quoꝗ l h attribuemus.Ex quatuor autem fæpedictis punctis
nō nifi fex eliciunt cōbinatiōes quas enumerauimus,uerū igit̃ eft ꝗd ꝓpofuimus.

I X.

Circulum in eadem fui parte duos habere polos eft impoffibile.

Quilibet circulus ex fphæra ipfum continente duas fcindit partes,quarū unā
fupra ,reliquā autē infra fe relinquit.Dico itaꝗ, ꝗ in neutra illarum duos circuli
unius polos reperire eft poffibile.Nam fi ita opinaberis,cōtinuemus ipfos cū cen=
tro circuli per duas rectas lineas,erit igitur utraꝗ earum orthogonalis ad fuperfi
ciem circuli per huius.cunꝗ ab uno puncto fcilicet centro circuli exurgant,mē=
tietur nobis undecimi Euclidis,quod eft inconueniens.Non ergo uerum eft ꝗd
putabas.Sonat autem hæc conclufio de polis in fuperficie unius fphæræ fignādis,
nō enim me latet eundē circulū in multis contineri poffe fphæris fe fecantibus,in
quarum fuperficiebus licebit ex eadem etiam parte circuli multos affignare po=
los.quotquot autem fignaueris polos,huiufmodi una linea recta orthogonaliter
à centro circuli egrediens,omnes eos complectetur .

I 2   Si lineam

point *H*, which we may not yet call a center. Then let the polar lines *KA*, *KB*, *KG*, and *KD* be drawn; similarly, [let] the radii of the sphere, *ZA*, *ZB*, *ZG*, and *ZD*, as well as lines *AH*, *BH*, *GH*, and *DH* [be drawn]. It was sufficient, however, to draw three of each of these instead of four of each. Therefore, since the four triangles, which share in vertex *K* [and] whose bases are the four radii of the sphere, [have] sides [respectively] equal [to each other] by the definition of a sphere and a pole, then, because line *KZ* is common [to the four triangles], they are similar by [Euclid] I.8. After we have learned that their angles at *K* are equal, the [argument] is to be turned toward the four triangles to which the mentioned angles [at *K*] are also common but [whose] four bases meet at point *H*. Since the two sides of each [triangle], enclosing the four mentioned equal angles respectively, are equal by the definition of a pole, and since line *KH* is common [to the four triangles], then by [Euclid] I.4 the triangles will have equal bases. Therefore, since more than two equal lines pass from point *H* to the circumference of circle *ABGD*, then, by [Euclid] III.9, *H* will be the center of the mentioned circle, and it is on line *KH*, [which] has no fixed limits. Since the center of the sphere is on this [line], we conclude by the preceding [theorem] that the other of the four points — namely, pole *L* — will be found on it.

Next, let two points, pole *K* and the center *H* of the small circle, be placed on line *KH*. Then by Th. 4 above, line *KH* will be perpendicular to the surface of circle *ABGD*. Thus, by Ths. 3 and 5 above, the other two points — namely, the center of the sphere and pole *L* — will be found on the limitless line *KH*.

But if we were to place the two poles *K* and *L* on line *KL*, then to prove our conclusion we will connect the four points *A*, *B*, *G*, and *D* with pole *L* by [polar] lines as the figure shows. Since these are equal because the definition of a pole so requires, and similarly, since the four polar lines drawn from pole *K* are also equal to each other, and since line *KL* is common to the four triangles *AKL*, *BKL*, *GKL*, and *DKL*, then their four angles at pole *K* will be equal. Again, because

these angles are equal, and because the upper polar lines are equal, and because line *KH* was taken [to be] common, from [Euclid] I.4 we prove the four bases *AH*, *BH*, *GH*, and *DH* to be equal. Therefore, by [Euclid] III.9, *H* will be the center of circle *ABGD*. From this and from the aforementioned, the center of the sphere will necessarily be found on line *KL*.

Finally, on line *ZH*, as is concluded from the preceding [theorem], the two poles of circle *ABG[D]* are necessarily found. Moreover, the center of the circle, *H*, as well as its pole *K*, will be included on line *ZL* because it may be proven to be none other than line *KZ*. Similarly, what we said about line *KH* we may apply also to line *LH*. Now, from these four often-mentioned points, no fewer than six combinations are produced, which we have enumerated. Therefore what we proposed is true.

Theorem 9

It is impossible for a circle to have two poles in the same direction.

If any circle divides the sphere containing it into two parts, of which one is above [the circle] and the other is left below it, then in neither of these [parts] is it possible to find two poles of that one circle.

Suppose you were to conjecture thus. Let us connect those [two poles on the same side] with the center of the circle by two straight lines. Then each of these [lines] will be perpendicular to the surface of the circle by Th. [4] above. And since they emerge from one point — namely, the center of the circle — then XI.[not given] of Euclid is lying to us, which is not fitting. Therefore what you considered is not true.

Moreover, this conclusion concerns poles which are marked on the surface of [only] one sphere, for it is apparent to me that the same circle can be contained in many spheres intersecting each other, on whose surfaces one can assign many poles still from the same side of the circle. But, however many poles you were to mark, [only] this one straight line that passes perpendicularly from the center of the circle includes them all.

X.

Si lineam polarem circulus quiſpiam diametro ſphæræ potentia=
liter ſubduplam habuerit, ipſe circulus maior erit.

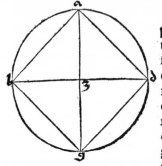

Sit a polus circuli in ſphæra ſignati, cuius linea
polaris a d quadratum habeat ſubduplū quadra=
to diametri ſphæræ. Dico, cp circulus ille erit ma=
ior. Ducatur enim ab a polo per centrum circuli
dicti, quod ſit 3 linea a 3, quæ continuata occur=
rat ſuperficiei ſphæræ in puncto g, erit itacp a g
diameter ſphæræ, nam per quartā & tertiam huius
ipſa incedit per centrū ſphæræ. Intelligatur dein=
ceps ſuperficies plana tranſiens per lineas a d & a
g, ſecando ſphæram. fiet autem cōmunis ſectio cir
cumferētia, circuli quidem ex prima huius, magni
uero per diffinitionem, habebit enim & centrum & diametrum ſphæræ. diameter
autem circuli in ſphæra ſignati ſit linea d 3 b, & linea polaris ſecunda g d. Quo
niam igitur quadratum a g duplum eſt quadrato a d per hypotheſim. quadratū
autem a g duobus quadratis linearum a d & d g æquale eſt per penultimā pri=
mi, cp angulus a d g in ſemicirculo rectus ſit, erit quadratum a d æquale quadra
to d g, & ideo linea lineæ æqualis. utercp autem angulorum a 3 d & g 3 d eſt re=
ctus per 4. huius & diffinitionem lineæ perpendicularis ad ſuperficiem, linea igit
d 3 commūni, erit per penultimā primi & cōmunem ſcientiā quadratū a 3 æqua=
le quadrato 3 g, & ideo coſta a 3 coſtæ 3 d æqualis. eſt autem a g diameter ſphæ
ræ, ut ſupra declarauimus, neceſſario igitur 3 centrum circuli ſignati erit centrum
ſphæræ: per diffinitionem itacp circulus ille maior eſt. Quandocūcp ergo linea
polaris circuli cuiuſdam potentialiter ſubdupla eſt diametro ſphæræ, aut æqualis
coſtæ quadrati magno circulo ſphæræ inſcriptibilis, ipſe circulus maior eſt, quod
expectabas declarandum.

X I.

Omnis circulus maior in ſphæra lineam polarem utrancp habet po
tentialiter ſubduplam diametro ſphæræ, æqualemcp lateri quadrati,
quod ipſi circulo magno inſcribitur. Vnde manifeſtum eſt arcum cir
culi magni, à polo ad circumferentiam circuli ſignati demiſſum, eſſe
quadrantem circumferentiæ ſuæ.

In figura præcedentis addo duas lineas a b & b g, intelligendo 3 centrum &
lineam b 3 d diametrum circuli in ſphæra ſignati. Dico, cp utracp linearū a b &
b g polarium potentialiter ſubdupla eſt diametro ſphæræ, & æqualis lateri qua=
drati, magno circulo ipſius ſphæræ inſcriptibilis. Erit enim per 4. huius unuſquiſ=
cp quatuor angulorum apud 3 rectus, quatuor itacp trianguli, quibus uertex com
munis eſt punctus 3, bina latera habentes æqualia, ſemidiametros ſcilicet ſphæræ,
per 4. primi baſes habebunt æquales. unuſquiſcp autem angulorum totalium; qui
apud puncta a b g d ſunt, rectus declaratur ex 3 o. tertij. per diffinitionem igitur
a b g d quadratum eſt, inſcriptum quidem circulo maiori a b g d : ſiccp ſecun=
da pars theorematis oſtenſa eſt, unde & per penultimā primi confirmabimus pri=
mam partē

Theorem 10

If some circle has a polar line [whose second] power is half the [second] power of the diameter of the sphere, the circle will be a great circle.

If *A* is a pole of a circle inscribed in a sphere [and if] the square of its polar line *AD* is half the square of the diameter of the sphere, then this circle will be a great circle.

Let line *AZ* be drawn from pole *A* through the center, *Z*, of the mentioned circle, and, when [*AZ*] is extended, let it meet the surface of the sphere at point *G*. Then *AG* will be the diameter of the sphere, for by Ths. 4 and 3 above this [line *AG*] passes through the center of the sphere. Then the planar surface passing through lines *AD* and *AG* is known to intersect the sphere. Furthermore, from Th. 1 above, the line of intersection will be the circumference of a circle — indeed, by definition, [that] of a great [circle], for it will have the center and the diameter of the sphere. Moreover, let the diameter of the circle inscribed in the sphere be line *DZB* and let a second polar line be *GD*. Therefore, because the square of *AG* is double the square of *AD* by hypothesis, and because the square of *AG* is equal to the [sum of the] squares of lines *AD* and *DG* by the penultimate theorem of [Euclid] I since ∠ *ADG* in the semicircle is a right angle, then the square of *AD* will be equal to the square of *DG*. Thus line *AD* is equal to line *DG*. Furthermore, each of the angles *AZD* and *GZD* is a right angle by Th. 4 above and the definition of a line perpendicular to a surface. Therefore, since line *DZ* is common [to the two right triangles *AZD* and *GZD*], the square of *AZ* will be equal to the square of *ZG* by the penultimate theorem of [Euclid] I and the axioms; and thus side *AZ* is equal to side *ZD*. Furthermore, *AG* is the diameter of the sphere, as we stated above, and so the center of the inscribed circle, *Z*, is

necessarily the center of the sphere. Thus by definition this circle is a great [circle]. Therefore, whenever the [second] power of a polar line of any circle is half the [second] power of the diameter of the sphere, or [whenever the polar line] is equal to the side of a square that could be inscribed within a great circle of a sphere, that circle is a great circle. Q.E.D.

Theorem 11

Every great circle in a sphere has the [second] power of each polar line [equal to] half the [second] power of the diameter of the sphere, and [the polar line] is equal to the side of the square which is inscribed within the great circle. Hence it is apparent that an arc of a great circle drawn from the pole to the circumference of a[nother] inscribed [great] circle is a quadrant of its circumference.

In the preceding figure I add two lines *AB* and *BG*, with *Z* understood to be the center and line *BZD* the diameter of the circle inscribed in the sphere. Then the [second] power of each of the polar lines *AB* and *BG* is half the [second] power of the diameter of the sphere, and [each polar line] is equal to the side of the square that can be inscribed in the great circle of that sphere.

By Th. 4 above, every one of the four angles at *Z* will be a right angle. Thus the four triangles, to which the vertex at point *Z* is common and which each have two sides respectively equal — namely, the radii of the sphere — will have equal bases by [Euclid] I.4. Moreover, every one of the [combined] angles, which are at points *A*, *B*, *G*, and *D*, is declared a right angle by [Euclid] III.30. Thus by definition *ABGD* is a square inscribed within the great circle *ABGD*. And thus the second part of the theorem is shown. From this and from the penultimate theorem of [Euclid] I, we may prove the first part.

mam partem, Corollariū autē nemo dubitabit poſtǫ ex tertij quatuor cordas
æquales, coſtas ſcilicet quadrati prædicti, quatuor arcus æquales abſcindere de
clarabimus,

XII.

Si ab aliquo puncto ſuperficiei ſphæræ duo quadrantes duarum
circumferentiarum magnarum egredientes, ad eundem arcum circuli
magni terminentur, punctus ille erit polus circuli, ad cuius arcum di
cti quadrantes terminantur.

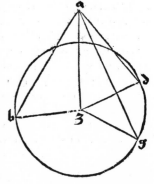

Sit punctus a in ſuperficie ſphæræ ſignatus, à
quo egrediantur duo quadrantes circumferentia
rum magnarum àngulariter coniuncti, qui ſint a
b & a c, & terminentur ad arcum circuli magni
b c. Dico, ǫ a ſit polus circuli magni, cuius erat
arcus ille b c. Producam enim per centrum ſphæ
ræ 3 diametrum a 3 g, complendo duas ſemicir
cumferentias magnas a b g & a c g. item binas cordas duorum quadrantum di
ctorum & duorum quadrantum reſiduorum, qui ſunt g b & g c. duas poſtremo
ſemidiametros circuli b c, quæ ſint 3 b & 3 c. Quoniam igitur duo trianguli a 3
c & c 3 g bina latera habent æqualia, duaſǫ baſes ſuas a c & c g per præceden
tem æquales, erunt per 8. primi duo eorum anguli apud 3 æquales : quare duæ li
neæ a 3 & 3 c ppendiculariter ſibi inuicem inſiſtunt. Non aliter declarabitur line
am a 3 perpendiculariter inſiſtere lineæ 3 b. per 4. igitur undecimi linea a 3 ppen
diculariter incidit ſuperficiei complectenti duas lineas 3 b & 3 c, quæ quidem ſu
perficies eſt ipſe circulus magnus, quem ſupra notauimus, & ideo per 5. huius pun
ctus a erit polus circuli b c, quod erat demonſtrandum.

XIII.

Si à puncto ſuperficiei ſphæricæ ad circumferentiam circuli cuiuſ
cuncǫ in ſphæra ſignati, plures ǫ duæ æquales rectæ lineæ deſcende
rint, punctus ille dicti circuli polus habebitur.

Sit punctus a in ſuperficie ſphærica notatus, à
quo ad circumferentiã circuli b g d plures ǫ duæ
æquales rectæ lineæ deſcendat, quæ ſint uerbi gra
tia tres a b, a g & a d. Dico, ǫ a ſit polus circu
li b g d. Demittatur enim ab a puncto ad ſuper
ficiem circuli b g d, per 11. undecimi perpendicu
laris a 3, cuius pedi puncto ſcilicet 3 tria puncta b
g d copulabimus per tres lineas 3 b, 3 g & 3 d,
claudendo tres angulos a 3 b, a 3 g & a 3 d, quo
rum angulos apud 3 punctum rectos eſſe oportet,
conuerſa diffinitione lineæ perpendicularis ad ſu
perficiem. cum itaǫ tres lineæ æquales ad circum
ferentiam circuli deſcendentes, illos rectos ſubten
dant angulos, per penultimam primi & cōmunes ſcientias linea a 3 communi ex
iſtente, tribus dictis triangulis tres lineæ 3 b, 3 g & 3 d æquales habebuntur : qua

I 3 re per

Furthermore, no one will doubt the corollary after we declare that by [Euclid] III.[not given] the four equal chords — namely, the sides of the previously mentioned square — cut off four equal arcs.

Theorem 12

If two quadrants of two great circles, emerging from some point on the surface of a sphere, are terminated at the same arc of a[nother] great circle, that point [from which they emerged] will be the pole of the circle at whose arc the mentioned quadrants are terminated.

Let point *A* be marked on the surface of a sphere, and let two quadrants *AB* and *AC* of great circles, joined together at an angle, emerge from that [point] and be terminated at arc *BC* of [another] great circle; then *A* is the pole of the great circle whose arc is *BC*.

Let diameter *AZG* be drawn through the center, *Z*, of the sphere to form two great semicircles *ABG* and *ACG*. Similarly, let two pairs of chords be drawn, [one pair] for the two quadrants [*AB* and *AC*] and [the other pair] for the other two quadrants, which are *GB* and *GC*. Finally, let two radii *ZB* and *ZC* of circle *BC* be drawn. Therefore, because the two triangles *AZC* and *CZG* each have two respectively equal sides, and since their bases *AC* and *CG* are equal by the preceding [theorem], then by [Euclid] I.8 their two angles at *Z* will be equal. Therefore the two lines *AZ* and *ZC* rest perpendicularly on each other. It may be explained in the same way that line *AZ* rests perpendicularly on line *ZB*. Thus by [Euclid] XI.4 line *AZ* meets the surface encompassing the two lines *ZB* and *ZC* perpendicularly. This [surface] is indeed the great circle that we mentioned above. Therefore, by Th. 5 above, point *A* will be the pole of circle *BC*. Q.E.D.

Theorem 13

If more than two equal straight lines descend from a point on the surface of a sphere to the circumference of any circle inscribed in the sphere, that point will be the pole of the given circle.

If point *A* is marked on the surface of a sphere, and if from this point more than two equal straight lines — for example, *AB*, *AG*, and *AD* — descend to the circumference of circle *BGD*, then *A* is the pole of circle *BGD*.

Let perpendicular *AZ* be drawn, by [Euclid] XI.11, from point *A* to the surface of circle *BGD*. We will join three points *B*, *G*, and *D* to the bottom of the perpendicular, namely to point *Z*, by three lines *ZB*, *ZG*, and *ZD* to form three [tri]angles *AZB*, *AZG*, and *AZD*. Their angles at point *Z* must be right angles by the converse of the definition of a line perpendicular to a surface. Therefore, since the three equal lines that descend to the circumference of the circle subtend those right angles, and because, by the penultimate theorem of [Euclid] I and the axioms, line *AZ* is common to the three mentioned triangles, then the three lines *ZB*, *ZG*, and *ZD* are equal. Then,

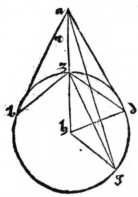

re per 9.tertij 3 erit centrum circuli b g d, & ideo
per 5.huius punctum a esse polū circuli b g d con
fiteberis, quod libuit attingere. Assumpsimus autē
in hoc processu punctum 3 cadere intra circulum
cuius rei certitudinē accipies hoc pacto. Non enim
potest punctus 3 esse in circumferentia circuli, si em
ita fuerit sententia quidem aduersarij, á centro cir=
culi, quod sit h, in nulla linearū 3 b, 3 g & 3 d exi
stente, ducantur tres semidiametri h 3, h g & h d,
erit itaq́p per octauam primi angulus 3 h d æqua=
lis angulo 3 h g, pars toti, quod est impossibile. Si=
mile inconueniens concludemus aduersario putan
ti punctum 3 extra circulum cadere. huiusmodi au
tem impossibilibus interemptis, reliquū est, ut pun=
ctus 3 in superficie circuli b g d reperiatur.

XIIII.

Omnes duæ superficies planæ in puncto uno communicantes, in li
nea quocꝗ recta per punctum ipsum incedente, si indefinitæ extendan
tur cōmunicabunt. hæc autem linea communis earū sectio habebitur.

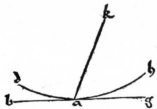

Sit punctus a communis duabus planis super
ficiebus. Dico, ꝙ ipsæ superficies, si indefinitæ ex=
tendantur, in linea per a punctum transeunte, cō
municabunt, & in ea linea se intersecabunt. Intelli
gat enim tertia superficies plana duas prædictas
in puncto a communi secans, fiatꝗ sectio cōmu=
nis superficiei tertiæ cum altera propositarum, li=
nea recta b a g per 3.undecimi, quæ si fuerit, etiam in reliqua constabit prima
pars theorematis. Si uero non sit communis sectio huius tertiæ superficiei, & reli
quæ duarum propositarum linea d a h , educaturꝗ á puncto a in superficie ter=
tia linea a k , argumento igitur 13.primi duo anguli d a k & k a h duobus re=
ctis æquiualent. similiter duo anguli b a k & k a g duos ualet rectos, quare duo
recti minores erunt duobus rectis, quod est impossibile. quo interempto, primam
propositionis partem astruemus. Similiter probabimus, ꝙ duæ superficies illæ in
linea communi prædicta se intersecabunt. Si enim non intersecent se in ea, intelli=
gatur tertia superficies secans prædictas duas, non quidem in linea communi iam
dicta, sed in alia. secet autem & hæc tertia superficies lineam in qua communicāt
duæ propositæ superficies in puncto a , á quo educatur linea a k in superficie ter=
tia, quo facto, duos angulos rectos duobus angulis rectis minores esse conclude=
mus, quod est impossibile. Non possunt igitur duæ propositæ superficies nō com=
municare in linea recta, & in ea se intersecare, quod erat lucubrandum.

X V.

Per duo puncta in superficie sphæræ signata, circulum magnum
producere.

In superficie sphæræ notentur duo puncta a & b, per quæ si producendum ue=
lis circulum magnum super puncto a tanꝗ polo, secundum quantitatem costæ
quadrati

by [Euclid] III.9, Z will be the center of circle BGD, and thus by Th. 5 above you will acknowledge point A to be the pole of circle BGD. Q.E.D.

Now we assumed in this procedure that point Z fell within the circle. You will prove the validity of this [assumption] as follows. Point Z cannot be on the circumference of the circle. For if it were so — indeed the contrary opinion — then, from the center, H, of the circle, being on none of the lines ZB, ZG, and ZD, let three radii HZ, HG, and HD be drawn. Then by [Euclid] I.8 ∠ ZHD will equal ∠ ZHG, [or] a part [will be] equal to a whole, which is impossible. In a similar way we may prove the inappropriateness of the contrary contention that point Z falls outside the circle. With these impossibilities eliminated, it remains that point Z is found on the surface of circle BGD.

Theorem 14

Every two planar surfaces that have one point in common will also have the straight line passing through that point in common, if they are extended to an unlimited [size]. Furthermore, this line is their line of intersection.

If point A is common to two planar surfaces, then these surfaces, if extended indefinitely, will have the line passing through point A in common and will intersect each other on this line.

Let a third planar surface be understood, intersecting the two aforementioned at common point A, and let the line of intersection of the third surface with one of the given [surfaces] be a straight line BAG by [Euclid] XI.3. If this [line] is also the [line of inter-

section] on the other [surface], the first part of the theorem will be established. However, if not, let line DAH be the line of intersection of the third surface and the other of the two given [surfaces]. Let line AK be drawn from point A on the third surface. Therefore, by the reasoning of [Euclid] I.13, the two angles DAK and KAH are equivalent to two right angles. Similarly the two angles BAK and KAG equal two right angles. Therefore two right angles will be smaller than two right angles, which is impossible. With this eliminated, we affirm the first part of the proposition.

Similarly, we may prove that these two surfaces will intersect each other on the aforementioned common line. For if they do not intersect each other on that [line], let a third surface be understood, intersecting the abovementioned two [surfaces], not on the common line just noted but on another. Furthermore, let this third surface also intersect the line that the two given surfaces have in common, at point A. From [point A] let line AK be drawn on the third surface. When this is done, we will conclude that two right angles are smaller than two right angles, which is impossible. Therefore the two given surfaces must have the straight line [BAG] in common and must intersect each other on it. Q.E.D.

Theorem 15

A great circle is determined through two points marked on the surface of a sphere.

On the surface of a sphere let there be two points A and B through which, [let us suppose,] you wish a great circle to be drawn on point A as a pole with the length of side

quadrati magno circulo fphæræ iſcriptibilis,quæ
ſit a d,deſcribe circulum e d g . ſecundum eandē
quoꝗ quantitatem,ſcilicet ſecundum lineā d b k,
æqualem ipſi a d ſuper polo k circulum e g k de
ſcribas,hos duos circulos ſe inuicem ſecare ex præ
miſſa liquet, ꝗ centrum ſphæræ cōmune habeant.
ſecent igitur ſe circumferentiæ eorum in punctis
e & g ,quorum alterum uidelicet g duobus pun=
ctis a & b. copulabo per lineas a g & b g , quas
neceſſe eſt eſſe æquales,&utranꝗ earum æqualem
eſſe lineæ a d aut b k coſtæ ſcilicet quadrati ma=
gni. duæ enim polares lineæ a d & a g æquales
ſunt,ponebatur autem b k æqualis ipſi a d, qua=
re& a g æquabitur lineæ b k, cui etiam b k æ=

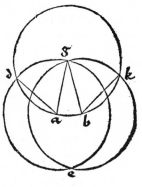

qualis exiſtit:ſunt enim b g & b k lineæ polares,ab eodem polo unius circuli de=
ſcendentes.circulus igiꞇ deſcriptus ſuper g puncto tanꝗ polo, trāſibit per utrūꝗ
punctorum a & b ,magnus autem erit per 8.huius, ꝗ linea polaris ſua g a æqua
lis ſit coſtæ quadrati magni.Quo igitur pacto ,ꝓoſitum efficere oporteat, ſatis
oſtendimus.

XVI.

Circulus magnus per unum polum alterius circuli incedens , per re
liquum quoꝗ tranſibit polum.

Circulus magnus a b g d per polum a circuli
b e d tranſeat.Dico, ꝗ tranſibit etiam per reliquū
eius polum.Continuemus enim polum a cum cen
tro 3 circuli magni prædicti, quod & ſphæræ ipſi
commune eſt per lineam a 3 ,quæ producta ampli
us,donec ſuperficiei occurret ſphæricæ per 4 aut 7.
huius,reliquum polum circuli b e d offendet, qui
ſit h .Si itaꝗ linea 3 h pars ſcilicet lineæ a h fue=
rit in ſuperficie circuli magni a b g d,uerū enun=
ciabat ꝓoſitio:ſi uero non, lineæ rectæ a h pars
erit in plano,& pars in ſublimi,quod eſt impoſſibi
le per primam undecimi.Circulus igitur a b g d,
per polum a circuli b e g tranſiēs, reliquum eius
polum præterire non poterit,quod libuit declarare.

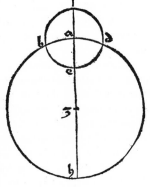

XVII.

Circulus magnus in ſphæra per polum alterius circuli tranſiens,
eum per æqualia & orthogonaliter ſecabit.

Sit circulus magnus a b g d, cuius centrum 3 tranſiens per polum a
circuli b e d. Dico, ꝗ circulus a b g d ſecabit circulum b e d per æqualia
&orthogonaliter.Ducatur enim à polo a ad centrum 3 circuli a b g d: quod
& ipſi ſphæræ eſt commune:linea a 3 ,quæ cum ſit in ſuperficie circuli a b g d,
&tran=

AD of the square that can be inscribed in the great circle of the sphere [as polar line]; draw circle *EDG*. With the same length — namely, line *DBK*,* equal to *AD* — upon pole *K*† you may draw circle *EGK*. It is clear, from the preceding [theorem], that these two circles intersect each other, because they have the center of the sphere in common. Therefore their circumferences intersect each other at points *E* and *G*, one of which — namely, *G* — is joined to the two points *A* and *B* by lines *AG* and *BG*. These [lines] are necessarily equal, and each of them is necessarily equal to line *AD*, or *BK* — namely, the side of the great square [i.e., of the square that can be inscribed in the great circle]. For the two polar lines *AD* and *AG* are equal. Moreover, *BK* was taken equal to *AD*. And therefore *AG* will be equal to line *BK*, to which *BK*‡ is equal, for *BG* and *BK* are polar lines, descending from the same pole of one circle. Therefore the circle drawn around point *G* as pole will pass through each of the points *A* and *B*. By Th. 8 above it will be a great circle because its polar line *GA* is equal to the side of the great square. Therefore we have shown sufficiently how one should prove the proposition.

Theorem 16

A great circle passing through one pole of another circle will also pass through the other pole [of that circle].

If a great circle *ABGD* passes through pole *A* of circle *BED*, then it will also pass through its other pole.

We will join pole *A* with the center, *Z*, of the previously mentioned great circle, which [center] is also common to the sphere, by line *AZ*. When this [line] is extended further until it meets the surface of the sphere, then, by Th. 4 or Th. 7 above, it strikes the other pole of circle *BED*, which is *H*. Thus, if line *ZH* — namely, a portion of line *AH* — is on the surface of great circle *ABGD*, the proposition is declared to be true. However, if not, part of straight line *AH* will be on the plane and part [will be] above, which is impossible by [Euclid] XI.1. Therefore circle *ABGD* passing through pole *A* of circle *BEG*§ cannot by-pass the other pole of this [circle]. Q.E.D.

Theorem 17

A great circle in a sphere, passing through the pole of another circle, will bisect [the other] orthogonally.

If great circle *ABGD*, whose center is *Z*, passes through pole *A* of circle *BED*, then circle *ABGD* bisects circle *BED* orthogonally.

Let line *AZ* be drawn from pole *A* to the center, *Z*, of circle *ABGD*, which [center] is also common to the sphere. Since [line *AZ*] is on the surface of circle *ABGD*,

*For *DBK* read *BK*.
†For *K* read *B*.
‡For *BK* read *BG*.
§For *BEG* read *BED*.

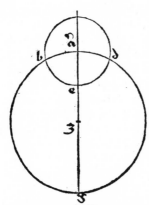

& tranfeat per centrum circuli b e d,ex diffinitio
ne quidē circuli magni, fi b e d circulus magnus
fuerit, aut per feptimam huius, fi minor, neceffe
quoque erit,ut circulus a b g d per centrum cir=
culi b e d tranfeat.quare fecabit eum per æqualia.
Præterea quoniam linea a 3 ex polo a ad centrū
circuli b e d defcendit (fiue illic quiefcat, fiue ul=
tra porrigatur,nihil refert) erit ipfa per quartam
huius pępēdicularis ad fuperficiem circuli b e d.
quare per 18.undecimi fuperficies circuli a b g d
per ipfam lineam a 3 incedens,orthogonalis erit
ad fuperficiem circuli b e d,quæ fuere lucubrāda.

XVIII.

Si quis in fphæra circulus alium per æqualia & orthogonaliter fe=
cuerit, ipfe magnus erit,& per polos eius quem fecat tranfibit.

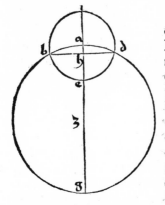

Hæc conuertit præmiffam.Circulus a b g d fe
cet circulum b e d p æqualia & orthogonaliter.
Dico,ꝗ circulus a b g d magnus eft.Sit enim cō=
munis eorum fectio linea b d, quam oportet effe
diametrum circuli b e d,quemadmodum ex hypo
thefi trahitur,cuius punctus h fit centrum circuli
b e d,à puncto autem h in fuperficie circuli a b g
d egrediatur orthogonalis ad diametrū b d, quæ
etiam orthogonalis erit ad fuperficiem circuli b e
d, ex diffinitione fuperficiei orthogonaliter fupra
fuperficiem erectæ & quarta undecimi . quare per
quintam huius orthogonalis illa tranfibit per po=
los circuli b e d,unde & circulus a b g d ortho=
gonalem prædictam continens per polos huiufmodi incedet.Item p 3 huius ipfa
orthogonalis tranfibit per centrum fphæræ.fi igitur utrinꝗ ad fuperficiem fphæ=
ræ aut circumferentiam circuli a b g d porrecta fuerit,ipfa erit diameter: fphæ=
ræ quidem per diffinitionem,circuli autem a b g d,ꝗ per centrum eius tranfeat
aut per corollarium primæ tertij.fecat enim cordam b d per æqualia & orthogo
naliter:p diffinitionē igiſ circulus a b g d magnus eft,quæ oportuit explanare.

XIX.

Omnes circuli magni in fphæra per æqualia fe inuicem fecant.

Commune enim omnibus circulis magnis in fphæra diffinitur centrum fphæ=
ræ.quicunꝗ igitur duo circuli in hoc uno puncto participāt,& in linea recta pun
ctum ipfum recipiente,tanꝗ fectione communi participabunt 12. huius confir=
mante.hæc autem fectio communis utriufꝗ circulorum erit diameter,utrunꝗ eo
rū p æqua fcindēs,quare & ipfi p æqua fe fcindent,ꝗd noftra enūciabat cōclufio.

XX.

Omnis circulus magnus per polos circuli in fphæra magni ad quē
ipfe erectus eft tranfibit.

Linea

and since it passes through the center of circle *BED* from the definition of a great circle if *BED* is a great circle or by Th. 7 above if [*BED*] is a small [circle], then it will be necessary that circle *ABGD* also pass through the center of circle *BED*. Therefore [*ABGD*] bisects [*BED*].

Furthermore, since line *AZ* descends from pole *A* to the center of circle *BED* (it makes no difference whether it stops there or extends beyond), by Th. 4 above it will be perpendicular to the surface of circle *BED*. Therefore by [Euclid] XI.18 the surface of circle *ABGD* passing through line *AZ* will be orthogonal to the surface of circle *BED*. Q.E.D.

Theorem 18

If any circle in a sphere bisects another orthogonally, it will be a great circle and will pass through the poles of [that circle] which it intersects.

This is the converse of the preceding [theorem]. If circle *ABGD* bisects circle *BED* orthogonally, then circle *ABGD* is a great circle.

Let the line of intersection of these be line *BD*, which must be the diameter of circle *BED*, as is concluded from the hypothesis. Let point *H* [of line *BD*] be the center of circle *BED*. Moreover, from point *H* on the surface of circle *ABGD*, let a [line] perpendicular to diameter *BD* be drawn. This [line] will also be perpendicular to the surface of circle *BED*, from the definition of a surface erected perpendicularly upon [another] surface and by [Euclid] XI.4. Therefore, by Th. 5 above, this perpendicular will pass through the poles of circle *BED*, and hence circle *ABGD*, containing the pre-

viously mentioned perpendicular, passes through these poles. Similarly, by Th. 3 above, the perpendicular itself will pass through the center of the sphere. Therefore, if [that perpendicular] is extended on both sides to the surface of the sphere or to the circumference of circle *ABGD*, it will be the diameter [both] of the sphere, by definition, and of circle *ABGD* because it passes through the center [of *ABGD*] or because of the corollary of III.1 for it bisects chord *BD* perpendicularly. Then by definition circle *ABGD* is a great circle. Q.E.D.

Theorem 19

All great circles in a sphere bisect each other.

It is defined that the center of the sphere is common to all the great circles in the sphere. Therefore whichever two circles have this one point in common will also have in common, as the line of intersection, the straight line having this point, by the confirmation of Th. 12 above. Furthermore, this line of intersection will be the diameter of each of the circles, [hence] bisecting each of them. Therefore they bisect each other, which our conclusion stated.

Theorem 20

Every great circle will pass through the poles of [another] great circle, upon which the former is erected perpendicularly,* in the sphere.

Erectus, meaning "set upright," of course, implies perpendicularity; however, it seems advisable to be explicit for the sake of the proof, even if the wording itself is redundant.

Linea nanꝗ à centro eorum communi (omnes enim circuli magni in centro ſphæræ participant) ad ſectionem communem in altero eorum orthogonaliter egrediens ad reliquum, erit orthogonalis per diffinitionem ſupeficiei ſupra ſuperficiem erectæ & quartam undecimi, quare per quintam huius ad polos eius perueniet, cunꝗ ipſa ſuperficiem circuli erecti non egrediatur, neceſſario ipſe circulus erectus circuli ſubſtrati polos complectetur, & hoc decuit confirmare.

XXI.

Quicunꝗ circulus magnus alium in ſphæra minorem circulum orthogonaliter ſecat, ipſum quoꝗ per æqua partietur. & ſi per æqualia ſcindet, orthogonaliter eum ſecabit. Vnde polos circuli minoris præterire non poterit circulus ipſe magnus.

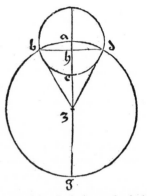

Secet enim circulus magnus a b g d, cuius centrum 3 circulum minorem b e d orthogonaliter ſecundum lineam b d. Dico, ꝗ ſecabit eũ p æqualia, quem ſi per æqua ſecuerit, orthogonaliter quoꝗ ſecuiſſe prædicabo. Diuiſa enim ſectione cõmuni b d per æqualia in puncto h, ducatur à centro ſphæræ circuli magni a b g d, quod eſt 3, ad punctum h linea 3 h, quæ ex diffinitione ſphæræ & octaua primi ductis duabus ſemidiametris 3 b & 3 d, ppendicularis erit ad ſectionem communem b d. quare per diffinitionem ſuperficiei erectæ ad ſuperficiem & quartam undecimi linea 3 h orthogonalis erit ad ſuperficiem circuli b e d, & ideo per corollarium primæ huius punctus h erit centrum circuli b e d, & linea b d diameter eius, quæ cum ſecet circulum b e d per æqualia, eundem quoꝗ circulus a b g d ſecabit per æqualia, quod aſſeruit prima pars theorematis. Secet deniꝗ eũ per æqualia, linea d b diametro circuli minoris exiſtente, in qua punctus h centrum eius accipiatur, quod cum centro 3 circuli magni & ſphæræ ipſius copulabimus per lineam 3 h, quam ex ſecunda huius ppendicularem eſſe oportet ad ſuperficiem circuli b e d. quare & per 18. undecimi ſuperficies circuli a b g d ad ipſam orthogonalis erit. Circulus igitur a b g d circulũ b e d orthogonaliter ſecabit, quod erat exponendũ. Neceſſe autẽ eſt, ex utraꝗ hypotheſi & eis quæ in pceſſu præſenti recitauimus, lineam 3 h utrinꝗ ſuperficiei ſphæricæ occurrentem circuli b e d polos inuenire, cũꝗ ipſa ſuperficiem circuli a b g d egredi non poſſit, ipſe circulus a b g d maior polos circuli minoris non præteribit, quod polliccebatur corollarium.

XXII.

Omnes circuli in ſphæra, quibus eidem ſui poli ſibi inuicem æquè diſtant: & ſi fuerint æquediſtantes quotlibet circuli, duos polos habebunt communes. Vnde patet, quotlibet circulorũ in ſphæra æquediſtantiũ centra cum polis ſuis in una reperiri linea recta.

Sint tres aut plures circuli quotlibet a b g d, & e h in ſphæra una, quibus duo poli k & l cõmunes exiſtant. Dico ꝗ ipſi inter ſe æquediſtabunt, qui ſi ponantur

A line from their common center (for all great circles have the center of the sphere in common), passing along [the surface of] one of these [circles] perpendicularly to the line of intersection, will be perpendicular to the other [surface] by the definition of a surface erected perpendicularly upon [another] surface and by [Euclid] XI.4. Therefore, by Th. 5 above, [the line] reaches to the poles [of the other]. And since [the line] does not leave the surface of the erected circle, this erected circle necessarily encompasses the poles of the underlying circle. Q.E.D.

Theorem 21

Whichever great circle intersects another small circle in the sphere orthogonally will also bisect [that small circle], and if [a great circle] bisects [a small circle], it intersects [the small circle] orthogonally. Hence the great circle cannot by-pass the poles of the small circle.

If great circle *ABGD*, with center *Z*, intersects small circle *BED* orthogonally along line *BD*, then it bisects [the small circle]; if it bisects [the small circle], one may state beforehand that it will also have intersected [it] orthogonally.

When the line of intersection, *BD*, is bisected at point *H*, let line *ZH* be drawn from the center, *Z*, of the sphere of great circle *ABGD* to point H. From the definition of a sphere and [Euclid] I.8, when two radii *ZB* and *ZD* are drawn, [line *ZH*] will be perpendicular to the line of intersection *BD*. Therefore, by the definition of a surface erected perpendicularly upon [another] surface and by [Euclid] XI.4, line *ZH* will be perpendicular to the surface of circle *BED*, and therefore, by the corollary of Th. 1 above, point *H*

will be the center of circle *BED* and line *BD* its diameter. Since [line *ZH*] bisects circle *BED*, circle *ABGD* will also bisect the same circle, which the first part of the theorem claimed.

Finally, let line *BD*, since it is the diameter of the small circle, bisect [the small circle]. On [line *BD*] point H is understood to be its center, which we will join with the center, *Z*, of [both] the great circle and its sphere by line *ZH*. From Th. 2 above [line *ZH*] must be perpendicular to the surface of circle *BED*. Therefore by [Euclid] XI.18 the surface of circle *ABGD* will be orthogonal to [the surface of circle *BED*]. Thus circle *ABGD* will intersect circle *BED* orthogonally. Q.E.D.

Furthermore, it is necessary, from each hypothesis [stated above] and from those [theorems] which we cited in the present procedure, for line *ZH*, meeting the surface of the sphere on both sides, to encounter the poles of circle *BED*. And since [line *ZH*] cannot leave the surface of circle *ABGD*, the great circle *ABGD* will not by-pass the poles of the small circle, which corollary was promised.

Theorem 22

All circles in a sphere which have the same poles are parallel to each other; and if any number of circles are parallel, they will have the same two poles. Hence it is clear that the centers, as well as the poles, of any number of parallel circles in a sphere are found on one straight line.

If three, or any number more, circles *AB*, *GD*, and *EH* are in one sphere [and] the two poles *K* and *L* are common to them [all], then they will parallel each other; [or] if they are given

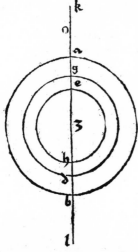

æquediſtantes in polis duobus cõmunicabunt. Con
tinuabo em duos polos k & l per lineam k l, quæ ꝑ
7. huius tranſibit ꝑ centrum ſphæræ, centraꝗ omni
um circulorum æquediſtantium, quare per quartã
huius uniuerſis circulis orthogonaliter incidet, &
ideo per undecimi circuli dicti inter ſe æquediſtabunt, quod enunciabat prima pars theorematis no
ſtri. Sed ponamus eos æquediſtantes, ꝓducamuſꝗ
à centro ſphæræ ad centrum unius eorum lineam re
ctam, quæ per 2. huius ꝑpendicularis erit ad ſuperfi
ciem circuli, ad cuius centrum ipſa ꝓducitur. Hæc
autem linea utrinꝗ ad ſupeficiem ſphæræ extenſa,
reliquis omnibus circulis æquediſtantibus ꝑpendiculariter incidet ex hypotheſi & undecimi. quare
per corollarium primæ huius per centra ſinguloru
tranſibit, & ideo ex 5. huius per polos omnium ince
det, cunꝗ polos in ſuperficie ſphæræ dutaxat ſigna

re ſoleamus, ipſa autem ſæpedicta linea in ſuperficie ſphæræ duo tantum offendit
puncta, conſtat puncta huiuſmodi eſſe polos omnibus circulis æquediſtantibus
communes, quod aſſerebat ſecunda pars ꝓpoſitionis. Corollarium autem huius
ex corollario primæ, & 7. huius confirmabitur.

XXIII.

Omnes duo circuli magni ex circulis in ſphæra æquediſtantibus,
per quorum polos incedunt, arcus abſumunt ſimiles.

Sint duo circuli æquediſtantes a b & g d in
ſphæra una, cõmunem ex præcedenti habentes po
lum 3, per quem & ſibi oppoſitum polum incedat
duo circuli magni, quorum arcus duntaxat in figu
ra hac poſuiſſe ſatis eſt, qui ſint 3 a & 3 g, ſecan-
tes circumferentias circulorum æquediſtantium,
circuli quidem a b in punctis a & b, circuli uero
d g in punctis g & d. Dico, ꝗ duo arcus a b &
g d ſint ſimiles. Egrediatur enim à polo 3 ad po
lum ſibi oppoſitum linea recta, quam conſtat eſſe
cõmunem ſectionem duorum circuloru magnoru

in qua neceſſe eſt reperiri duo centra circulorum a b & g d ex corollario præcedentis. ſit igitur h centrum circuli a b, & e centrum circuli g d, à quibus binæ
educantur ſemidiametri ad quatuor puncta ſectionum, quæ ſint h a, & h b quidem circuli a b, e g autẽ & e d circuli g d. oportet autẽ has ſemidiametros, eſſe
in cõmunibus ſectionibus duorum circuloru magnorum & ipſorum æquediſtantium, ꝗ puncta eas terminantia in utriſꝗ circulis exiſtant. Quoniã itaꝗ circulus
a g 3 ſecat duos circulos æquediſtantes a b & g d, erunt per undecimi duæ li
neæ a h & g e ſectionis cõmunis æquediſtantes, ſimiliter duæ lineæ b h & d e
æquediſtabunt. duæ igitur lineæ a h & b h angulariter coniunctæ æquediſtant
duabus g e & d e angulariter coniunctis, & ideo per undecimi angulus a h b,
æqualis erit angulo g e d, utriuſꝗ ergo eorum ad quatuor rectos una eſt ꝓportio, quæ quidem, ut ex ultima ſexti trahitur, eſt tanꝗ utriuſꝗ duorum arcuu a b
& g d

parallel, they will share in the two poles.

The two poles K and L are joined by line KL, which, by Th. 7 above, will pass through the center of the sphere and through the centers of all parallel circles. Therefore, by Th. 4 above, KL will cut through all the circles perpendicularly, and thus by [Euclid] XI.[not given] the mentioned circles will be parallel to each other, which the first part of our theorem declared.

But let us take them [to be] parallel, and from the center of the sphere to the center of one of these [circles] let us draw a straight line, which, by Th. 2 above, will be perpendicular to the surface of the circle to whose center it is drawn. This line, extended on both sides to the surface of the sphere, will cut perpendicularly through all other parallel circles, from the hypothesis and [Euclid] XI.[not given.] Therefore, by the corollary of Th. 1 above, it will pass through the centers of them individually, and thus from Th. 5 above it will proceed through all the poles. And since we are accustomed to mark the poles only on the surface of a sphere, this often-mentioned line hits only the two points on the surface of the sphere. It is established that these points are the poles common to all the parallel circles, which the second part of the proposition stated.

Furthermore, the corollary of this [theorem] may be confirmed from the corollary of Th. 1 and from Th. 7 above.

Theorem 23

In a sphere any two great circles intercept similar arcs from parallel circles through whose poles they pass.

In one sphere let AB and GD be two paral-lel circles, sharing pole Z by the preceding theorem. Through [pole Z] and through the opposite pole let two great circles pass, of which it is sufficient to have put only the arcs, which are ZA and ZG, in this figure, [these great circles] intersecting the circumferences of the parallel circles—circle AB at points A and B and circle DG at points G and D. Then the two arcs AB and GD are similar.

Let a straight line be drawn from pole Z to the pole opposite it. This [straight line] is established to be the line of intersection of the two great circles, [and] on that [line] the two centers of circles AB and GD are necessarily found, by the corollary of the preceding theorem. Therefore let H be the center of circle AB and E the center of circle GD. From these [centers] two radii each are drawn to the four points of the intersections [i.e., of each small circumference with each great circumference], which [radii] are HA and HB of circle AB, while EG and ED [are radii] of circle GD. Furthermore, these radii must be on the lines of intersection of the two great circles [with] the parallel [circles] because the points terminating them exist in both sets [of intersecting circles]. Therefore, since circle AGZ intersects the two parallel circles AB and GD, then by [Euclid] XI.[not given] the two lines AH and GE of the line[s] of intersection will be parallel. Similarly, the two lines BH and DE will be parallel. Therefore the two lines AH and BH, joined at an angle, are parallel to the two lines GE and DE, also joined at an angle, and thus by [Euclid] XI.[not given] $\angle AHB$ will equal $\angle GED$. Hence there is one ratio of each of them to four right angles. Indeed, this [ratio], as it is understood from the last theorem of [Euclid] VI, is as that of each of the two arcs AB

& g d ad fuam circumferentiam:proportio itacʒ arcus a b ad fuā circumferen=
tiam eſt ut arcus g d ad fuam.per diffinitionem igitur duo arcus a b & g d du
obus circulis magnis intercepti funt fimiles,quod oportuit explanare.Non aliter
‚pcedemus,fi plures cʒ duo æquediſtantes circuli nobis offerantur,nifi cp media,
quibus freti fumus,quoties res ipfa popofcerit ingeminemus.

<h3 style="text-align:center">XXIIII.</h3>

Si fuper duas diametros duorum circulorum æqualium eri=
gantur duæ portiones unius circuli æquales,aut duorum circulorum
æqualium,ex ipfis autem portionibus accipiātur duo arcus æquales,
quorum utercʒ minor fit dimidio arcu portionis fuæ,itemcʒ ex circū=
ferentijs circulorum arcus æquales conterminales quidē arcubus por=
tionum acceptis,lineæ rectæ continuantes extrema arcuum acceptorū
æquales erunt.& fi lineæ ipfæ fint æquales , arcus autem portionum
æquales,minores tamen dimidijs arcubus fuarum portionum,arcus
quocʒ circulorum æquales erunt.

Sint duo circuli a b g & d e h æquales,à quo
rum diametris a g & d h exurgant duæ portiōes
æquales a k g & d l h unius circuli,aut duorū cir=
culorum æqualium,orthogonales ad circulos fub=
iacentes,quarum portionum uertices fint puncta
m & n ,diuidentia arcus portionū per æqualia, ac=
cipiaturcʒ ex una earū arcus a k minor arcu a m,
ex reliqua uero d l minor arcu d n .fed & duo ar=
cus a b & d e circulorum fubſtratorū æquales ſta=
tuanť.‚pductas itacʒ lineas rectas b k & e l æqua
les,concludam hac argumentatione. A duobus pū
ctis k & l duas ppendiculares k r & l s, demitto
ad fectiones cōmunes circulorum & portionum,
punctacʒ b & e pedibus ppendicularium demiſſa
rum & centris circulorum copulabo,b quidem per
lineas b r & b 3,e autem per lineas e s & e x,du
ascʒ cordas k g & l h in portionibus ipfis ponā.
Quia itacʒ duo arcus a k & d l æquales funt,erūt
duo anguli k g r & l h s æquales,utercʒ autem an
gulorum apud puncta r & s rectus eſt,latus autem
k g trianguli k g r æquatur lateri l h trianguli l
h s ‚ppter d uos arcus k g & l h æquales.quare per

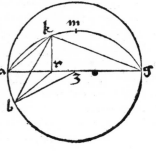

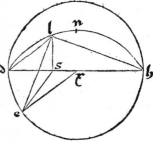

26. primi duo latera refidua trianguli k g r duobus lateribus reliquis trian=
guli l h s æqualia erunt, r g quidem ipfi s h & k r lateri l s. demptis ergo cir=
culorum æqualibus femidiametris,relinquetur r 3 æqualis ipfi s x,eſt autem b 3
æqualis ipfi e x .funt enim ex hypotheſi duo circuli æquales,quorum ipfæ femidi
ametri habentur,angulus autem b 3 r æquatur angulo e x d ‚ppter arcus a b
& d e ,quos hypotheſis æquales fubiecit,ultima fexti cooperante:quare per quar
tam primi bafis b r trianguli b 3 r æquabitur bafi e s trianguli e x s. Cum au
tem utracʒ linearum k r & l s demiſſa fit perpendiculariter ad fectionem cōmu=

K 2   nem fu=

and *GD* to its [respective] circumference. Therefore the ratio of arc *AB* to its [respective] circumference is as that of arc *GD* to its [respective circumference.] Thus, by definition, the two arcs *AB* and *GD*, intercepted by the two great circles, are similar. Q.E.D.

We proceed no differently if more than two parallel circles are given us, except that we repeat the method by which we reasoned as many times as the situation requires it.

Theorem 24

If, upon two diameters of two equal circles, two equal portions of one circle or of two equal circles are erected perpendicularly, and if from these portions two equal arcs are taken of which each is less than half the arc of its portion, and similarly if from the circumferences of the circles arcs are taken, equal and, indeed, adjacent to the intercepted arcs of the portions, then straight lines connecting the extremes of the intercepted arcs will be equal. And if these lines are equal and the arcs [intercepted] from the portions are equal yet less than half the arcs of their [respective] portions, the arcs of the circles will be equal also.

Let the two circles *ABG* and *DEH* be equal, and on their diameters *AG* and *DH* let two equal portions *AKG* and *DLH*, of one circle or of two equal circles, be erected, orthogonal to the circles [*ABG* and *DEH*] sitting beneath. Of these portions the summits are points *M* and *N*, bisecting the arcs of the portions. Let arc *AK*, smaller than arc *AM*, be intercepted from one of these [portions] while *DL*, less than arc *DN*, is [intercepted]

from the other. Furthermore, let two arcs *AB* and *DE* of the underlying circles be established equal. Therefore I may conclude that the straight lines *BK* and *EL*, which have been drawn, are equal by the [following] reasoning.

From the two points *K* and *L* two perpendiculars *KR* and *LS* are dropped to the lines of intersection of the circles and the portions, and the points *B* and *E* are joined to the bases of the perpendiculars that were drawn and to the centers of the circles — *B* by lines *BR* and *BZ*, *E* by lines *ES* and *EX*. Let the two chords *KG* and *LH* be placed in the portions.

Therefore, because the two arcs *AK* and *DL* are equal, the two angles *KGR* and *LHS* will be equal. Furthermore, each of the angles at points *R* and *S* is a right angle, and in addition side *KG* of △ *KGR* is equal to side *LH* of △ *LHS* because the two arcs *KG* and *LH* are equal. Therefore by [Euclid] I.26 the two other sides of △ *KGR* will be equal to the two other sides of △ *LHS* — *RG* [equal] to *SH* and *KR* [equal] to side *LS*. Thus, when the equal radii of the circles are subtracted, *RZ* is left equal to *SX* and *BZ* is equal to *EX*. For from the hypothesis the two circles, of which these are radii, are equal. Furthermore, ∠ *BZR* is equal to ∠ *EXD* because of arcs *AB* and *DE*, which the hypothesis gives equal, by the assistance of the last theorem of [Euclid] VI. Therefore by [Euclid] I.4 base *BR* of △ *BZR* will equal base *ES* of △ *EXS*. Furthermore, since each of the lines *KR* and *LS* was drawn perpendicularly to the line of intersection

nem superficierum, ex diffinitione superficiei ad superficiem erectæ & quarta undecimi, erit utracp earum perpendicularis ad superficiem circuli sibi substrati, &
ideo per diffinctionem perpendicularis ad omnem lineam in substrata superficie si
bi conterminalem. duo igitur anguli b r k & e s l recti sunt, quos cum circundēt
æqualia latera quemadmodū deductum est, erunt per 4. primi duæ lineæ b k & e l
rectis angulis oppositæ æquales, quod enūciabat theorematis nostri prima pars.
Qd̕ si eas lineas ponendo æquales uelimus ostendere æqualitatem arcuum a b &
d e, reliquis ut antehac manentibus, hoc pacto ratiocinabimur, ꝓpter duas lineas
b k & k r æquales duabus e l & l s. anguloscp apud r & s rectos penultima
primi & cōmunibus scientijs confirmantibus, linea b r æqualis erit lineæ e s, ba
sis uidelicet trianguli b r 3 basi trianguli e s x, quorum etiam bina latera sunt
æqualia, quemadmodū in priori processu explanauimus, quare per 8. primi duo
eorum anguli b 3 r & e x s in centris suorum circulorū quiescentes æquabuntur,
& ideo per tertij duos arcus a b & d e æquales esse oportebit, quod̕ ꝓmittebat
secunda pars ꝓpositionis. Aduertendum tamen, cp si ex erectis portioibus arcus
æquales absumpserimus, non minores, sed maiores dimidijs arcubus suarum portionum cæteris ut antea manentibus, eandem per omnia passionem demonstrabi
mus, syllogismo etiam non mutato. Præterea si erexeris semicirculos æquales, idē
accidet, uerum semicirculis erectis, aut portionibus minoribus semicirculo ꝑpendiculares demittendæ occurrent diametris circulorum substratorū, forma igitur
superioris argumentationis haud mutabitur. Si uero portiones semicirculo maio
res statuerimus, possibile erit ex eis abscindere arcus æquales adeo paruos, ut per-
pendiculareʒ supradictæ non offendant diametros substratorum circulorum, sed
occurrant eis extra circulos suos prolongatis, aut fortasse fient diametris conterminales. Postremo quod de duobus substratis circulis diximus, ad unicum applicare poterimus, siue supra diametrum eius unam portionem, siue plures erexeri-
mus. Igitur secundum hæc uariabis parumper figurationem, ꝓcessum autem demonstratum, si supradicta satis tenueris, facile comparabis.

X X V.

Si ex portione supra diametrum circuli erecta arcum minorem dimidio arcu portionis absumpseris, omnium linearum rectarū ab eius
puncto terminali ad circumferentiam circuli substrati demissarū corda arcus absumpti erit breuissima, corda autem residui arcus portiōis
omnium longissima, reliquæ uero, quo breuissime sunt uiciniores, eo
sunt remotioribus breuiores, æqualiter autem à breuissima remotæ,
æquales habebuntur.

Sit circulus a b g d super centro 3, cuius diameter a 3 d sit corda portionis
a h d orthogonaliter insistentis circulo dicto, sumaturcp arcus a h minor dimidio arcu portionis, à cuius puncto h ad circumferentiam circuli a b g d demittant lineæ rectæ, h a quidem corda arcus h a, & h d corda arcus residui, itemcp
quotlibet aliæ lineæ rectæ, quarū duæ h b & h g inæqualiter à breuissima distan
tes, duæ uero h b & h k æqualiter ab ea remotæ, demonstrationi perficiendæ suf
ficient. Dico, cp corda a h omnium demissarum linearum est breuissima, & corda
d h omnium longissima, linea autem h b breuior linea h g, quod sic accipias. A
puncto h ad diametrum a d perpendicularis descendat h l, cuius pedi puncto
                                                                  scilicet

of the surfaces, then, from the definition of a surface erected perpendicularly upon [another] surface and [by Euclid] XI.4, each of these [lines *KR* and *LS*] will be perpendicular to the surface of the circle beneath it, and thus by definition each will be perpendicular to every line adjacent to it on the underlying surface. Therefore the two angles *BRK* and *ESL* are right angles. Since, as was deduced, equal sides surround these [right angles], then by [Euclid] I.4 the two lines *BK* and *EL* opposite the right angles will be equal, which the first part of our theorem declared.

But if, when it is given that these lines are equal while the other [conditions] remain as before, we wish to show the equality of the arcs *AB* and *DE*, we will reason in this [following] way. Because the two lines *BK* and *KR* are equal to the two [lines] *EL* and *LS*, and because the angles at *R* and *S* are right angles by the penultimate theorem of [Euclid] I and by the confirmation of the axioms, then line *BR* will be equal to line *ES* — namely, the base of △ *BRZ* to the base of △ *ESX*. Of [these triangles] two sides each are [respectively] equal, as we explained in the earlier procedure. Therefore by [Euclid] I.8 their two angles *BZR* and *EXS*, resting in the centers of their [respective] circles, will be equal, and thus by [Euclid] III.[not given] the two arcs *AB* and *DE* must be equal, which the second part of the proposition asserts.

It should be noted, however, that if, from the erected portions, we had intercepted equal arcs, not smaller than, but greater than half the arcs of their portions, while the rest remained as before, we would show the same proof throughout, because the syllogism is still not changed. Furthermore, if you had erected equal semicircles, the same happens; indeed, when semicircles or portions smaller than a semicircle are erected, the perpendiculars that are dropped will meet the diameters of the underlying circles. Therefore the form in the above argument will not be changed at all. However, if we were to set the portions larger than a semicircle, it would be possible to intercept from these [portions] equal arcs so small that the above-mentioned perpendiculars do not hit the diameters of the underlying circles but meet [the diameters] extended outside their [respective] circles or perhaps [meet] the diameters [at their end points on the circumference].

Finally, what we said concerning the two underlying circles we can apply to each one whether we had erected one portion upon the diameter [of the circle] or several [portions]. Therefore you will change the construction, the procedure, and the proof a little in accordance with these [situations]; if you have understood the above-mentioned sufficiently, you will provide [any changes] with ease.

Theorem 25

If from a portion [of a circle] erected upon the diameter of [another] circle, you were to intercept an arc that is less than half the arc of the portion, then, of all the straight lines drawn from the end point of this [arc] to the circumference of the underlying circle, the chord of the intercepted arc will be the shortest and the chord of the remaining arc of the portion will be the longest of all, while of the rest [of the chords those] closer to the shortest are shorter than [those that] are farther away from [the shortest], and those equally distant from the shortest will be equal [to each other].

Let *ABGD* be a circle, upon center *Z*, whose diameter *AZD* is the chord of [circular] portion *AHD* that rests upon the mentioned circle [*ABGD*] perpendicularly, and let *AH* be an intercepted arc, smaller than half the arc of the portion. From point *H* of this [arc] to the circumference of circle *ABGD*, let [two] straight lines be drawn — *HA*, the chord of arc *HA*, and *HD*, the chord of the other arc — as well as any number of other straight lines, of which the two, *HB* and *HG*, [that are] unequal distances from the shortest and the two, *HB* and *HK*, [that are] equally distant from [the shortest] are sufficient for completion of the proof. Then, of all [these] drawn lines, chord *AH* is the shortest and chord *DH* the longest of all, and line *HB* is shorter than line *HG*, which you will consider as follows.

From point *H* to the diameter *AD*, let a perpendicular *HL* be dropped, to whose base — namely, point

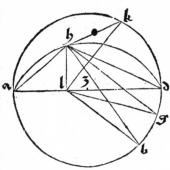

scilicet l duo puncta b & g connectantur per li‑
neas b l & g l.oportet autem punctum l distare
à centro z uersus a punctum, quoniã ex primi
huius casus a l minor est casu l d ,ppter latus a
h trianguli a h d ,minus latere h d ,quod comi‑
tatur hypothesim.puncto igitur l præter centrũ
circuli a b g d signato,erit per   tertij linea l a
omnium ab eo puncto ad circumferentiam circu
li egredientium linearum breuissima, l d autem
omnium longissima, item l b breuior linea l g.
Cum autem h l in superficie alterius duarum su‑
perficierum orthogonaliter se secantium,sectioni
earum communi orthogonaliter insistat,erit ipsa
per diffinitionem superficiei ad superficiem erectæ,& quartam undecimi ortho‑
gonalis ad reliquam,& ideo per diffinitionem lineæ erectæ ad superficiẽ ,ipsa erit
orthogonalis ad omnem lineam sibi in reliqua superficie conterminalem , quare
omnes anguli,quos ambiunt lineæ à puncto l egredientes,cum ipsa linea h l re‑
cti habebuntur,si igitur quatuor triangulos in latere h l communicantes nota‑
ueris,quorum reliqua latera sunt lineæ ab h puncto demissæ,& ab l puncto egre
dientes per penultimã primi & cõmunes animi conceptiones confiteberis lineam
h a omnium demissarũ linearum esse breuissimam,& h d longissimam, h b ue‑
ro breuiorẽ linea h g, & linea h k æqualem h b,quæ pollicebamur demonstran
da. Huiusmodi passionem demonstrabimus etiam de semicirculo erecto supra di
ametrum circuli,non aliter q̃ de portione quam minorem posuimus semicirculo,
simile præterea accidit portioni maiori semicirculo,uerũ perpendicularis h l nõ
semper residebit in diametro circuli substrati,nam possibile est abscindere arcũ ex
huiusmodi portione adeo paruum, q̃ perpendicularis dicta non occurrat diame‑
tro,nisi extra circulum prolongetur , aut fortasse cõcurret cum ea in termino suo,
quod cũ euenerit,utemur & tertij in processu demonstratiuo,ubi pridem addu
ximus   tertij,reliqua uero omnia prorsus repetemus.

XXVI.

Circulus magnus in sphæra transiens per polos duorum circulorũ
se secantium,diuidit arcus separatos eorum per æqualia.& si diuiserit
arcus eorum separatos per æqualia,transibit per polos eorum. q̃ si ar
cus unius eorum per æqua partiatur,per polum alterius eorum transẽ
undo,ipse quocõ arcus separatos reliqui diuidet per æqualia , & per
polos amborum incedet.

Sint duo circuli in sphæra a b g & e b g qualescunqᷓ secantes se in punctis
b & g,per quorum polos incedat circulus magnus a e z d ,secans binos arcus se
paratos circulorum a b g & e b g in punctis a,e, z & d .Dico,q̃ arcus a b æ‑
qualis est arcui a g ,& b z æqualis g z.item duo arcus b e & e g inuicem, duoᷓcõ
arcus b d & d g sibi æquales erunt.Polus nanqᷓ circuli a b g sit k,& polus cir‑
culi e d g sit punctus h ,à quo ad duo puncta b & g protrahantur duæ rectæ h
b & h g,quas oportet esse æquales, h polo circuli e d g existente.Quoniam igi
tur circulus a e z d transit per polos amborum circulorum, ipse secabit utrunqᷓ

K 3   eorum

L — let the two points B and G be joined by lines BL and GL. Moreover, point L must be away from the center Z in the direction of point A because, from Th. I[44] above, segment AL is shorter than segment LD since side AH of △ AHD is shorter than side HD, which follows from the hypothesis. Therefore, since point L is marked to one side of the center of circle $ABGD$, then by [Euclid] III.[not given] line LA will be the shortest of all the lines passing from that point [L] to the circumference of the circle, while LD will be the longest of all, and likewise LB will be shorter than line LG. Moreover, since HL, on the surface of one of the two surfaces intersecting each other orthogonally, rests perpendicularly on their line of intersection, [HL] will be perpendicular to the other [surface] by the definition of a surface erected perpendicularly upon [another] surface and by [Euclid] XI.4. And therefore, by the definition of a line erected perpendicularly upon a surface, [HL] will be perpendicular to every line adjacent to itself on that other surface. Therefore all the angles, which the lines passing from point L form with line HL, will be right angles. Thus, if you have noted the four triangles which have line HL in common but whose other sides are the lines drawn from point H and the lines emerging from point L, then, from the penultimate theorem of [Euclid] I and the axioms, you will acknowledge line HA to be the shortest of all the lines drawn and HD the longest, while HB will be shorter than line HG and line HK will equal HB, which we promised would be proven.

We will also show the proof concerning a semicircle erected upon the diameter of the circle in the same way as [we showed] that concerning the portion which we took smaller than a semicircle. Furthermore, it happens in a similar manner for a portion larger than a semicircle; however, perpendicular HL will not always rest on the diameter of the underlying circle, for it is possible to intercept from this portion an arc so small that the mentioned perpendicular will not meet the diameter unless [the diameter] is extended outside the circle; or perhaps the [perpendicular] meets with [the diameter] at the end point [of the diameter]. When this happens, we use [Euclid] III.[not given] in the proving procedure, where before we brought in [Euclid] III.[not given]; however, we may recall all the rest directly.

Theorem 26

A great circle in a sphere passing through the poles of two circles intersecting each other bisects their severed arcs; and if [the great circle] bisects their severed arcs, it will pass through their poles. But if the arc of one of them is bisected by the [great circle's] passage through the poles of the other of them, [the great circle] will bisect the severed arcs of the other and will pass through the poles of both.

Let the two circles in the sphere be ABG and EBG, which intersect each other in any way at points B and G and through whose poles a great circle $AEZD$ passes, intersecting both the severed arcs of each of the circles ABG and EBG at points A, E, Z, and D. Then arc AB is equal to arc AG and BZ is equal to GZ. Similarly, the two arcs BE and EG and the two arcs BD and DG will be equal to each other.

Let the pole of circle ABG be K, and let the pole of circle EDG be point H. From this [point H] to the two points B and G, let two straight [lines] HB and HG be extended, which must be equal because H is a pole of circle EDG. Therefore, because circle $AEZD$ passes through the poles of both circles, it will bisect each of them

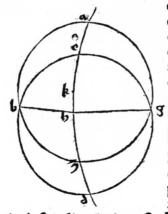

eorum per æqualia & orthogonaliter ex huius.
quare portio a e ʒ erigitur supra diametrum circuli a b g ,ex qua abſcinditͬ arcus ʒ h minor dimi
dio arcu portionis, à cuius termino h duæ rectæ
æquales ad circumferentiam circuli a b g deſcen
dunt,ſit igitur per huius arcus b ʒ æqualis arcui
ʒ g .cunǫ uterǫ arcuum a b ʒ & a g ʒ ſit ſemicir
cumferentia circuli,ǫ circulus a e d tranſiens per
polum circuli a b g ,ſecet eum per æqualia , demptis utrobiǫ æqualibus arcubus,manebit arcus a
b æqualis arcui a g .Non aliter oſtendemus arcū
e b æqualem eſſe arcui e g ,ſi portionem e h d ſu
pra diametrū circuli e b g erectam eſſe intellexe-
rimus,erit em̄ arcus e k minor dimidio arcu por-
tionis ſuæ, lineæǫ à puncto k ad duo puncta b & g demiſſæ ,æquales proba-
buntur, k polo circuli a b g exiſtente.Hinc etiam reliqui duo arcus b d & d g
æquales oſtendentur,primæ igitur parti theorematis aſſentire compellimur , ſed
pone circulum a e ʒ d diuidere arcus ſeparatos eorum per æqualia in punctis a
e ʒ & d .Dico,ǫ tranſibit per polos eorum.Si enim non, tranſeat alius circulus
magnus per polos eorum,quod quidem poſſibile eſt per & huius,qui per iam de
monſtrata tranſibit etiam per quatuor puncta a ,e, ʒ & d .ſecabunt ergo ſe duo
circuli magni in ſphæra in iſtis quatuor punctis,quare per huius unuſquiſǫ arcuum a e ,a ʒ ,e ʒ & e d erit ſemicircumferentia,omnes autem ſemicircumferentiæ unius circuli æquales phibentur,pars igitur æqualis erit toti,quod eſt im-
poſſibile.Non poteſt ergo circulus alius tranſire per polos huiuſmodi,niſi ille qui
dicta quatuor puncta tranſcurrit.Cum itaǫ per omnia duo puncta in ſuperficie
ſphæræ ſignata tranſire poſſit circulus magnus,ut ex huius didicimus,& nō po
teſt tranſire per duos polos k & h alius circulus præter eum,qui per quatuor pun
cta a ,e ,ʒ ,d incedit,manifeſtum eſt,ǫ circulus magnus per quatuor dicta puncta tranſiens,duos polos præterire non poterit ,qui tandem ex huius per reliquos duos polos incedere cogitur,& illud ,pponebat pars ſecunda.Poſtremo ſi cir
culus a e ʒ d ſecuerit,uerbi gratia,arcum b a g per mediū in puncto a ,& trāſie
rit per polos alterius ſe ſecantium circulorum.Dico,ǫ ſecabit & reliquū p æqualia per polos amborum tranſeundo.Si enim non,ſecet eos alius quiſpiam p æqua
lia,aut tranſeat per polos eorum,ille igitur per polos amborum incedens, ſecabit
arcus ſeparatos eorum per æqualia .comunicabunt itaǫ duæ circumferentiæ circulorum magnorum in tribus punctis, uidelicet duobus polis alterius eorum &
puncto a ,eritǫ unuſquiſǫ arcuum inter duo quælibet puncta comprehenſus,
ſemicircumferentia per huius,unde ſequitur partem ſuo toti æqualem eſſe,quod
eſt impoſſibile.Non igitur ſtabit hæc hypotheſis,niſi circulus ille tranſeat per polos amborū,& ſecet utriuſǫ arcus ſeparatos per æqualia,quæ fuere demōſtranda.

XXVII.

Omnes circuli ſecundum æquales lineas polares deſcripti in ſphæra una ,aut diuerſis æqualibus tamen,æquales exiſtunt. & ſi circuli æ-
quales fuerint,lineas eorum polares æquari neceſſe eſt.

Si lineæ polares lateri quadrati magni æquales habeantur, conſtabit per 8.
huius

perpendicularly by Th. [17] above. Thus, portion *AEZ* is erected perpendicularly upon the diameter of circle *ABG*, [and] from this [portion], arc *ZH*, which is less than half the arc of the portion, is intercepted. From end point *H* [of arc *ZH*], two equal straight [lines] descend to the circumference of circle *ABG*. Thus by Th. [24] above, arc *BZ* is equal to arc *ZG*. And since each of the arcs *ABZ* and *AGZ* is half the circumference of the circle because circle *AED*, passing through the pole of circle *ABG*, bisects the circle, then, when the equal arcs [*BZ* and *ZG*] are subtracted on both sides [i.e., from each half-circumference], arc *AB* is left equal to arc *AG*. We show no differently that arc *EB* is equal to arc *EG*, if we understand that portion *EHD* is erected perpendicularly upon the diameter of circle *EBG*. For arc *EK* will be less than half the arc of its portion, and the lines drawn from point *K* to the two points *B* and *G* will be proven equal because *K* is the pole of circle *ABG*. Hence the other two arcs *BD* and *DG* will also be shown equal. Therefore we are compelled to agree to the first part of the theorem.

Now take circle *AEZD* to bisect the severed arcs of the [circles] at points *A*, *E*, *Z*, and *D*. Then [the great circle] will pass through the poles [of the circles]. For if not, let another great circle pass through the poles [of the circles], which is possibly by Ths. [15 and 17] above. This [other great circle], by [those things] already proven, will pass through the four points *A*, *E*, *Z*, and *D*, and therefore the two great circles in the sphere will intersect each other at these four points. Therefore by Th. [19] above, each one of the arcs *AE*, *AZ*, *EZ*, and *ED* will be a semicircumference; moreover, all semicircumferences of one circle are held to be equal. Therefore the part will be equal to the whole, which is impossible. Thus another circle cannot pass through these poles except

that one which runs through the four mentioned points. Therefore, since a great circle can pass through [any] two points marked on the surface of a sphere, as we learned from Th. [15] above, and since no other circle can pass through the two poles *K* and *H* besides the one that passes through the four points *A*, *E*, *Z*, and *D*, then it is obvious that the great circle passing through the four mentioned points cannot by-pass the two poles. Finally, by Th. [16] above this [great circle] is known to pass through the other two poles, and that [is what] the second part proposed.

Finally, if circle *AEZD* were to bisect arc *BAG*, for example, at point *A* and pass through the poles of the other of the circles intersecting each other, then it will also bisect the other [circle] by its passage through the poles of both. For if not, let some other [great circle] bisect them or pass through their poles. Then the one which passes through the poles of both will bisect the severed arcs of the [circles]. Therefore the two circumferences of the great circles will have three points in common — namely, the two poles of the other of the [circles] and point *A* — and each of the arcs will be included between any two [of these] points and will [also] be semicircumferences by Th. [19]. Hence it follows that the part is equal to its whole, which is impossible. Therefore this hypothesis will not stand unless that [one great] circle crosses through the poles of both and bisects the severed arcs of both [circles]. Q.E.D.

Theorem 27

All circles drawn in one sphere or in different but equal [spheres] according to equal polar lines are equal. And if circles are equal, their polar lines are necessarily equal.

If the polar lines are equal to the side of the great square [that can be inscribed in a great circle], then, by Th. 8 above and by definition, it will be established

huius & diffinitionem:omnes circulos & magnos
& æquales inuicem esse.Sed ponamus eas minores
aut maiores huiusmodi latere quadrati magni, sint
cp a g & d h secundum quas duo circuli b m g &
e n h in sphæra una,aut diuersis æqualibus tamen
describantur,quos dico esse æquales. Transeant em
per polos eorum qui sint a & d duo circuli magni
a b g,cuius centrum 3 ,& d e h super centro s, q
per 15.huius secabūt descriptos circulos per æqua=
lia & orthogonaliter,sintcp communes sectiones li
neæ b g & e h ,descendant demum à polis a & d
aliæ duæ polares lineæ a b & d e pdictis duabus
æquales,à centris autem circulorum binæ egrediā=
tur semidiametri 3 b & g 3 circuli a b g, s e au=
tem & s h circuli d e h ,erit igitur ex hypothesi &
tertij arcus a g æqualis arcui d h ,itemcp arcus
a b æqualis arcui d e ,quare per communem scien
tiam totus arcus b g toti arcui e h æqualis habe=
bitur,& ideo per tertij corda b g, quæ est etiam
diameter circuli b g m æqualis erit cordæ e h ,
quæ est diameter circuli e h ,quare per diffinitionē
circulorum æqualium patebit prima pars theore=
matis.Conuertendo autem processum iam recita=
tum,facile concludemus æqualitatem linearum po
lariū,si prius circulos ipsos subiecerimus æquales,
erunt enim duæ eorum diametri b g & e h æqua=
les per conuersionem diffinitionis circulorū æqua=
lium ,unde & per tertij arcus b g & e h,quorum ipsæ sunt cordæ,æquales inue
nientur.quare per cōmunem scientiam arcus eorū dimidij a g scilicet & d h nō
erūt inæquales,& ideo per tertij cordæ suæ a g & d,quæ & polares lineæ sunt,
æquales demonstrabuntur.utrancp igitur theorematis partem firmauimus, quod
quidem studiosus expectabat discipulus.

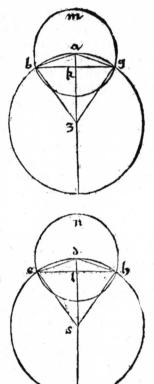

XXVIII.

Omnes minores circuli æquales à centro sphæræ eos continentis
æqualiter distant,& si ab eo centro æqualiter distiterint circuli mino=
res quotlibet,ipsos æquales esse conuincemus.

Sint duo circuli minores a b & g d .aut plures quotlibet in
sphæra æquales.Dico,cp ipsi æqualiter distent à centro sphæ=
ræ,qui si fuerint æquedistantes à centro sphæræ,æquales neces
sario habebuntur.Primam partem sic confirmabimus. A cen=
tro sphæræ quod sit 3 ad centra duorū circulorum h & k edu=
cantur duæ lineæ 3 h & 3 k, duæcp semidiametri sphæræ 3 b
& 3 g,quarum duos terminos connectemus per lineas h b &
k g,eritcp per huius utracp linearum 3 h & 3 k perpendicu=
laris ad superficiem circuli cui incidit,& ideo per diffinitionē
perpendicularis ad semidiametrum sibi conterminalem,quare
utercp angulorum 3 h b & 3 k g rectus habetur .oportet autē
& semi

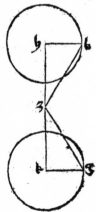

that all the circles are both great circles and equal to one another.

But now let us take [the polar lines] to be smaller or larger than the side of a great square, and let [the lines] be *AG* and *DH*, according to which let two circles *BMG* and *ENH* be described in one sphere or in different but equal [spheres]; then these [two circles] are equal.

Let two great circles, *ABG* with center *Z* and *DEH* around center *S*, pass through their poles, which are *A* and *D*. Then by Th. 15 above, these [great circles] bisect the described circles orthogonally, and the lines of intersection are lines *BG* and *EH*. Next, from poles *A* and *D* let two other polar lines *AB* and *DE*, equal to the aforementioned two, be dropped. Furthermore, from the centers of the circles let two radii each emerge, *ZB* and *GZ* of circle *ABG* and *SE* and *SH* of circle *DEH*. Therefore, from the hypothesis and [Euclid] III.[not given], arc *AG* will be equal to arc *DH*, and similarly arc *AB* will be equal to arc *DE*. Hence by axiom, the total arc *BG* will be equal to the total arc *EH*, and thus by [Euclid] III.[not given] chord *BG*, which is also the diameter of circle *BGM*, will be equal to chord *EH*, which is the diameter of circle *EH*. Then by the definition of equal circles, the first part of the theorem will be clear.

By a reversal of this procedure just recited, we will conclude with ease the equality of the polar lines. If first we submit that the circles themselves are equal, certainly their two diameters, *BG* and *EH*, will be equal by the converse of the definition of equal circles. From this and from [Euclid] III.[not given],

arcs *BG* and *EH*, of which [the diameters] are the chords, will be found to be equal. Hence by axiom their half-arcs — namely, *AG* and *DH* — will not be unequal, and thus their chords *AG* and *D*,* which are polar lines, will be proven equal. Thus we have shown each part of the theorem, which, of course, the eager student was expecting.

Theorem 28

All equal small circles are equally distant from the center of the sphere containing them. And if any number of small circles are equally distant from that center, we will prove them to be equal.

If two, or any number more, small circles, *AB* and *GD*, within a sphere are equal, then they are equally distant from the center of the sphere; and if they are equidistant from the center of a sphere, they are necessarily equal.

We shall prove the first part thus. From the center of the sphere, which is *Z*, to the centers, *H* and *K*, of the two circles, let two lines *ZH* and *ZK* be drawn, as well as two radii of the sphere, *ZB* and *ZG*, whose two ends we shall join [to *H* and *K*] by lines *HB* and *KG*. By Th. [2] above each of the lines *ZH* and *ZK* will be perpendicular to the surface of the circle to which it falls, and thus by definition [each of the lines] is perpendicular to the radius adjacent to it. Hence each of the angles *ZHB* and *ZKG* is a right angle. Moreover, both radii *HB* and *KG* must

*For *D* read *DH*.

&semidiametros h b & k g esse aequales ppter circulos eorum, quos hypothesis
aequales tradidit. per penultimam igitur primi & communes scientias erit 3 h linea aequalis 3 k, unde & per diffinitionem circuli dicti aequaliter à cētro sphaerae
distabunt, quod intendebat prima pars ppositionis. Sed ponatur duo circuli praedicti aequaliter distantes à centro sphaerae, quamobrem oportebit duas lineas 3 h
& 3 k esse aequales, unde ex medijs praeassumptis duas semidiametros h b & k g
aequales esse constabit, diametris ergo circulorum aequalibus existentibus, ipsi circuli per diffinitionem circuloꝝ aequalium aequales habebuntur, quod asseruit secunda pars theorematis. Poterimus autem & has mutuas passiones demonstrare
de circulis diuersarum sphaerarum aequalium tamen, non aliter quàm de circulis
unius sphaerae.

XIX.

    Si duos circulos minores in sphaera aequales & aequedistantes, circulus quidam secet magnus per polos eorum non transiens, arcus ex
eis coalternos abscindet aequales. circulus praeterea magnus duobus
praedictis aequedistans, arcus circuli inclinati interceptos duobus circulis aequedistantibus minoribus per aequalia secabit.

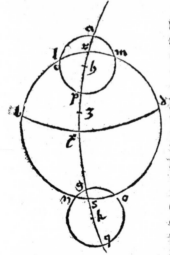

Ex duobus circulis minoribus a l p & g n q in
sphaera una aequalibus & aequedistantibus, circulus
magnus l b o d per polos eorum non incedens, secet arcus coalternos l a m & n g o, quibus quidē
circulis aequediftet circulus magnus b x d, secans
duos arcus l n & m o circuli l b o d in duobus
punctis b & d. Dico, φ arcus l a m aequalis est ar
cui n g o, & arcus l p m arcui n q o aequalis, &
φ circulus magnus b x d aequedistans duobus minoribus, duos arcus l n & m o per aequa scindet in
punctis b & d. Duos enim polos per huius communes esse oportet, tribus dictis circulis aequedistantibus, qui sint h & k, polus autem circuli l b o
d sit punctus 3 . per duos itaqꝫ polos h & 3 incedat
circulus a h 3 k huius edocente, qui transibit etiā
per punctum k ex huius. Cum itaqꝫ arcus r s sit
semicircumferentia per huius, arcus φ h k sit me

dietas circumferentiae, φ duo poli h & k diametrum sphaerae terminent, quemad
modum ex huius trahitur, ablato arcu communi h s, relinquetur arcus h r aequalis arcui k s. ex autem huius duo arcus k g & h a aequales sunt, quare & residuus a r residuo g s aequalis habebitur. Supra diametrum autem circuli l b o
d erectae sunt duae portiones aequales, ex quibus sumuntur arcus aequales r h & s
k, quorum uterqꝫ minor est dimidio arcu portionis suae, lineae φ à punctis h & k
ad duo puncta m & o protensae, aequales sunt per huius, quare per huius arcus a m aequabitur arcui s o, arcus autem r m aequalis est arcui l r, similiter ar
cus s o arcui s n, cum circulus a h 3 k per polos horum circulorum se secantiū
transeat. Rursus quoniam supra diametros duorum circulorum a l p & g n q,
erectae sunt duae portiones aequales a p & g q, ex qbus absumpti sunt arcus aequales a r & g s, quorum uterqꝫ minor est dimidio arcu portionis suae, lineae uero à
                                                                punctis

be equal because of their circles, which the hypothesis gave to be equal. Then, by the penultimate theorem of [Euclid] I and the axioms, line ZH will be equal to line ZK. From this and by definition, the mentioned circles will be equally distant from the center of the sphere, which the first part of the proposition maintained.

But now let the two aforementioned circles be placed equally distant from the center of the sphere; whereupon the two lines ZH and ZK must be equal. Hence, by the previously applied methods, it will be established that the two radii HB and KG are equal. Therefore, since the diameters of the circles are equal, the circles themselves will be equal by the definition of equal circles, which the second part of the theorem avers.

We can also show the corresponding proofs concerning circles of different but equal spheres no differently from those concerning circles of one sphere.

Theorem 29

If a particular great circle, not passing through their poles, intersects two equal and parallel small circles in a sphere, it intercepts arcs from them [which are] reciprocally equal. Furthermore, a great circle parallel to the two aforementioned [small circles] will bisect the arcs of the inclined circle [that were] intercepted by the two parallel small circles.

From the two equal and parallel small circles ALP and GNQ in one sphere, let a great circle LBOD, not passing through their poles, cut reciprocal arcs LAM and NGO. Let a great circle BXD, intersecting the two arcs LN and MO of circle LBOD at the two points B and D, be parallel to those [two small circles]. Then arc LAM is equal to arc NGO,

and arc LPM is equal to arc NQO; [furthermore] great circle BXD, parallel to the two small circles, bisects the two arcs LN and MO at points B and D.

The two poles must be shared by the three mentioned parallel circles by Th. [22] above. Let these be H and K. Moreover, the pole of circle LBOD is point Z. Then, by the instruction of Th. [15] above, let circle AHZK pass through the two poles H and Z, [and] by Th. [16] above this [circle] will also pass through point K. Then, since arc RS is a semicircumference by Th. [19] above, and since arc HK is [also] half of the circumference because the two poles H and K terminate the diameter of the sphere as is concluded from Ths. [9 and 18 and the definition of poles — see Th. 22], then, when the common arc HS is subtracted [from both], arc HR will be left equal to arc KS. Moreover, from Th. [23] above the two arcs KG and HA are equal. Therefore the remaining [arc] AR will be equal to the [other] remaining [arc] GS. Now upon the diameter of circle LBOD are erected two portions, from which equal arcs RH and SK are taken, each of which is smaller than half the arc of its portion. The lines drawn from points H and K to the two points M and O are equal by Th. [24] above. Therefore, by Th. [25] above, arc AM will be equal to arc SO. Moreover, arc RM is equal to arc LR. Similarly arc SO is equal to arc SN since circle AHZK passes through the poles of those [two] circles that intersect each other.

Again, because upon the diameters of the two circles ALP and GNQ were erected two equal portions AP and GQ, from which equal arcs AR and GS were intercepted, each of which is less than half the arc of its portion, then the lines extended from

punctis r & s ad puncta circulorum substratorum m scilicet & o protensæ sunt
æquales,propter arcus r m & s o,quos æquales nuperrime conclusimus,erit per
huius arcus a m æqualis arcui g o,arcus autem a m æquaturarcui a l,& ar
cus g o arcui g n propter circulum magnum a h 3 h,per polos circulorum se
secantium transeuntem.quare totus arcus l a m æquabitur toti arcui n g o, reli
quusq̃ l p m reliquo n q o non erit inæqualis.constat ergo prima pars propo=
sitionis nostræ.Postremo unusquisq̃ quatuor arcuum r b, b s, s d & d r est qua
drans circumferentiæ,quandoquidem circulus a h 3 k per polos duorum circu=
lorum b l o d & b x d magnorum transit,à quibus singulatim quadrantibus,si
dempseris quatuor arcus æquales l r,r m,n s & s o,relinquentur quatuor ar=
cus l b, b n,o d & d m inter se æquales.utrunq̃ igitur arcuum l n & m o cir=
culus magnus b x d duobus minoribus æquedistans,per æqualia scindit, quod
secundo loco demonstrandum extitit.

XXX.

Omnis anguli sphæralis ad quatuor rectos eam esse proportionẽ,
quam basis eius ad circumferentiam suam.

Si ultimam sexti Euclidis satis uidisti,non latebit te nostri theorematis demõ
stratio,per habitudines enim æquemultiplicium usitato iuuabis te syllogismo.

XXXI.

Cuncti sphærales anguli,siue in sphæra una,siue diuersis, quorum
bases sunt similes,sibi inuicem æquabuntur.

Erit enim basibus similibus ad suas circumferẽtias una proportio per cõuer
sionem diffinitionis arcuum similium,anguloꝝ autem ad quatuor rectos præce
dens,eam cõclusit ꝓportionem,quam baſſum ad suas circumferentias,erit igitur
omnium angulorum ad quatuor rectos una proportio,per 7.ergo quinti anguli
ipsi æquales erunt,quod oportuit declarare.

XXXII.

Omnium angulorum æqualium bases inueniri similes.

Æqualiũ nanq̃ angulorum ad quatuor rectos eandem constat esse propor=
tionem,quæ quidem per huius est,ut baſium ad suas circumferentias.basiũ igĩ
huiusmodi angulorum ad suas circumferentias eadem habetur proportio, quare
per diffinitionem similiũ arcuum bases ipsæ similes conuincent,quod est ꝓpositũ.

XXXIII.

Quorum dissimiles sunt bases,angulos inæquales esse,angulorum
quoq̃ inæqualium dissimiles reperiri bases.

Hæc ex contrario subiecti præmissæ & ante præmissæ contrarium infert pas
sionum suarum aduersarium ducendo ad impossibile.Si enim bases dissimiles ad=
miserit,angulos autem æquales,erunt ex præmissa bases similes,igitur similes &
dissimiles esse confitebitur easdem bases,quod est inconueniens.Quòd si angulis
inæqualibus existentibus bases putauerit similes,sequitur ex antepræmissa & an=
gulos esse æquales.quare eosdem angulos æquales inuicem & inæquales affirma=
bit,quod,quoniam repugnantiam parturit,esse non potest.Destructis autem im=
possibilibus,ueritatẽ ꝓpositionis inferemus. L Iuxta

points *R* and *S* to points *M* and *O* on the underlying circles are equal because of arcs *RM* and *SO*, which we concluded not long ago to be equal. By Th. [11] above arc *AM* will be equal to arc *GO*. Moreover, arc *AM* equals arc *AL* and arc *GO* equals arc *GN* because great circle *AHZH*** passes through the poles of the intersecting circles. Therefore the total arc *LAM* will equal the total arc *NGO*, and the remaining [arc] *LPM* will not be unequal to the other arc *NQO*. Therefore the first part of our proposition is established.

Finally, each one of the four arcs *RB*, *BS*, *SD*, and *DR* is a quadrant of the circumference, since circle *AHZK* passes through the poles of the two great circles *BLOD* and *BXD*. If you subtract the four equal arcs *LR*, *RM*, *NS*, and *SO* from the quadrants, one at a time, the four arcs *LB*, *BN*, *OD*, and *DM* will be left equal to each other. Therefore the great circle *BXD*, parallel to the two small [circles], bisects each of the arcs *LN* and *MO*, which was to be shown in the second part [of the theorem].

Theorem 30

The ratio of every spherical angle to four right angles is as that of its base [i.e., the arc subtending the angle] to its circumference.

If you have observed the last theorem of VI of Euclid sufficiently, the proof of our theorem will not hide from you, for you will assist yourself through the conditions by the customary syllogism of equimultiples [because they are integral multiples of the same number].

Theorem 31

All spherical angles, whether in one sphere or in different [spheres], whose bases are similar will be equal to each other.

There will be [only] one ratio for the similar bases [relative] to their circumferences, by a reversal of the definition of similar

arcs, and the preceding [theorem] concluded that the ratio of the angles to four right angles is as that of the bases to their circumferences. Therefore there will be [only] one ratio of all the angles to four right angles. Thus by [Euclid] V.7 these angles will be equal. Q.E.D.

Theorem 32

The bases of all equal angles are similar.

It is established that the ratio, which is [known] through Th. [30] above, of the equal angles to four right angles is the same as [the ratio] of the bases to their circumferences. Therefore the ratio of the bases of these angles to their circumferences will be the same [and known]. Hence by the definition of similar arcs, these bases are proven similar, which was proposed.

Theorem 33

The angles whose bases are dissimilar are unequal, and dissimilar bases are found from unequal angles.

This [theorem], by the opposite of the substance of the [two] preceding theorems, introduces the contrary of their proofs in order to lead to the impossibility of [these contrary suggestions].

If it were asserted that the bases are dissimilar while the angles are equal, then from the preceding [theorem] the bases will be similar. Therefore it is acknowledged that the same bases are both similar and dissimilar, which is contradictory. But if it were considered that the bases are similar when the angles are unequal, it follows from the theorem before the preceding [theorem] that the angles are equal. Therefore one will assert that the same angles are equal and unequal to each other, which, because it teems with incompatibility, cannot be. With these impossibilities eliminated, we infer the truth of the proposition.

*For *AHZH* read *AHZK*.

XXXIIII.

Iuxta punctum quotlibet in superficie sphæræ signatum, angulo proposito æquum angulum statuere.

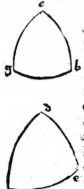

Sit angulus propositus b a g, cui iuxta pūctum d in sphæra quacunq; æqualem constituere libet. Angulo b a g substerno ba sim suam b g, deinde super puncto d secundum distantiam quan tamlibet describo circulum, ex cuius circumferentia abscindo arcum e 3 similem arcui b g, quod per 23 primi & ultimam sexti facile comparabis, duoc; puncta eius terminalia e & 3 puncto d per duos arcus circulorum magnorum copulabo, qui sint d e & d 3, quos dico apud punctum d continere angulum æqualem angulo b a g, cuius ratio propter similitudinem duarum basium b g & e 3 ex huius manifesta apparet. Facilius tamen id effici es in sphæra una, aut diuersis æqualibus tamen. Super puncto eñ d secundum distantiam a b circumduces circulum, ex cuius circumferentia arcum e 3 abscindes æqualem arcui b g, reliqua ut antehac disponendo. Erunt enim duo circuli, ex quarum circumferentijs bases ab scinduntur, per huius æquales. cunq; bases ipsæ sint æquales, erunt etiam similes, per igitur huius syllogismum conficies.

XXXV.

Omnium duorum triangulorum sphæralium, quorum cuncta latera unius cunctis lateribus alterius æqualia sunt, oēs angulos æquis oppositos lateirbus æquales esse.

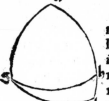

Omnes trianguli, de quibus futurum habebimus sermonem, in superficie unius sphæræ, aut duarum uel pluriū æqualium tamen, ex arcubus circulorum magnorum constantes intelligemus. Sint itaq; duo trianguli a b g, & d e 3, quorum trina latera sunt æqualia, latus quidem b g unius lateri e 3 alterius, & reliqua reliquis. Dico, φ anguli æqualibus oppositi lateribus sunt æquales, angulus uidelicet a angulo d, & reliqui suis relatiuis. Si enim utriq; arcuum angulos a & d ambientium fuerint quadrantes, erunt duo puncta a & d poli circulorum b g & e 3 per huius, quare per huius duo anguli a & d æquales erunt, φ bases eorum sint similes, imo æquales. Si uero bini arcus dictos angulos conti nētes fuerint æquales, minores tamē quadrāte diuisim, descriptis duobus circulis super polis a & d secundum distantias æquales a g & e 3, circuli ipsi æquales erunt per hu

ius, arcus autem eorum ad bina puncta b g & e 3 terminari (necesse enim est eos ad pūcta dicta terminari propter æqualitatem binorum ar cuum à punctis a & d descendentium) arcus inq; illi æquales erunt, quoniam ha bebunt cordas æquales, ppter arcus b g & e 3 eis conterminales, quos quidem hypothesis subiecit æquales, quare per 21. huius anguli a & d æquales habebun tur. Qd si bini arcus, qui ambiunt duos angulos a & d, inæquales fuerint, sint mi nores eorum duo arcus a g & d 3, superq; punctis a & d factis polis secundum

distantias

Theorem 34

[At] any indicated point on the surface of a sphere, an angle equal to a given angle may be constructed.

Let ∠ *BAG* be given. One wishes to construct an [angle] equal to it [at] point *D* on any sphere.

Under ∠ *BAG* draw its base *BG*. Then, around point *D* with any length as radius, describe a circle, from whose circumference intercept arc *EZ* similar to arc *BG*. Through [Euclid] I.23 and the last [theorem] of [Euclid] VI, you will arrange this with ease. Join its two end points *E* and *Z* to point *D* by two arcs, *DE* and *DZ*, of great circles. These, then, enclose an angle at point *D* equal to ∠ *BAG*. The reason for this is obvious from Th. [31] above because of the two similar bases *BG* and *EZ*.

You can accomplish this just as easily in one sphere or in different but equal [spheres]. For upon point *D*, with the length *AB* [as polar line], you will circumscribe a circle from which you intercept an arc *EZ*, equal to arc *BG*, to arrange the rest as before. For the two circles, from whose circumferences the bases are cut, will be equal by Th. [27] above. And since these bases are equal, they will also be similar. Therefore by Th. [31] above you will complete the syllogism.

Theorem 35

For every pair of spherical triangles, in which all the sides of one are [respectively] equal to all the sides of the other, all the angles opposite the equal sides are equal.

We shall understand that all triangles, concerning which we will have future discourse, are established on the surface of one sphere or of two or more equal [spheres] from the arcs of great circles. Then, let *ABG* and *DEZ* be two triangles whose three sides are [respectively] equal — indeed, side *BG* of the one to side *EZ* of the other and the others [of the one] to the others [of the second]. Then the angles opposite the equal sides are equal — namely, ∠ *A* is equal to ∠ *D* and the other [angles are equal] to their respective angles.

If both of the arcs enclosing angles *A* and *D* are quadrants, the two points *A* and *D* will be poles of circles *BG* and *EZ* by Th. [12] above. Thus by Th. [31] above the two angles *A* and *D* will be equal because their bases are similar — indeed, are equal.

However, if the two arcs of each [triangle] that enclose the mentioned angles are equal yet less than a quadrant division, then, when two circles are drawn upon poles *A* and *D* with the equal lengths *AG* and *EZ** [as polar lines], these circles will be equal by Th. [27] above. Furthermore, their arcs, terminated at the pairs of points *B,G* [in one] and *E,Z* [in the other] (for they must be terminated at the mentioned points because of the [respective] equality of the two arcs of each [triangle] that descend from points *A* and *D*) — those very arcs — will be equal, since they will have equal chords because arcs *BG* and *EZ*, which the hypothesis gave to be equal, are adjacent to those [arcs on the described circle]. Therefore, by Th. 21† above, angles *A* and *D* will be equal.

But if the two pairs of arcs which [respectively] enclose the two angles *A* and *D* were unequal, let the two arcs *AG* and *DZ* be the smaller of them, and let circles *GH* and *ZK* be described around points *A* and *D* as poles with the

*For *EZ* read *DZ*.
†For 21 read 31.

diſtantias æquales a g & d ʒ deſcribantur circuli g h & ʒ k ,quos conſtat eſſe
æquales per ·· huius,ſupra quorum diametros erectæ ſunt duæ portiones æqua=
les,ex quibus quidé portionibus accepti ſunt duo arcus h b & k e æquales,quo=
rum uterq̓ minor eſt dimidio arcu portionis ſuæ.b g autem recta linea,ſi produ
cta fuerit,æqualis eſt ipſi e ʒ ,propter arcus b g & e ʒ æquales,erit igitur per
huius arcus g h æqualis arcui ʒ k . illi autem duo arcus, cum ſint baſes duorum
angulorum a & d,per huius angulos ſuos afferent æquales. Quemadmodum
autem circa angulos a & d proceſſimus,circa reliquos quoq̓ faciemus,& hæc in
ſtituimus declaranda.

X X X ꝟ I.

Omniũ duorũ triangulorum,quorum duo latera unius duobus la
teribus alterius ſunt æqualia,angulusq̓ unius dictis lateribus cõtétus
angulo alterius,baſis quoq̓ unius baſim alterius æquabit, reliqui de=
mum anguli unius reliquis angulis alterius,quiſq̓ uidelicet ſuo rela=
tiuo æquabuntur.

Trianguli a b g latus a b lateri d e trianguli d e ʒ æ
quale habeatur,latus uero a g æquale lateri d ʒ ,& angulus
a æqualis angulo d .Dico, q̓ latus b g lateri e ʒ æquabit.
angulus etiam b angulo e ,& angulus g angulo ʒ æquabũ
tur.Si enim utrunq̓ laterum a b & a g fuerit quadrans cir=
cumferentiæ,ſimiliter utrunq̓ laterum d e & d ʒ ,erũt duo
puncta a & d poli circulorum b g & e ʒ per huius,cũq̓
poſiti ſint anguli a & d æquales,erunt per huius baſes eo
rum ſimiles,arcus ſcilicet b g & e ʒ ,& ideo æquales,q̓ cir=
culi eorum æquales habeantur.Si uero bini arcus,qui duos angulos a & d ambi
unt,inter ſe fuerint æquales,non tamen quartæ circulorum,neceſſe eſt circulos de
ſcriptos ſuper polis a & d ſecundum diſtantias a g & d
ʒ tranſire per puncta b & e,eruntq̓ ex hypotheſi & hu
ius arcus horum circulorum,quos latera triangulorũ in-
tercipiunt ſimiles,& ideo æquales,q̓ circuli eorum huius confirmante æquales habeantur.unde etiam eorũ cor=
das æquari oportet,quæ quidem cordæ arcubus b g & e
ʒ lateribus ſcilicet triangulorum ꝓpoſitorum cõmunes
ſunt,quare & arcus ipſi æquales erunt.Poſtremo duorum
arcuum a b & a g alter altero maior ſit arcus,uerbi gra
tia, a b maior arcu a g ,itemq̓ d e maior d ʒ .deſcribã
tur itaq̓ ſuper duobus polis a & d ,ſecundum diſtantias æquales a g & d ʒ cir
culi g h & ʒ k ,quos huius æquales arguit.ex portionibus autem æqualibus ſu
pra diametros eoꝛ erectis duo arcus h b & k e abſumuntur æquales,quoꝛ uter
q̓ minor eſt dimidio arcu portionis ſuæ.eſt autẽ arcus g h ſimilis arcui ʒ k ex
hypotheſi & huius,& ideo æqualis eidem,q̓ circuli ſui æquales exiſtant, quare
per huius linea b g ,ſi producta fuerit,æquabitur lineæ e ʒ .unde & arcus b g
& e ʒ ,quorum ipſæ ſunt cordæ,æquales reperientur.Tres igitur arcus triangulũ
a b g claudentes,tribus triangulum d e ʒ ambientibus,æquales habentur,angu
lusq̓ a unius angulo d alterius æqualis,unde & per præcedentem reliquos angu
los unius reliquis duobus angulis alterius æquales uideri neceſſe eſt , pro quibus
hactenus fatigati ſumus.

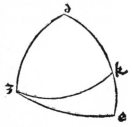

equal lengths *AG* and *DZ* [as polar lines]. It is established that these [circles] are equal by Th. [27] above. Upon their diameters two equal portions are erected. From these portions two equal arcs *HB* and *KE* are intercepted, each less than half the arc of its portion. Moreover, straight line *BG*, if it is drawn, is equal to *EZ* because arcs *BG* and *EZ* are equal. Then by Th. [24] above arc *GH* will be equal to arc *ZK*. Furthermore, these two arcs, since they are the bases of the two angles *A* and *D*, have their angles equal by Th. [31] above.

Just as we proceeded concerning angles *A* and *D*, so also we can accomplish concerning the others and we have [then] established those statements [above].

Theorem 36

Of [any] two triangles in which two sides of the one are equal to two sides of the other and in which the angle contained by the mentioned sides of the one is equal to the [corresponding] angle of the other, then the base of the one will be equal to the base of the other and, finally, the other angles of the one will be equal to the other angles of the other — that is, each one to its corresponding [angle].

If side *AB* of △ *ABG* is equal to side *DE* of △ *DEZ*, side *AG* is equal to side *DZ*, and ∠ *A* is equal to ∠ *D*, then side *BG* will equal side *EZ*, ∠ *B* will also equal ∠ *E*, and ∠ *G* will be equal to ∠ *Z*.

If each of the sides *AB* and *AG*, as well as each of the sides *DE* and *DZ*, is a quadrant of a circumference, then the two points *A* and *D* will be poles of the circles *BG* and *EZ* by Th. [12] above. And since the angles *A* and *D* are given equal, by Th. [32] above their bases will be similar — namely, arcs *BG* and *EZ* — and therefore [*BG* and *EZ* will be] equal because their circles are equal.

However, if the two pairs of arcs which respectively enclose the two angles *A* and *D* are equal to each other but are not quarters of circles, then it is necessary that the circles described around poles *A* and *D*, with the lengths *AG* and *DZ* [as polar lines], pass through points *B* and *E*. From the hypothesis and Th. [32] above, the arcs of these circles, which [arcs] the sides of the triangles intercept, will be similar, and therefore [those arcs will be] equal because their circles are equal by the confirmation of Th. [27] above. Hence their chords must also be equal. These chords are shared by arcs *BG* and *EZ* — namely, the sides of the given triangles — and therefore the arcs [*BG* and *EZ*] themselves will be equal.

Finally, of the two arcs *AB* and *AG*, let one be greater than the other — for example, let arc *AB* be greater than arc *AG*, and similarly let *DE* be greater than *DZ*. Then let circles *GH* and *ZK* be drawn around the two poles *A* and *D*, with the equal lengths *AG* and *DZ* [as polar lines], and these [circles] Th. [27] above concludes to be equal. Moreover, from equal portions that are erected upon the diameters of these [circles], two equal arcs *HB* and *KE* are taken, each of which is less than half the arc of its portion. Moreover, arc *GH* is similar to arc *ZK* from the hypothesis and Th. [32] above, and therefore [*GH*] is equal to the same [*ZK*] because their circles are equal. Thus, by Th. [24] above, line *BG*, if it is drawn, will be equal to line *EZ*. Hence arcs *BG* and *EZ*, of which the [lines *BG* and *EZ*] are chords, will be found [to be] equal. Therefore the three arcs enclosing △ *ABG* will be equal to the three [arcs] enclosing △ *DEZ*, and ∠ *A* of one will be equal to ∠ *D* of the other. Hence by the preceding, the rest of the angles of the one necessarily appear equal to the two remaining angles of the other. And this has wearied us enough.

XXXVII.

Omnis trianguli duo quælibet latera tertio reliquo funt longiora.

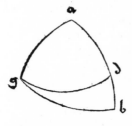

Duo latera a g & g b trianguli a b g collecta dico
effe longiora latere a b .Si enim latus a g æquale fuerit
lateri a b ,aut longius eo,planum eft duo latera a g & g
b congregata fuperare latus a b .Si uero minus eo fue‹
rit fecundum diftantiam a g, fuper polo a defcribatur
circulus, cuius arcus g d latus triâguli a b offendat in
puncto d .Supra diámetrum itacӡ circuli g d erecta eft
portio circuli,ex qua fumptus eft arcus d b minor dimi
dio arcu portionis fuæ,quare per huius linea b d re‹
cta,fi producta fuerit,breuior eft linea b g,unde & arcus b d breuior arcu b g.
adiectis igitur utrobicӡ æqualibus a g & a d arcubus,erit per communem fcien
tiam aggregatum ex duobus arcubus a g,g b maius aggregato ex duobus ar‹
cubus a d, d b fcilicet arcu a d,quod libuit abfoluere,Non aliter de duobus reli‹
quis quibufcuncӡ ad tertium collatis procedemus.

XXXVIII.

Si à duobus pûctis terminalibus unius lateris trianguli duo arcus
exierint,intra angulum ipfum concurrentes,erunt ipfi collecti mino‹
res duobus reliquis trianguli lateribus.

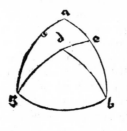

A duobus pûctis b & g terminâtibus latus b g triâ‹
guli a b g exeant duo arcus b d & g d ,intra triangu‹
lum confluentes in puncto d .Dico,qͻ duo arcus b d, d
g congregati,minores funt duobus arcubus a b & a g
coniunctis.Protendatur enim arcus g d in e punctum
lateris a b , eruntcӡ per præcedentem duo arcus g a, a
e longiores arcu g e.quare adiecto communi e b fient
duo arcus g a, a b longiores duobus g e & e b . item
duo arcus d e, e b longiores funt arcu d b ,facto igi‹
tur d g communi:erunt duo arcus g e, e b longiores duobus g d, d b ,fed erât
duo arcus g a, a b longiores duobus g e, e b ,multo igitur longiores habebun
tur duobus arcubus g d & d b côiunctis.unde & illi uiceuerfa minores iftis, qͻd
uoluimus pertractare. Côftat autem ex dictis,qͻ fi ab aliquo puncto terminali la
teris trianguli arcus,producatur latus fibi oppofitum fecans,erunt duo arcus fcili
cet productus,& ex latere trianguli refectus minores duobus trianguli lateribus.

XXXIX.

Cuiuslibet trianguli fphæralis tria latera duobus femicirculis effe
minora.

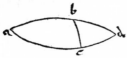

Producantur duo latera a b & a c ut concurrât in d .
erit itacӡ b c minor duobus arcubus b d & c d ,adie‹
ctis itacӡ communibus a b & a c,tres arcus a b, a c &
b minores erunt duobus femicirculis a b d & a c d.

X L.

Cuiuslibet trianguli duos æquales angulos habentis ,duo quocӡ
ateral eos refpicientia æquari necefle eft. Habeat

Theorem 37

Any two sides of every triangle are longer than the third side.

If the two sides *AG* and *GB* of △ *ABG* are added, then they are longer than side *AB*.

If side *AG* were equal to side *AB* or longer than it, it is obvious that the sum [of] the two sides *AG* and *GB* exceeds side *AB*.

However, if side *AG* were less than [*AB*], let a circle be described around pole *A* with the length *AG* [as polar line]. Arc *GD* of that [circle] hits the side *AB* of the triangle at point *D*. Then upon the diameter of circle *GD* a portion of a circle is erected, from which arc *DB* is taken less than half the arc of its portion. Therefore, by Th. [25] above, straight line *BD*, if it is drawn, is shorter than line *BG*. Hence, arc *BD* is shorter than arc *BG*. Therefore, when the equal arcs *AG* and *AD* are added on each side, by axiom the sum of the two arcs *AG* plus *GB* will be greater than the sum of the two arcs *AD* plus *DB* — namely, arc *AD*,* which one wished to prove. We will proceed no differently concerning whichever two others combined [are compared] to a third.

Theorem 38

If from the two end points of one side of a triangle two arcs emerge, meeting within the [tri]angle, their sum will be less than the other two sides of the triangle.

If, from the two end points *B* and *G* of side *BG* of △ *ABG*, two arcs *BD* and *GD* emerge, meeting at point *D* within the triangle, then the sum of the two arcs *BD* plus *DG* is less than the sum of the two arcs *AB* and *AG*.

Let arc *GD* be extended to point *E* on side *AB*. By the preceding [theorem] the two arcs

GA plus *AE* will be longer than arc *GE*. Therefore, when *EB* is added to both, the two arcs *GA* plus *AB* are made longer than the two *GE* plus *EB*. Similarly, the two arcs *DE* plus *EB* are longer than arc *DB*. Then, when *DG* is made common [i.e., is added to both], the two arcs *GE* plus *EB* will be longer than the two arcs *GD* plus *DB*. But the two arcs *GA* plus *AB* were longer than the two arcs *GE* plus *EB*; therefore [*GA* plus *AB*] will be much longer than the sum of the two arcs *GD* plus *DB*. In any combination these [constructed arcs are] smaller than those [of the given triangle], which we wanted to prove.

Moreover, it is established from [those things] mentioned that if, from some end point of the sides of a triangle, an arc is drawn intersecting the side opposite it, the two arcs — namely, the one drawn and the one cut short from the side of the triangle — will be smaller than the two sides of the triangle.

Theorem 39

The three sides of any spherical triangle are smaller than two semicircles.

Let the two sides *AB* and *AC* be drawn to meet at *D*. Then *BC* will be smaller than the two arcs *BD* and *CD*. Therefore, when the common [arcs] *AB* and *AC* are added [to *BC*], the three arcs *AB*, *AC*, and *B*† will be smaller than the two semicircles *ABD* and *ACD*.

Theorem 40

In any triangle that has two equal angles, the two sides opposite those [angles] are necessarily equal.

*For *AD* read *AB*.
†For *B* read *BC*.

Habeat nanqʒ triangulus a b g duos angulos b &
g æquales. Dico, qʒ latus a b æquabiť lateri a g. Si enim
non, alterum altero maius erit, sitʨ a g longius ipso a b
ex quo abſcindam arcum g d æqualem arcui a b produ
cendo arcum b d. Duo itaqʒ trianguli a b g & b g d bi
na latera habent æqualia, angulos'ʨ hiſce contentos late
ribus æquales, angulum uidelicet a b g æqualem angu⸗
lo b g d ex hypotheſi. quare per huius angulus d
b g æqualis erit angulo a g b, quem hypotheſis ſubiecit
æqualem angulo a b g. unde & angulus d b g angulo
eidē a b g æqualis habebiť: pars ſcilitet toti, ʠd eſt impoſſibile. Nō erit ergo alteʀ
altero maius, & ideo æqualia inuicē relinquenť, quod expectabas demonſtrandū.

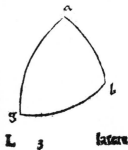

X L I.

Duo latera cuiuſuis trianguli æqualia angulos æquales ſubtendere
oportet.

Hæc conuertendo præmiſſam ex paſſione eius ſubiectū inferre ſuum pollice⸗
tur. Triangulus ergo a b g duo latera a b & a g habeat æqualia. Dico qʒ angu
lus g æqualis erit angulo b. Non enim alter eoʀ altero maior eſſe poteſt. quod ſi
forſitan ita arbitreris, ſit angulus g maior angulo b, fiat'ʨ per huius iuxta ter
minum g arcus b g angulus b g d æqualis angulo a b g, producto arcu g d.
Trianguli itaqʒ d b g duo anguli d b g & d g b æquales ſunt, quare per præce
dentem duo eius latera d b & d g ſibi inuicem æquabuntur. adiecto'ʨ cōi arcu
a d erit aggregatū ex duobus arcubus a d, d g æquale arcui a b, qui ponebatur
æqualis arcui a g. Vnde & aggregatum ex duobus arcubus a d, d g æquabitur
arcui a g, quod eſt impoſſibile, repugnans huius. Deceptus igitur affirmabas
alterum altero maius eſſe, quare propoſitioni noſtræ aſſentire compelleris. Hanc
autem & præcedentem oſtenſiue potuimus demonſtrare, uerum breuiores in om⸗
ni opere uias eligendi fuit conſilium.

X L I I.

In omni triangulo ſphærali maiorem angulū
longius ſubtendit latus.

Angulus enim b trianguli a b g maior occurrat an
gulo g. Dico qʒ latus a g longius eſt latere a b. Iuxta
punctum enim b arcus b g fiat angulus d b g æqualis
angulo a g b. erit igitur per huius arcus d b æqualis
arcui d g. adiecto'ʨ arcui a d cōi erit arcus a g æqua⸗
lis aggregato ex duobus arcubus a d, d b. quod quidē aggregatū per huius ma
ius eſt latere a b. unde & latus a g latere a b lōgius habebiť, ʠd placuit addiſcere.

X L I I I.

Longiori demū lateri trianguli cuiuslibet ma
iorē opponi angulum.

Sit triangulus a b g latus a g longius latere ſuo a
b habens. Dico qʒ angulus a b g maior erit angulo a
g b. Nō enim poteſt eſſe æqualis ei, ſic enim per huius
latera a b & a g conuincerentur æqualia, ſed neqʒ mi
nor eo poteſt haberi. ſic enim angulus a g b maior eſſet
angulo a b g, & ideo ex præcedenti latus a b longius

L 3 latere

If △ *ABG* has two equal angles *B* and *G*, then side *AB* will be equal to side *AG*.

If not, one [side] will be longer than the other. Let *AG* be longer than *AB*. From [*AB*] let arc *GD*, equal to arc *AB*, be intercepted to produce arc *BD*. Then the two triangles *ABG* and *BGD* have two [respectively] equal sides each [*AB* = *GD*, *GB* = *GB*] and, by hypothesis, two equal angles included by these very sides — namely, ∠ *ABG* equal to ∠ *BGD*. Therefore, by Th. [31] above, ∠ *DBG* will be equal to ∠ *AGB*, which hypothesis gave equal to ∠ *ABG*. Hence ∠ *DBG** will be equal to the same *ABG*;† that is, the part will be equal to the total, which is impossible. Therefore the one [side] will not be longer than the other, and thus they are left to be equal to each other. Q.E.D.

Theorem 41

Two equal sides of any triangle must subtend equal angles.

This [theorem] promises to argue its subject by a reversed proof of the preceding. Thus, if △ *ABG* has two equal sides *AB* and *AG*, then ∠ *G* will be equal to ∠ *B*.

If not, one of them must be larger than the other. But if perchance you were to consider in that way, then let ∠ *G* be greater than ∠ *B*. By Th. [34] above, when arc *GD* is drawn at end *G* of arc *BG*, ∠ *BGD* is made equal to ∠ *ABG*. Then the two angles *DBG* and *DGB* of △ *DBG* are equal. Therefore, by the preceding [theorem], its two sides *DB* and *DG* will be equal to each other. When the common arc *AD* is added, the sum of the two arcs *AD* plus *DG* will be equal to arc *AB*, which was given equal to arc *AG*. Hence the sum of

the two arcs *AD* plus *DG* will be equal to arc *AG*, which is impossible, opposing Th. [37] above. Thus you have proven false [the possibility] that one is larger than the other; hence you will be compelled to agree with our theorem. Moreover, this and the preceding we could have proven declaratively; however, it is advisable to select the shortest routes in all work.

Theorem 42

In every spherical triangle the side subtending the larger angle is the longer [side].

If ∠ *B* of △ *ABG* is larger than ∠ *G*, then side *AG* is longer than side *AB*.

At point *B* of arc *BG* let an ∠ *DBG* be constructed equal to ∠ *AGB*. Then, by Th. [32] above, arc *DB* will be equal to arc *DG*. When the shared arc *AD* is added, arc *AG* will be equal to the sum of the two arcs *AD* plus *DB*. By Th. [37] above that sum [*AD* + *DB* = *AG*] is greater than side *AB*. Hence side *AG* will be longer than side *AB*. Q.E.D.

Theorem 43

The greater angle of any triangle is opposite the longer side.

If △ *ABG* has side *AG* longer than side *AB*, then ∠ *ABG* will be larger than ∠ *AGB*.

[Angle *ABG*] cannot be equal to [∠ *AGB*], for then, by Th. [32] above, sides *AB* and *AG* will be proven equal. Nor can [∠ *ABG*] be less than the other, for then ∠ *AGB* would be larger than ∠ *ABG*, and therefore, from the preceding [theorem], side *AB* will be longer

*For *DBG* read *DGB*.
†For *ABG* read *AGB*.

latere a g,quorum cum utrunḡ contrarium sit hypothesi, destructis ipsis, relinquitur ueritas theorematis nostri,quam deducere sperabamus.

XLIIII.

A punƈto in arcu circuli magni signato, orthogonalem arcum circuli magni educere.

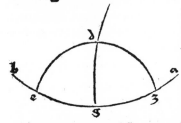

Sit arcus huiusmodi a b,à cuius punƈto g libet educere orthogonalē.Super punƈto g tanḡ
polo secundum distantiam quantamlibet mino
rem tamen diametro sphæræ describatur circulus e d 3,oportet autem arcum e d,e 3 esse semicircumferentiam,circulus enim magnus a b
per polos circuli iam descripti e d 3 transit,sece
tur ergo arcus e d 3 per medium in punƈto d,
per duóḡ puncta d &'g arcus circuli magni dirigente　huius producatur, qui
sit d g,hunc dico esse orthogonalem ad arcum b g,erit enim per　huius angulus a g d æqualis angulo b g d.per diffinitionem igitur arcus d g orthogonaliter insidebit arcui a b,quod libuit efficere.Qd̄ si polus circuli a b datus fuerit,
facilius operabimur,ipsum nanḡ punƈto g copulabimus per arcum circuli magni,qui orthogonalis erit ad arcum a b,circulo illius per polū alterius trāseunte.

XLV.

Circuli in sphæra nobis propositi polum inuenire.

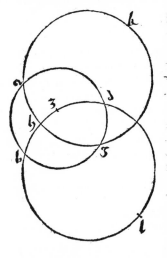

Sit circulus a b g d in sphæra signatus,siue maior siue minor existat,cuius polū reperire libet. Describe circulum,ut libet,secantem circulum a b g d
propositum,quod facile uidebiꞇ,si super aliquo pun
ƈto circunferentiæ a b g d secundum quantitatem
minorem diametro circuli a b g d circulum circū
duxeris,qui sit a g k,utrunḡ autem arcuum a b
g & a h g per medium scindas,hunc quidē in pun
ƈto b,illum uero in punƈto h,& per duo puncta b
& h circulum magnum b h 3 d producas　huius
docente,cuius arcum b h d, quem resecat circulus
a b g d propositus per medium partiaris in punƈto 3,quē oportet esse polum circuli propositi,Nā
per　partem　huius circulus b h 3 d transiꞇ ꝑ po
los circuli a b g d.oportet autem polum æqualiter
distare à punƈtis b & d in circumferentia circuli a
b g d signatis,quod profecto nulli punƈto arcus b
h d conuenit præter ipsum punctum 3,quod facil
lime ostenditur.punctus igitur 3 est polus circuli a b g d quæsitus.Reliquus au
tem polus inuenietur,si reliquus arcus circuli b h 3 d per medium diuisus fuerit,
nam punctus mediæ diuisionis erit polus ille,cuius demonstrationem ut antehac
fabricabimus .

XLVI.

A punƈto in superficie sphæræ signato ad arcum circuli magni pun-
ƈtum ipsum non includentis,perpendicularem demittere.

Sit pun

than side *AG*. Because each of these [possibilities] is contrary to the hypothesis, then, since they have been eliminated, the truth of our theorem remains. Q.E.D.

Theorem 44

From a point marked on the arc of a great circle, an orthogonal arc of [another] great circle may be drawn.

Let this arc be *AB*, from whose point *G* it is desired to draw an orthogonal. Upon point *G* as a pole with any length less than the diameter of the sphere [as polar line], let circle *EDZ* be described. Moreover, arc *ED* plus [arc] *EZ** must be a semicircumference, for great circle *AB* passes through the poles of circle *EDZ*, just described. Therefore arc *EDZ* is bisected at point *D*. Through the two points *D* and *G* let an arc, *DG*, of a great circle be drawn by the directions of Th. [15] above. This [arc], then, is orthogonal to arc *BG*, for by Th. [31] above ∠ *AGD* will be equal to ∠ *BGD*. Therefore by definition arc *DG* rests orthogonally on arc *AB*. Q.E.D.

But if the pole of circle *AB* is given, we will work more easily, for we will join that [pole] to point *G* by the arc of the great circle, which will be orthogonal to arc *AB* because the circle of the one will pass through the pole of the other.

Theorem 45

To find the pole of a circle, given to us, in a sphere.

Let circle *ABGD*, inscribed in a sphere, be either a great circle or a small circle, whose pole it is desired to find. Draw any desired circle, intersecting the given circle *ABGD*. This will seem easy if you would circumscribe the circle, *AGK*, around some point on circumference *ABGD*, with a length less than the diameter of circle *ABGD* [as radius]. Moreover, bisect each of the arcs *ABG* and *AHG*, the one at point *B*, the other at point *H*. By the instruction of Th. [15] above, through the two points *B* and *H*, draw a great circle *BHZD*, whose arc *BHD*, which the given circle *ABGD* intercepts, you bisect at point *Z*. This [point *Z*] is necessarily the pole of the given circle. For, by the [third] part of Th. [21] above, circle *BHZD* crosses through the poles of circle *ABGD*. Moreover, the pole must be equally distant from points *B* and *D* marked on the circumference of circle *ABGD*. Indeed, this [description] fits no point of arc *BHD* except point *Z* itself, which is very easily shown. Thus point *Z* is the desired pole of circle *ABGD*. Moreover, the other pole may be found if the other arc of circle *BHZD* is bisected, for the point of bisection will be that pole. We can formulate the proof of this as above.

Theorem 46

To draw a perpendicular from a point marked on the surface of a sphere to the arc of a great circle that does not incorporate that point.

*For *EZ* read *DZ*.

Sit punctus a extra arcū b g signatus, à quo ad arcum ipsum demittere libet perpendicularem. Per præcedētem inueniatur polus circuli, cuius est arcus b g, qui sit 3, ducaturꜭ per duo puncta 3 & a circulus magnus, quemadmodū huius docuit, cuius arcus 3 d occurrat arcui b g si possibile est, aut ipsi quantum oportet prolongato in puncto d. Dico ꝙ arcus 3 d est perpendicularis ad arcum b g. Circulus enim 3 a d per polum 3 circuli b g transiēs orthogonalis ad eum est ex huius, quod & arcubus ipsorum circulorum accidere necesse est. quicquid enim de inclinatione aut erectione arcuū ad arcus dicitur, non nisi ex habitudine circulorum suorū trahitur. Forsitan illud te non satiat, pone igitur arcum h d quantumlibet æqualem arcui d k, duoꜭ pū cta h & k puncto 3 polo scilicet circuli b g connectas per duos arcus circuloru magnorum qui sint 3 h & 3 k, ut docuit huius. duo itaꜭ trianguli 3 d h & 3 d k terna latera habent æqualia, quare per huius angulus 3 d h æquabitur an gulo 3 d k, & ideo arcus 3 d propter quod & a d perpendicularis est ad arcum b g, quod uolebas declarandum. Aliter idem efficies. Super puncto a facto polo describe circulum, cuius circumferentia secet arcū b g, si possibile est, aut eum prolongatum quoad satis est in duobus punctis d & e, diuidaturꜭ ar= cus d e per medium in puncto h, qui continue= tur cum puncto a per arcum circuli magni a h, quem oportet esse perpēdicularem ad arcum b g. producam enim duos arcus circulorum magnoꝝ a d & a e. duo igitur trianguli a h d & a h e tri na latera habent æqualia, latere a h communi existente, quare per huius angu lus a h d æqualis est angulo a h e, & ideo per diffinitionem arcus a h perpendi cularis est ad arcum b g, quod libuit attingere.

XLVII.

In triangulis sphæralibus educto latere uno, contingit angulum ex trinsecum alteri intrinsecorum sibi oppositorum nunc esse æqualem, nunc uero maiorem, interdum etiam minorem eo.

Duorum enim circulorum ad se inclinatoꝝ semicircumferentiæ coincidant in punctis a & b, abscindaturꜭ ex una earum arcus a g mi= nor quadrante, à cuius termino g ad arcum a b substratum perpēdicularis arcus circuli magni descendat g d, erit itaꜭ trianguli a g d angu= lus a d g rectus maior angulo acuto d a g, an gulo scilicet inclinationis. est autem angulus a æqualis angulo b, quod constabit per huius descripto circulo secundū distantiam quantamlibet minore diametro sphæræ super polo a, & ideo angulus a d g extrinsecus ad triāgulum b g d ma= ior est intrinseco angulo b sibi opposito. Rursus cū sit arcus a g minor quadran te, maior autem arcu a d per huius, erit & arcus a d minor quadrante, & ideo

minor

Let *A*, outside arc *BG*, be the indicated point, from which it is desired to draw a perpendicular to that arc [*BG*]. By the preceding [theorem] the pole of the circle to which arc *BG* belongs may be determined. Let that [pole] be *Z*. Through the two points *Z* and *A*, as Th. [15] above taught, let a great circle be drawn, whose arc *ZD* meets arc *BG*, if possible, or its length extended, if necessary, at point *D*. Then arc *ZD* is perpendicular to arc *BG*.

Circle *ZAD*, passing through pole *Z* of circle *BG*, is orthogonal to [*BG*] from Th. [17] above. This [orthogonality] necessarily occurs with the arcs of these circles also, for whatever is said concerning the inclination or perpendicular erection of arcs is understood only from the condition of their respective circles.

Perhaps this does not satisfy you; then take arc *HD* of any size [that is] equal to arc *DK*, and connect the two points *H* and *K* to point *Z* — namely, the pole of circle *BG* — by two arcs *ZH* and *ZK* of the great circle, as Th. [15] above teaches. Then the two triangles *ZDH* and *ZDK* each have three sides respectively equal. Therefore, by Th. [35] above, ∠ *ZDH* will be equal to ∠ *ZDK*; and hence arc *ZD*, and because of that [arc] *AD*, is perpendicular to arc *BG*. Q.E.D.

You may do the same thing in another way. Around point *A* as a pole, describe a circle whose circumference intersects arc *BG*, if possible, or [*BG*] extended as much as is necessary, at the two points *D* and *E*. Let arc *DE* be bisected at point *H*, which is joined with

point *A* by arc *AH* of the great circle. [Arc *AH*] is necessarily perpendicular to arc *BG*. Let two arcs *AD* and *AE* of the great circle be drawn. Then the two triangles *AHD* and *AHE* each have three sides respectively equal since side *AH* is common. Thus, by Th. [35] above, ∠ *AHD* is equal to ∠ *AHE*, and therefore by definition arc *AH* is perpendicular to arc *BG*. Q.E.D.

Theorem 47

In spherical triangles, when one side is extended, it happens that the exterior angle is sometimes equal to one of the interior [angles] opposite itself, sometimes larger [than it], and sometimes even smaller than it.

Let the semicircumferences of two circles [that are] inclined toward each other meet at points *A* and *B*, and from one of them let arc *AG*, less than a quadrant, be intercepted. From the end [point] *G* of that [arc *AG*] let a perpendicular arc *GD* of a great circle descend to the underlying arc *AB*. Therefore ∠ *ADG* of △ *AGD* will be a right angle, greater than acute ∠ *DAG* — namely, the angle of inclination. Moreover, ∠ *A* is equal to ∠ *B*, [a fact] which will be established by Th. [31?] above when a circle is described around pole *A* with any length less than the diameter of the sphere [as polar line]. Therefore ∠ *ADG*, exterior to △ *BGD*, is greater than the interior ∠ *B* opposite it.

Again, when arc *AG* is less than a quadrant but greater than arc *AD* by Th. [42] above, arc *AD* will also be less than a quadrant and therefore

minor arcu d g, ex quo abſcindatur ei æqualis arcus d e. productus itaꝗ arcus
g e per   huius, æqualis erit arcui a g, quare per   huius angulus a e g æqua
lis erit angulo e a g, illud tamen ſola   huius inferre potuit, eſt autem angulus
e a g æqualis angulo b. quare angulus extrinſecus a e g æquabitur angulo b
intrinſeco ſibi oppoſito. Poſtremo ſignetur punctus quilibet in arcu e b, qui ſit 3
copulandus g puncto per arcum 3 g, qui profecto ſuperabit arcum e g, & ideo
arcum a g ſibi æqualem, per   igitur huius angulus 3 a g æqualis angulo b ma
ior erit angulo a 3 g, & uiceuerſa angulus a 3 g extrinſecus ad triangulum b 3
g minor erit angulo b intrinſeco ſibi oppoſito. quæ ꝓpoſuimus lucubrare.

XLVIII.

Si fuerit angulus extrinſecus æqualis alteri intrinſecorum ei oppo=
ſitorum, aggregatum ex lateribus reliquum angulum intrinſecum ei
oppoſitū ambientibus, æquabit ſemicircumferentiæ. ſi uero maior in
trinſeco ſibi oppoſito, erit aggregatū huiuſmodi minus, & ſi minor,
ipſum maius.

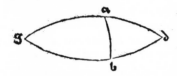

        Sit triangulus a b g, cuius latere g b ꝓducto
ad partē puncti b, fiat angulus extrinſecus æqua
lis intrinſeco ſibi oppoſito angulo a g b. Dico ꝙ
duo latera g a & a b collecta æquabunt ſemicir
cumferentiā. ſi uero maior fuerit angulo g, ipſa
duo latera minora habebuntur ſemicircumferen-
tiā, & ſi minor maiora. Protendantur enim duo arcus g a & g b donec concur
rent in puncto d. Cum itaꝗ angulus a b d æqualis fuerit angulo g, erit ipſe eti
am æqualis angulo d. quare per   huius arcus a b æqualis erit arcui a d, adhibi
toꝗ arcu communi a g, erunt duo arcus b a & a g collecti æquales arcui g d,
qui eſt ſemicircumferentiā. & hoc erat primum. Quòd ſi angulus a g d maior fu
erit angulo g, erit & ipſe maior angulo d. quare per   huius arcus a d ſuperabit
arcum a b. adiectoꝗ cōi a g, erunt duo arcus b a & a g cōiuncti minores ar
cu g d, quem conſtat eſſe ſemicircūferentiam. aperta igitur eſt pars ſecunda. Po
ſtremo ſi angulus a b d minor exiſtat angulo g, minor quoꝗ erit angulo d. ar
cus ergo a b per   allegatam maior inuenietur arcu a d. cōiꝗ ſociato arcu a g
erit aggregatum ex duobus arcubus b a, a g maius arcu g d, qui eſt circumfe
rentiæ dimidiū. Verum itaꝗ hoc enunciauimus theoremate. Conuerſam autē hu
ius medijs utendo conuerſis demonſtrare haud uidebitur difficile.

XLIX.

Omnis triangulus ſphæralis tres habet angulos duobus rectis ma
iores.

     Pleraſꝗ cōmuniter demonſtrare ſolemus paſſiones de triangulis & planis &
ſphæralibus, nonnullas autem differenter. Sphæralibus enim nō accidit tres angu
los ſuos duobus rectis æquales habere, quemadmodum planis. quo fit, ut cognitis
duobus angulis trianguli ſphæralis non pendeat inde tertij anguli cognitio. Ne
igitur circa hæc quempiā errare contingat, monimento theorematis huius caut
orem reddere libuit. Sit itaꝗ triangulus a b g ſphæralis. Dico ꝙ tres eius anguli
a, b, & g, maiores ſunt duobus rectis. Prolōgatis enim arcubus g a & g b, donec
in puncto d concurrent, erit angulus extrinſecus a b d per   huius aut æqualis
intrinſeco angulo b a g ſibi oppoſito, aut minor eo aut maior. Si æqualis adie
cto

less than arc *DG*,* from which arc *DE* is intercepted equal to [*AD*]. Then arc *GE*, [when it is] drawn, will be equal to arc *AG* by Th. [24] above; therefore, by Th. [35] above, ∠ *AEG* will be equal to ∠ *EAG*. That one could have inferred by Th. [36] above alone. Moreover, ∠ *EAG* is equal to ∠ *B*. Therefore exterior ∠ *AEG* will be equal to interior ∠ *B* opposite it.

Finally, let any point be marked on arc *EB*; let this be *Z*, [which is] to be joined to point *G* by arc *ZG*. This [arc *ZG*] will indeed exceed arc *EG*, and therefore [will exceed] arc *AG* equal to [*EG*]. Therefore, by Th. [43] above, ∠ *ZAG*, equal to ∠ *B*, will be greater than ∠ *AZG*, and conversely ∠ *AZG*, exterior to △ *BZG*, will be less than interior ∠ *B* opposite it. Q.E.D.

Theorem 48

If an exterior angle were equal to one of the interior angles opposite it, then the sum of the sides enclosing the other interior angle opposite that [exterior angle] will be equal to the semicircumference. However, if [the exterior angle] is larger than the interior [angle] opposite it, this sum will be less [than the semicircumference]; and if [the exterior angle] is smaller [than one of the interior angles opposite it], this [sum] will be greater [than the semicircumference].

If *ABG* is a triangle whose exterior angle, when side *GB* is extended beyond point *B*, is equal to interior ∠ *AGB* opposite it, then the sum of the two sides *GA* and *AB* will be equal to the semicircumference. However, if [the exterior angle] is larger than ∠ *G*, the [sum of the] two sides will be less than the semicircumference; and if [the exterior angle] is smaller, then [the sum of the two sides will be] greater [than the semicircumference].

Let the two arcs *GA* and *GB* be extended until they meet at point *D*. Therefore, when ∠ *ABD* is equal to ∠ *G*, it will also be equal to ∠ *D*. Therefore by Th. [40] above arc *AB* will be equal to arc *AD*, and when the common arc *AG* is added, the sum of the two arcs

BA and *AG* will be equal to arc *GD*, which is the semicircumference. And that was the first [part].

But if ∠ *AGD*† is greater than ∠ *G*, it will be greater than ∠ *D*. Therefore, by Th.[42] above, arc *AD* will exceed arc *AB*. When the common [arc] *AG* is added, the sum of the two arcs *BA* and *AG* will be less than arc *GD*, which was established to be the semicircumference. Therefore the second part is revealed.

Finally, if ∠ *ABD* is less than ∠ *G*, it will also be less than ∠ *D*. Therefore arc *AB*, by Th. [42 just] cited, is found [to be] greater than arc *AD*. And when the common arc *AG* is added, the sum of the two arcs *BA* plus *AG* will be greater than arc *GD*, which is half the circumference. Therefore we have declared the truth in this theorem. Moreover, by the use of the reverse of this method, it will seem not at all difficult to prove the converse.

Theorem 49

Every spherical triangle has three angles [whose sum is] greater than two right angles.

We usually show most proofs concerning triangles in general, both planar and spherical, but several [must be shown] specifically. For it happens that in spherical [triangles] the three angles are not equal to two right angles, as [is true] in planar [triangles], which means that knowledge of the third angle of a spherical triangle does not follow when two angles are known. Therefore, lest anyone should chance to err about this fact, it might be desirable to offer [a word of] caution through the written reminder of this theorem.

Thus, if *ABG* is a spherical triangle, then its three angles, *A*, *B*, and *G*, are greater than two right angles.

When arcs *GA* and *GB* are extended until they meet at point *D*, exterior ∠ *ABD* will be equal to or greater than or less than interior ∠ *BAG* opposite it by Th. [47] above. If [it is] equal,

*For *DG* read *DB*.
†For *AGD* read *ABD*.

çto cõmuni angulo a b g, erunt duo anguli a
b g & b a g æquales duobus a b d & a b g,
quos liquet æquales eße duobus recti s, quare &
duo anguli a b g & g a b æquabuntur duo=
bus recti s.& ideo tres anguli a,b,& g, duos ßu=
perabunt rectos. Si uero angulus a b d minor
fuerit angulo b a g, quoniã ipße cũ angulo a b
g duobus recti s æquatur, erunt duo anguli a b
g & b a g maiores duobus recti s. Tres igitur anguli triangui a b g multo ma
iores erunt duobus recti s. Quòd ßi angulus a b d maior fuerit angulo b a g, fi=
at iuxta punctũ b arcus a b, angulus a b e æqualis angulo b a g producto ar=
cu b e ßemicircũferentiæ d a g, occurrête in puncto e. Cum itacß angulus b a g
extrinßecus æqualis ßit intrinßeco angulo a b e triangui a b e, erũt duo arcus a e
& e b coniuncti per præcedentem æquales ßemicircũferentiæ g a d, ablatocß cõi
arcu a e, manebit arcus b e æqualis duobus arcubus a g & e d. maior itacß eßt
arcus b e ipßo e d arcu. quare per   huius angulus b d e maior angulo e b d
habebitur. Eßt autê angulus b d e æqualis angulo a g b, quare & angulus a g b
maior eßt angulo d b e. adiecti s igitur æqualibus angulis a b g & b a g, item a
b g & a b e, erunt tres anguli triangui propoßiti maiores tribus angulis a b g,
a b e & e b d, quos conßtat eße duobus recti s æquales, unde tres anguli dicti ma
iores erunt duobus recti s, quod oportuit demonßtrare.

L

Si fuerint duo latera unius triangui æqualia duobus lateribus al=
terius, angulorum autem hißce æqualibus lateribus contentorßum alte
ro altero maior, erit quocß baßis maiorem ßubtendens angulum unius
maior baßi alterius.

Duo triangui a b g & e d z bina latera habent æqualia,
a b quidem æquale d e, & a g æquale ipßi d z, angulus autem a
maior ßit angulo d. Dico cß baßis b g longior eßt baßi e z. Simi=
lem paßßionem 24.primi Euclidis de triangui s plani s concludit.
Modus autem demonßtrandi hanc & illam non eßt uarius.

L I.

Si fuerint duo latera unius triangui æqualia duobus
lateribus alterius, baßis autem unius maior baßi alterius,
erit & angulus longiorem baßim reßpiciens maior angu=
lo breuiorem reßpiciente baßim.

Haec conuertit præcedentem, & correßpondêt 25. primi Euclidis . Quemad=
modum autem illa ex 4.& 24.primi Euclidis ad impoßßibile demõßtratur, ita hæc
ex   huius & præcedenti ad inconueniens ducendo aduerßarium, ueritatem pro=
batur habere neceßßariam.

LII.

Omnium duorum triangulorum binos angulos æquales haben=
tium, duocß latera æqualibus ßubiecta angulis æqualia, reliqua quocß

M   latera

then, when the common ∠ *ABG* is added, the two angles *ABG* and *BAG* will be equal to the two [angles] *ABD* and *ABG*, which are obviously equal to two right [angles]. Therefore the two angles *ABG* and *GAB* will also be equal to two right [angles]. Thus the three angles *A*, *B*, and *G* will exceed two right [angles].

However, if ∠ *ABD* is less than ∠ *BAG*, then, since it, plus ∠ *ABG*, equals two right [angles], the two angles *ABG* and *BAG* will be greater than two right [angles]. Therefore the three angles of △ *ABG* will be larger by far than two right [angles].

But if ∠ *ABD* is greater than ∠ *BAG*, then let ∠ *ABE*, next to point *B* of arc *AB*, be made equal to ∠ *BAG* when arc *BE* is extended to the semicircumference *DAG*, meeting [it] at point *E*. Then since exterior ∠ *BAG* is equal to interior ∠ *ABE* of △ *ABE*, the two arcs *AE* and *EB* combined will be equal to the semicircumference *GAD* by the preceding [theorem]. And when the common arc *AE* is taken away, arc *BE* will be left equal to the two arcs *AG* and *ED*. Thus arc *BE* is greater than arc *ED*. Therefore, by Th. [43] above, ∠ *BDE* is greater than ∠ *EBD*. Moreover, ∠ *BDE* is equal to ∠ *AGB*; therefore ∠ *AGB* is greater than ∠ *DBE*. Thus, when equal angles *ABG* and *BAG* are added, [and] similarly [when] *ABG* and *ABE* [are added], the three angles of the given triangle will be greater than the three angles *ABG*, *ABE*, and *EBD*, which are established to be equal to two right [angles]. Hence the three mentioned angles will be greater than two right [angles]. Q.E.D.

Theorem 50

If two sides of one triangle are equal to two sides of another but one of the angles contained by these equal sides is greater than the other, then the base subtending the greater angle of the one will also be larger than the base of the other.

If the two triangles *ABG* and *EDZ* each have two sides respectively equal — *AB* equal to *DE* and *AG* equal to *DZ* — but if ∠ *A* is greater than ∠ *D*, then base *BG* is longer than base *EZ*.

I.24 of Euclid argues the same proof concerning planar triangles. Moreover, the method of proving this [theorem] and that [one] is no different.

Theorem 51

If two sides of one triangle are equal to two sides of another but the base of one is longer than the base of the other, then the angle opposite the longer base will be greater than the angle opposite the shorter base.

This is the converse of the preceding [theorem] and corresponds to I.25 of Euclid. Moreover, just as the latter is shown from I.4 and I.24 of Euclid through [elimination of] the impossible, so this is proven to have the necessary truth from Th. [36] above and the preceding, through consideration of the contrary [statement].

Theorem 52

In every pair of triangles having two pairs of angles respectively equal and having the pair of [corresponding] sides included by the equal angles [also] equal, the other sides also

latera æquos angulos respicientia sunt æqualia, & angulus reliquus an
gulo reliquo.

Sint duo trianguli a b g & d e 3 , quorum duo anguli b &
g æquales sint duobus angulis e & 3 , latuścꝫ b g æquale la
teri e 3 .Dico cp latus a b æquabitur lateri e d ,& latus a g late
ri d 3 ,similiter angulus a angulo d æqualis habebitur. Si enim
non fuerint duo latera a b & d e æqualia,erit alterum altero ma
ius.ex maiori igitur eorum quod sit,uerbi gratia, d e ,abscindatur
arcus e h æqualis a b producendo arcum 3 h .sequitur itacꝫ per

huius angulum e 3 h æqualem esse angulo a g b,qui poneba
tur æqualis angulo d 3 e ,angulus igitur e 3 h æqualis erit angu
lo d 3 e pars toti,quod est impossibile.Non potest igitur alterũ
duorum laterum a b & d e altero maius esse,quare & æqualia ha
bebuntur.Similiter probabis duo latera a g & d 3 esse æqualia,
angulum postremo a æqualem angulo d conuincet huius,aut
quæ decuit stabilire.

LIII.

**Omnium duorum triangulorum binos angulos habentium æqua
les,duocꝫ latera æqualibus opposita angulis æqualia,latera uero reli
quos æquales respicientia angulos coniuncta non æqualia semicircũ
ferentiæ,bina latera reliqua erunt æqualia, & angulus reliquus angu
lo reliquo .**

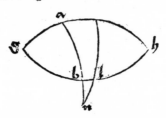

Sint duo anguli b & g trianguli a b g æ
quales duobus angulis e & 3 trianguli d e 3 ,
latuścꝫ a b æquale lateri d e : duo autem late
ra a g & d 3 coniuncta semircircumferentiæ
non æqualia.Dico cp latus a g æquale erit la
teri d 3 ,latuścꝫ b g æquale lateri e 3 , & an
gulus a æqualis angulo d ,Producam enim du
os arcus g b & g a donec concurrent in pun
cto h ,scindamcꝫ ex semicircumferentia h b g arcum h l æqualem arcui e 3 .&
ex semicircumferentia h a g arcum h k æqualem arcui d 3 ,necesse est autem
punctum k aliud esse cp punctum a ,cp duo arcus a g & d 3 non æquentur semi
circumferentiæ,continuabócꝫ duos arcus a b & k l donec concurrent in puncto
n ,erit autem per huius arcus b l æqualis arcui d e ,sed & angulus h l k æqua
lis angulo d e 3 ,itécꝫ angulus h k l æqualis angulo e d 3 .po
nebatur autẽ angulus d e 3 æqualis angulo a b g .quare & an

gulus h l k æqualis erit angulo a b g ,unde & eorũ contrapo
siti uidelicet n b l & n l b æquales erũt,quare per huius duo
arcus b n & n l æquales erũt.cuncꝫ duo arcus a b & k l sint æ
quales,est enim utercꝫ eorũ æqualis arcui d e ,erit totus arcus a
n æqualis toti arcui k n,& ideo per huius angulus n a k æqualis angulo n k
a ,residuícꝫ ex duobus rectis anguli scilicet b a g & h k l inuicem æquabuntur.
erat autem angulus h k l æqualis angulo e d 3 ,quare & angulus b a g æquabi
tur angulo e d 3 ,duo itacꝫ trianguli a b g & d e 3 duo latera a b & d e haben
tes æqualia,binoścꝫ angulos hisce lateribus insidentes æquales , per præmissam
itacꝫ quod reliquum est absoluemus.

Quicuncꝫ

[which are] opposite the [corresponding] equal angles are [respectively] equal and the remaining angle [of the one is equal to] the remaining angle [of the other].

If *ABG* and *DEZ* are two triangles, of which the two angles *B* and *G* are equal to the two angles *E* and *Z* and side *BG* is equal to side *EZ*, then side *AB* will be equal to side *ED*, side *AG* [will be equal] to side *DZ*, [and] similarly ∠ *A* will be equal to ∠ *D*.

If the two sides *AB* and *DE* were not equal, one would be greater than the other. Therefore, from the larger of them — *DE*, for example — let arc *EH*, equal to *AB*, be intercepted to form arc *ZH*. Then it follows, by Th. [31] above, that ∠ *EZH* is equal to ∠ *AGB*, which was given equal to ∠ *DZE*. Thus ∠ *EZH* will be equal to ∠ *DZE* — the part [will be equal] to the total, which is impossible. Hence, one of the two sides *AB* and *DE* cannot be larger than the other, and therefore they will be equal. Similarly you may prove that the two sides *AG* and *DZ* are equal. Finally, Th. [31] or Th. [35] above proves that ∠ *A* is equal to ∠ *D*, which it was appropriate to establish.

Theorem 53

In every pair of triangles having two pairs of angles respectively equal and [having] the pair of [corresponding] sides opposite [one of the pairs of] equal angles equal but the sum of the [corresponding] sides opposite the other [pair of] equal angles not equal to the semicircumference, the two other pairs of sides will be [respectively] equal and the remaining angle [will be equal] to the remaining angle.

If the two angles *B* and *G* of △ *ABG* are equal to the two angles *E* and *Z* of △ *DEZ*, and if side *AB* is equal to side *DE* but the two sides *AG* and *DZ* combined are not equal to a semicircumference, then side *AG* will be equal to side *DZ*, and side *BG* [will be equal] to side *EZ*, and ∠ *A* [will be] equal to ∠ *D*.

Let the two arcs *GB* and *GA* be extended until they meet at point *H*;* let arc *HL*, equal to arc *EZ*, be intercepted from the semicircle *HBG*, and [let] arc *HK*, equal to arc *DZ*, [be cut] from the semicircle *HAG*. Moreover, point *K* must be different from point *A* because the two arcs *AG* and *DZ* are not equal to a semicircumference. The two arcs *AB* and *KL* are extended until they meet at point *N*. Moreover, arc *BL*† will be equal to arc *DE* by Th. [36] above, and furthermore ∠ *HLK* will be equal to ∠ *DEZ*. Similarly, ∠ *HKL* will be equal to ∠ *EDZ*. In addition ∠ *DEZ* was given equal to ∠ *ABG*. Therefore ∠ *HLK* will be equal to ∠ *ABG*, and hence their opposite [vertical angles] — namely, *NBL* and *NLB* — will be equal. Then by Th. [32] above the two arcs *BN* and *NL* will be equal. And since the two arcs *AB* and *KL* are equal because each of them is equal to arc *DE*, then the total arc *AN* will be equal to the total arc *KN*. Thus by Th. [41] above ∠ *NAK* will be equal to ∠ *NKA*, and angles *BAG* and *HKL*, [which are] left from two right [angles], will be equal to each other. Moreover, ∠ *HKL* was equal to ∠ *EDZ*; therefore, ∠ *BAG* will be equal to ∠ *EDZ*. Thus the two triangles *ABG* and *DEZ* have two equal sides *AB* and *DE* and two pairs of [respectively] equal angles that rest on those equal sides. Therefore we may prove what is left by the preceding [theorem].

*In the figure, for arc *gah* read *gakh*.
†For *BL* read *KL*.

LIIII.

Quicunɋ duo trianguli trinos angulos habent æquales, & trina la
tera habebunt æqualia.

In eadem fphæra aut diuerfis æqualibus ta=
men fint duo trianguli a b g & d e ʒ, quorum
unius tres anguli tribus angulis alterius fingula
tim fint æquales. Dico ꝗ tria latera unius eorū
tribus lateribus alterius æqualia habebunꞇ late=
ra ꝗdē æquos angulos refpicientia ad fe cōferen
do. Producā em duos arcus a b & g b ad partē
puncti b, ponamꝗ arcū b h æqualē arcui d e, arcū aūt b k æ=
qualē e ʒ, & per duo puncta h k ducā arcū circuli magni, ꝗ cō
tinuabo utrinꝗ donec cū arcu a g utrinꝗ protenfo concurrat in
duobus punctis l & m. Cum itaꝗ duo latera triāguli h b k fint
æqualia duobus lateribus triāguli d e ʒ, & anguli hifce cōtenti late
ribus æquales, angulus uidelicet h b k æqualis angulo d e ʒ, erit
per huius & bafis h k unius bafi d ʒ alterius æqualis. angulus quoꝗ b k h æ=
qualis angulo d ʒ e, fed & angulus b h k æqualis angulo e d ʒ. duo autem an=
guli d & ʒ ponebantur æquales duobus a & g. quare angulus b k h extrinfecus
ad triangulum g l k æquabitur angulo a g b intrinfeco, & ideo per huius ag=
gregatum ex duobus arcubus g l & l k æquabitur femicircumferentiæ. Simili=
ter angulus g a b extrinfecus ad triangulum a l h, intrinfeco angulo b h k fiue
a h l æqualis. duos arcus a l & l h collectos æquari conuincet femicircumferen=
tiæ, demptis ergo communibus arcubus a l & l k, relinquetur arcus a g æqualis
arcui h k. Duo itaꝗ trianguli a b g & b h k duo latera a g & h k habēt æqua
lia, angulosꝗ binos ipfis infidentes lateribus æquales, quare per huius triangu=
lus a b g æquilaterus & æquiangulus erit triangulo h b k, qui pridem æquilate
rus & æquiangulus demonftrabatur triangulo d e ʒ, unde & duo trianguli a b g
& d e ʒ trina latera habebunt æqualia, quod erat abfoluendum. Ne autem fufpi=
ceris arcum a g utrinꝗ protenfum occurrere arcui h k in punctis h & k: fic em
caffaretur forma argumentationis noftræ, oftendemus id fieri non poffe. Nam fi
ita accideret, fieret arcus h g a k femicircumferentia per huius, ꝗ duæ circum
ferentiæ circulorum magnorum fe fecarent in duobus punctis h & k, fimiliter o=
portet arcum g a k effe femicircūferentiam. cūꝗ omnes femicircūferentiæ uni=
us circuli fint æquales, fieret pars æqualis toti, quod eft impoffibile. Idem fequereꞇ
fi arcus a g continuatus alteri duntaxat duorum punctorum h & k occurreret,
oportet igitur duo puncta l & m effe diuerfa â punctis k & h.

LV.

Omnes duos triangulos, quorum duo latera unius funt æqualia
duobus lateribus alterius, duoꝗ anguli eorum duobus æquis lateri=
bus oppofiti æquales, reliqui uero duo anguli eorum reliquis duo=
bus æquis lateribus oppofiti, aut ambo acuti, aut ambo obtufi, æquila
teros & æquiangulos effe neceffe eft.

Duo latera a b & a g trianguli a b g æqualia fint duobus lateribus d e &

Theorem 54

Whichever two triangles have three angles respectively equal will also have the three sides respectively equal.

If *ABG* and *DEZ*, in the same sphere or different but equal [spheres], are two triangles, of which the three angles of the one are respectively equal to the three angles of the other, then the three sides of one of them will be equal to the three sides of the other — indeed, the sides opposite the correspondingly equal angles.

Let the two arcs *AB* and *GB* be extended beyond point *B*, and let arc *BH* be taken equal to arc *DE* and arc *BK* equal to *EZ*. And, through the two points *H* [and] *K*, let an arc of a great circle be drawn, which will be continued on both sides until it meets with arc *AG*, extended on both sides, at the two points *L* and *M*. Therefore, since two sides of △ *HBK* are equal to two sides of △ *DEZ*, and since the angles contained by these equal sides are equal — namely, ∠ *HBK* [is] equal to ∠ *DEZ* — then by Th. [36] above base *HK* of the one will also be equal to base *DZ* of the other. In addition ∠ *BKH* will be equal to ∠ *DZE*, and furthermore ∠ *BHK* will be equal to ∠ *EDZ*. Moreover, the two angles *D* and *Z* were given equal to the two [angles] *A* and *G*. Therefore ∠ *BKH*, exterior to △ *GLK*, will be equal to interior ∠ *AGB*, and thus by Th. [48] above the sum of the two arcs *GL* and *LK* will be equal to the semicircumference. Similarly ∠ *GAB*, exterior to △ *ALH*, will be equal to interior ∠ *BHK*, or *AHL*; one may prove conclusively that the two arcs *AL* and *LH* combined are equal to the semicircumference. Therefore, when the common arcs *AL* and *LK* are subtracted, arc *AG* is left equal to arc *HK*. Thus the two triangles *ABG* and *BHK* have two equal sides *AG* and *HK* and the two respectively equal pairs of angles, resting upon these sides; therefore, by Th. [52] above, △ *ABG* will be equal-sided and equal-angled to △ *HBK*, which was already shown to be equal-sided and equal-angled to △ *DEZ*. Hence the two triangles *ABG* and *DEZ* will have three respectively equal sides. Q.E.D.

Moreover, lest you might suspect that arc *AG* extended on both sides meets arc *HK* at points *H* and *K* — for in that case the form of our reasoning is empty — we will show that it cannot happen. For if it were to happen that way, arc *HGAK* would be a semicircumference by Th. [19] above, because the two circumferences of the great circles would intersect each other at the two points *H* and *K*. Similarly arc *GAK* must be a semicircumference. And since all semicircumferences of one circle are equal, then a part would be equal to the whole, which is impossible. The same would follow if arc *AG* extended were to meet only one of the two points *H* and *K*. Therefore, the two points *L* and *M* must be different from points *K* and *H*.

Theorem 55

Every pair of triangles, of which two sides of one are equal to two sides of the other and [of which] the two [corresponding] angles opposite [one of the] pairs of equal sides are equal while the other two [corresponding] angles opposite their other two equal sides are either both acute or both obtuse, are necessarily equal-sided and equal-angled.

If the two sides *AB* and *AG* of △ *ABG* are equal to the two sides *DE* and

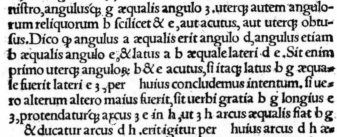

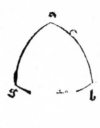

d 3 trianguli d e 3 ,dextrum uidelicet dextro & finiftrum fi=
niftro,angulusᷓ g æqualis angulo 3 .uterᷓ autem angulo=
rum reliquorum b fcilicet & e ,aut acutus, aut uterᷓ obtu=
fus.Dico ᵠ angulus a æqualis erit angulo d,angulus etiam
b æqualis angulo e,& latus a b æquale lateri d e .Sit enim
primo uterᷓ anguloᷓ b & e acutus,fi itaᷓ latus b g æqua=
le fuerit lateri e 3 ,per　huius concludemus intentum. fi ue=
ro alterum altero maius fuerit,fit uerbi gratia b g longius e
3 ,protendaturᷓ arcus 3 e in h ,ut 3 h arcus æqualis fiat b g
& ducatur arcus d h .erit igitur per　huius arcus d h æ=
qualis arcui a b ,& angulus d h 3 æqualis angulo a b g.
ponebatur autem arcus a b æqualis arcui d e,quare duo
arcus d h & d e fibi æquabuntur,& ideo per　huius an=
gulus d h e æqualis angulo d e h . cunᷓ angulus d h e
fit acutus propter angulum a b g ei æqualem acutũ , erit
& angulus d e h acutus,unde angulus d e 3 obtufus ha=
bebitur,quod eft contrariũ pofito.Nõ aliter procedemus,
fi aduerfarius arcum e 3 maiorem arbitretur arcu b g. Qᵈ fi pofuerimus utrũᷓ
anguloꝝ b & e obtufum,concludemus fimili argumentatione angulũ e effe acu=
tum.Volenti igitur contradicere propofitioni noftræ,concluditur eundem angu=
lum effe acutum & obtufum,quod cum effe nequeat,manifefta relinquetur ueritas
theorematis.

LVI.

Omnes trianguli rectanguli bina latera habentes æqualia , utraᷓ
autem latera rectos angulos ambientia minora quadrante diuifim,
æquianguli & æquilateri comprobantur.

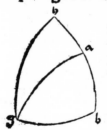

Sint duo trianguli a b g & d e 3 ,quorũ angli b & e funt
recti,utrunᷓ autem laterũ a b & b g trianguli a b g minus
quadrante,fimiliter utrunᷓ d e & e 3 minus quadrante, duo
ᷓ latera unius duobus reliqui lateribus quibufcũᷓ fint æqua
lia.Dico ᵠ reliquũ latus unius æquale erit reliquo lateri alte=
rius,& anguli reliqui unius angulis reliquis alterius : hoc eft,
ipfi duo trianguli æquilateri erunt & æquianguli. Si enim bi
na eoꝝ latera æqualia circa rectos fuerint angulos,per　hu=
ius conftabũt omnia.Si aũt latera huiufmodi æqualia angu=
los ambiant alios,fint uerbi gratia duo latera b a, a g æqua
lia duobus e d, d 3 ,dextrũ dextro & finiftrum finiftro com=

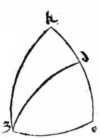

parando,producaturᷓ uterᷓ arcuum b a & e d, donec fi=
at quadrans, b a quidem ad h ,& e d ad k , erunt itaᷓ pun
cta h & k poli circulorum b g & e 3 ,à quibus demittant
duo quadrantes h g & k 3 ,quos æquales cõftat,cum in una
fphæra aut duabus æqualibus eos imaginari foleamus. eft
autẽ & arcus a h æqualis arcui d k. hi eᷚ duo funt comple=
menta duoꝝ arcuũ a b & d e,ᷡs æquales tradidit hypothefis.
per　igiᷓ huius duobus arcubus a g & d 3 æqualibus exiftentibus , erit angulus
h a g æqualis angulo k d3,unde & refidui ex duobus rectis anguli fcilicet b a g &
e d 3 nõ erũt inæquales, ponebanᷓ aũt & duo arcus b a, a g æquales duobus e d, d
3 .quare per　huius triangulos propofitos æquiangulos conuincet & æquilate=
ros,quæ fuere lucubranda.　　　　　　　　　　　　Finis tertij trianguloꝝ.

DZ of △ *DEZ* — namely, the right [side equal] to the right and the left to the left — and if ∠ *G* is equal to ∠ *Z*, and furthermore if each of the other angles — namely, *B* and *E* — is either acute or obtuse, then ∠ *A* will be equal to ∠ *D*, ∠ *B* [will] also be equal to ∠ *E*, and side *AB* [will be] equal to side *DE*.

First, let each of the angles *B* and *E* be acute; therefore, if side *BG* is equal to side *EZ*, we may conclude [what was] intended by Th. [35] above. But if the one [side] were longer than the other, let *BG*, for example, be longer than *EZ*. Let arc *ZE* be extended to *H* so that arc *ZH* is made equal to *BG*, and let arc *DH* be drawn. Then, by Th. [36] above, arc *DH* will be equal to arc *AB* and ∠ *DHZ* will be equal to ∠ *ABG*. Moreover arc *AB* was given equal to arc *DE*. Therefore the two arcs *DH* and *DE* will be equal to each other, and thus, by Th. [41] above, ∠ *DHE* will be equal to ∠ *DEH*. Since ∠ *DHE* is acute because ∠ *ABG*, equal to it, is acute, then ∠ *DEH* will also be acute; hence ∠ *DEZ* will be obtuse, which is contrary to [what was] given. [Therefore *BG* must be equal to *EZ*.] We would proceed no differently if the contrary that arc *EZ* is greater than arc *BG* were assumed.

But if we had taken each of the angles *B* and *E* [to be] obtuse, we would conclude by similar reasoning that ∠ *E* is acute. Therefore, because our proposition would say the opposite, it is concluded that the same angle is acute and obtuse, but, since this cannot be, the truth of the theorem is left revealed.

Theorem 56

All right triangles having two sides each respectively equal and [having] both sides that enclose the right angles less than a quadrant division are confirmed [to be] equal-angled and equal-sided.

If *ABG* and *DEZ* are two triangles whose angles *B* and *E* are right [angles], and if each of the sides *AB* and *BG* of △ *ABG*, as well as each of the sides *DE* and *EZ*, is less than a quadrant, and if two sides of the one are equal to any two sides of the other, then the remaining side of the one will be equal to the remaining side of the other, and the remaining angles of the one will be equal to the remaining angles of the other; that is, these two triangles will be equal-sided and equal-angled.

If the two sides surrounding the right angles [of each] of them are respectively equal, everything will be established by Th. [36] above. But if these equal sides surround angles other [than the right angles], let the two sides *BA* and *AG*, for example, be equal to the two *ED* and *DZ*, in a pairing of the right with the right and the left with the left. And let each of the arcs *BA* and *ED* be extended until it is a quadrant — indeed, *BA* [is extended] to *H* and *ED* to *K*. Then the points *H* and *K* will be poles of the circles *BG* and *EZ*, from which two quadrants *HG* and *KZ* are drawn. It is established that these [quadrants] are equal since we usually represent them in one sphere or in two equal [spheres]. Moreover, arc *AH* is equal to arc *DK*, for these two [arcs] are the complements of the two arcs *AB* and *DE*, which the hypothesis gave equal. Therefore, by Th. [35] above, since the two arcs *AG* and *DZ* are equal, ∠ *HAG* will be equal to ∠ *KDZ*. Hence the angles remaining from two right [angles] — namely, *BAG* and *EDZ* — will not be unequal. Moreover, the two arcs *BA* and *AG* were given equal to the two [arcs] *ED* and *DZ*. Therefore, by Th. [36] above, one may prove that the given triangles are equal-angled and equal-sided. Q.E.D.

The End of the Third [Book] of Triangles.

LIBER QVARTVS
TRIANGVLORVM.

I.

Si à polo circuli magni in fphæra ad circumferentiam ipfius, aut ar
cum eius arcum magnum demiferis, arcus ille demiffus erit quadrans
perpendicularis circumferentiæ, duos angulos fupra arcum, cui inci=
dit, rectos fecernens.

Sit circulus magnus in fphæra a b g, à cuius
polo 3 demittatur arcus circuli magni qui fit 3 b.
Dico cp arcus 3 b erit quadrans circumferentiæ
magnæ, & utercp angulop a b 3 & 3 b g rectus
erit. Producam enim lineam polarem circuli a b
g, quæ fit 3 b, quam per tertij huius oportet ef=
fe latus quadrati infcripti circulo magno. quatu=
or autem latera quadrati huiufmodi, cũ fint æqua
lia, quatuor abfcindunt æquales arcus per ter='
tij, ex circũferentia circuli, quorũ unus eft arcus 3
b, arcus igitur 3 b eft quadrans circuli. Præterea
circulus, cuius eft arcus 3 b, tranfit per polũ 3 cir
culi a b g, quare erectus eft ad eum per tertij huius, quod non poteft effe, nifi
utercp angulorũ a b 3 & 3 b g fit rectus. Sed fortaffe infirmam fufpicaris hãc ar=
gumentationem, defcribe igitur fuper polo b fecundũ quantitatem b 3 circulũ in
fphæra, cuius femicircumferentia fit a 3 g arcus, erit itacp ex eis, quæ fuper hoc
theoremate præfenti primũ diximus, utercp arcuũ a 3 & 3 g quadrans circũfren=
tiæ, quare per tertij huius duo anguli a b 3 & 3 b g æquales declarantur, per dif
finitionem igitur arcus 3 b eft perpendicularis ad circumferentiam circuli a b g,
quæ fuerunt explananda.

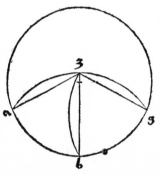

II.

Si ab aliquo puncto arcus circuli magni quadrans magnus ortho=
gonaliter egrediatur, terminus eius erit polus circuli à quo egredieba
tur quadrans ipfe, cum quo declarabitur punctum concurfus duorum
arcuum orthogonaliter à tertio arcu exurgentium, effe polum circuli
arcum ipfum continentis.

A puncto b arcus a b g circuli magni orthogonaliter egrediatur quadrans
circuli magni b 3. Dico cp punctus 3 terminus uidelicet quadrantis egreffi erit po
lus circuli a b g. Subtenfa enim quadranti dicto corda fua b 3, quam conftat ef
fe coftam quadrati magni, fecundũ eius quantitatem defcribat circulus magnus,
cuius femicircumferentia fit a 3 g .cuncp duo anguli a b 3 & 3 b g fint æquales
hypothefi id exigente, erunt per tertij huius duo arcus a 3 & 3 g fimiles, unde &
æquales, cp de eadem circumferentia exiftant, eft autem arcus a 3 g femicircumfe

M 3 rentia

FOURTH BOOK OF THE TRIANGLES

Theorem 1

If you were to drop a great arc from the pole of a great circle in a sphere to its circumference or to an arc of [the circumference], that lowered arc will be a quadrant perpendicular to the circumference, forming two right angles upon the arc to which it falls.

If *ABG*, in a sphere, is a great circle, from whose pole *Z* an arc *ZB* of [another] great circle is dropped, then arc *ZB* will be a quadrant of the great circumference and each of the angles *ABZ* and *ZBG* will be a right [angle].

Let a polar line *ZB* of circle *ABG* be extended. By Th. III.[11] above, this must be a side of the square inscribed in the great circle. Moreover, the four sides of this square, since they are equal, intercept four equal arcs from the circumference of the circle by [Euclid] III.[not given]. One of these [equal arcs] is arc *ZB*; therefore arc *ZB* is a quadrant of the circle. Furthermore, the circle of which *ZB* is an arc passes through pole *Z* of circle *ABG*; therefore it is perpendicular to [*ABG*] by Th. III.[17] above. This cannot be unless each of the angles *ABZ* and *ZBG* is a right [angle].

But perhaps you suspect that this argument is weak. Then in the sphere describe a circle upon pole *B* with length *BZ* [as polar line]. The semicircumference of this [circle] is arc *AZG*. Then, from those [theorems] which we mentioned first, above this, in the present theorem, each of the arcs *AZ* and *ZG* will be a quadrant of the circumference. Therefore by Th. III.[31] above the two angles *ABZ* and *ZBG* are declared equal. Then, by definition, arc *ZB* is perpendicular to the circumference of circle *ABG*. Q.E.D.

Theorem 2

If, from some point on the arc of a great circle, a great quadrant emerges orthogonally, its end [point] will be the pole of the circle from which the quadrant emerged; with this [fact], it may be stated that the meeting point of two arcs emerging orthogonally from a third arc is the pole of the circle containing that [third] arc.

If, from point *B* of arc *ABG* of a great circle, a quadrant *BZ* of [another] great circle emerges orthogonally, then point *Z* — namely, the end [point] of the outgoing quadrant — will be the pole of circle *ABG*.

Since its chord *BZ*, which is established to be the side of a great square, is subtended by the quadrant, let a great circle be described with the length of that [chord *BZ* as polar line]. The semicircumference of this [circle] is *AZG*. And because two angles *ABZ* and *ZBG* are equal since the hypothesis demands it, then, by Th. III.[32] above, the two arcs *AZ* and *ZG* are similar and, hence, are equal because they are from the same circumference. Moreover, arc *AZG* is a semicircumference

rentia circuli magni,quemadmodū trahitur ex   tertij huius,quare uterœ arcuū
a 3 & 3 g est medietas semicircumferentiæ,& ideo quadrans circūferentiæ totius
circuli magni.tres igitur arcus a 3, b 3 & g 3 æquales sunt quadrantes circumfe
rentiæ magnarū æqualium,quare per   tertij cordæ eoœ æquales declarantur . à
puncto itacœ 3 ad circumferentiam circuli a b g tribus æqualibus rectis descen=
dentibus   tertij huius punctū 3 polum circuli a b g declamabit, quod uoleba=
mus aperire. Corollariū autem sic constabit. In utrocœ arcuū orthogonaliū, quan
tum sat est protensoœ,necesse est inueniri polū circuli huiusmodi,quemadmodum
ex præsenti trahitur,aut igitur punctus coincidentiæ duoœ arcuū orthogonaliū
est polus,aut duo erunt poli unius circuli ex eadem parte,sed nullus circulus duos
ex eadem parte polos habet per   tertij.punctus igitur in quo confluūt dicti arcus
orthogonales,polus circuli, à cuius arcu egrediunt ipsi habebitur.

III.

In omni triangulo rectangulo latera rectum angulum continen=
tia ad quadrantem circūferentiæ & anguli eis oppositi ad rectum an=
gulum similes habebunt comparationes.

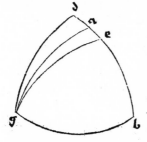

Volo dicere. Si angulus,qui uni ex lateribus re=
ctum angulū continentibus, opponitur recto, fuerit
æqualis,latus ipsum quadranti æquale erit.si maior re
cto,& ipsum latus quadrante maius. si minor, ipsum
minus quarta circūferentiæ.Versa demū uice.Si alte=
rum ex huiusmodi lateribus rectum ambientibus qua
drās existat,angulus ei oppositus erit rectus.si maius
quadrante,maior recto erit angulus.& si minus, mi=
nor.   Sit igitur exempli causa triangulus sphæralis
a b g,ex arcubus circulorū magnorū constitutus,angulū b rectum habens. Di=
co œ si angulus g rectus fuerit latus, a b erit quadrās.si uero recto maior,arcus a
b quadrantem superabit,& si minor recto extiterit, arcus a b quadrante minor
habebitur.Similiter si arcus a b quadranti æqualis occurrat, angulus g rectus
erit.si uero quadrante maius,& angulus g rectū superabit.& si minus quadrante,
angulū g recto minorem enunciabimus,quæ sic habebis.Sit primo angulus g re
ctus,uterœ igitur circuloœ,in quibus sunt duo arcus a g & a b, erectus est ad su=
perficiem circuli,cuius est latus tertium b g ,transibitœ per polum circuli b g.cū
autem hij duo arcus in puncto a coincidant,erit a polus circuli b g magni,qua
re per corollarium   tertij huius arcus a b respiciens angulū g ,erit quadrans cir
cumferentiæ.Sit deinde angulus a g b maior recto,& fiat iuxta punctū g arcus
b g angulus rectus b g e ,docente   tertij huius,pducendo arcū g e . erit itacœ e
polus circuli b g ,& ideo arcus e b quarta circuli,quare arcus a b angulū a g b
subtendens quadrantem superabit.Qd'si angulus a g b minor fuerit recto, sta=
tuatur angulus d b g rectus,eritœ,quemadmodū antea conclusum est, d polus
circuli b g ,& arcus d b quadrans circūferentiæ,quare arcus a b angulū a g b
subtendens minor quadrante.   Præterea ponamus arcū a b quartam circumfe
rentiæ,eritœ propter hoc a polus circuli b g ,& ideo arcus a g erectus ad arcū b
g ,angulus ergo a g b rectus habebitur.Sed intelligatur arcus a b maior qua=
drante,fiatœ arcus e b quarta circuli,& ideo e polus circuli b g ,arcus itacœ e g
erectus erit ad arcum b g ,& angulus e g b rectus, quare angulus a g b maior
recto

of a great circle, as is concluded from Th. III.[18] above. Therefore each of the arcs *AZ* and *ZG* is half of a semicircumference and hence is a quadrant of the total circumference of a great circle. Thus the three equal arcs *AZ*, *BZ*, and *GZ* are quadrants of equal great circumferences. Hence by [Euclid] III. [not given] their chords are declared equal. Then since three equal straight lines descend from point *Z* to the circumference of circle *ABG*, Th. III.[13] above will declare that point *Z* is the pole of circle *ABG*. Q.E.D.

Moreover, one may establish the corollary in this manner. On each of the orthogonal arcs, extended as much as is sufficient, the pole of this circle is necessarily found, as is concluded from the present [theorem]. Then either the point of coincidence of the two orthogonal arcs is the pole or there will be two poles of one circle on the same side. But no circle has two poles on the same side, by [Euclid] III.[not given, but see Th. III.9 above]. Therefore the point at which the mentioned orthogonal arcs meet will be the pole of the circle from whose arc [the two arcs] emerged.

Theorem 3

In every right triangle [the length of] the sides including the right angle compared to [the length of] a quadrant of the circumference and [the size of] the angles opposite these [sides] compared to [the size of] a right angle will have similar relationships.

[What] I mean [is this]. If an angle which is opposite one of the sides including the right angle were equal to a right angle, that side will be equal to a quadrant. If [that angle] is greater than a right [angle], the side will be greater than a quadrant. If [the angle] is less [than a right angle], that [side] will be less than a quarter circumference. Finally, the converse [is true]. If one of those sides surrounding the right [angle] is a quadrant, the angle opposite that [side] will be a right [angle]. If [one of those sides] is greater than a quadrant, the [opposite] angle will be greater than a right [angle], and if [the side] is less [than a quadrant], [the opposite angle] will be less [than a right angle].

Therefore, as an example, let *ABG* be a

spherical triangle, formed from the arcs of great circles and having a right ∠ *B*. Then, if ∠ *G* is a right [angle], side *AB* will be a quadrant. However, if [∠ *G*] is greater than a right [angle], arc *AB* will exceed a quadrant, and if [∠ *G*] is less than a right [angle], arc *AB* will be less than a quadrant. Similarly, if arc *AB* is equal to a quadrant, ∠ *G* will be a right [angle]. But if [*AB*] is greater than a quadrant, ∠ *G* will exceed a right [angle]; and if [*AB*] is less than a quadrant, we pronounce ∠ *G* [to be] less than a right [angle]. You will find these in this way.

First, let ∠ *G* be a right [angle]. Then each of the circles on which the two arcs *AG* and *AB* are [found] is perpendicular to the surface of the circle of which the third side *BG* is [an arc], and [each of the circles] will pass through the pole of circle *BG*. Moreover, since these two arcs meet at point *A*, *A* will be the pole of great circle *BG*. Therefore, by the corollary of Th. III.[11] above, arc *AB* opposite ∠ *G* will be a quadrant of the circumference.

Next, let ∠ *AGB* be greater than a right [angle], and, next to point *G* of arc *BG*, let a right ∠ *BGE* be made, by the teaching of Th. III.[34 or 44] above, to produce arc *GE*. Then *E* will be the pole of circle *BG*, and thus arc *EB* will be a quarter of the circle. Therefore arc *AB* subtending ∠ *AGB* will exceed a quadrant.

But if ∠ *AGB* is less than a right [angle], let right ∠ *DBG* be set up. Then, as was concluded before, *D* will be the pole of circle *BG*, and arc *DB* will be a quadrant of the circumference. Therefore arc *AB* subtending ∠ *AGB* will be less than a quadrant.

Furthermore, let us take arc *AB* to be a quarter of the circumference. Because of that, *A* will be the pole of circle *BG*, and thus arc *AG* is perpendicular to arc *BG*. Therefore ∠ *AGB* will be a right [angle].

But let arc *AB* be understood to be greater than a quadrant. Then let arc *EB* be made a quarter of the circle. Therefore *E* will be a pole of circle *BG*. Arc *EG* will then be perpendicular to arc *BG*, and ∠ *EGB* will be a right angle. Hence ∠ *AGB* will be greater than a right [angle].

recto.Qd'fi ftatuerimus arcũ a b minoréquadrante,prolongetur ipfe in direḍ
ctum ufcḡ ad d,donec b d fiat quarta circuli,ideocḡ d polus circuli b g.demiſ
fo igitur arcu d g,erit angulus d g b rectus,& angulus a g b minor recto,maḍ
nifeftam itacḡ effecimus theorematis noftri ueritatem.

IIII.

In omni triangulo rectangulo,fi fuerit alterum ex lateribus rectum
ambientibus quarta circuli,latus quocḡ rectum fubtend ns angulum
erit quarta circuli.fi uero fuerint latera rectum angulum continentia,
aut ambo maiora quadrante,aut ambo minora,erit latus rectũ fubten
dens angulum minus quarta circuli.cḡ fi alterum maius quadrante,&
alterum minus extiterit,latus rectum angulum refpiciens,maius quaḍ
drante pronunciabitur.

Habeat triangulus a b g angulũ b re
ctum,&latus a b quadrantem circumferen
tiæ.Dico cḡ latus a g rectum fubtendẽs an
gulum,erit quarta circuli. Erit enim a poḍ
lus circuli b g,quare per huius arcus a g
erit quarta circuli.Si uero utercḡ arcuũ a b
& b g minor quarta fuerit, erit arcus a g
minor quadrante.fiat enim arcus g e quar
ta circuli & arcus b d fimiliter,tranfeatcḡ p
duo puncta d & e arcus circuli magni d e,
arcus autem g a prolongetur,donec occur
ret arcui d e in puncto ȝ .Quoniã itacḡ anḍ
guli apud b funt recti, & arcus b d quarta
circuli erit d polus circuli g e,& ideo angu

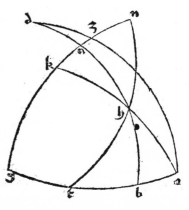

lus e rectus,fed & e g eft quarta circuli.quare punctus g eft polus circuli d e,&
ideo arcus g ȝ eft quadrans circumferentiæ,arcus igitur partialis a g minor qua
drante fiet.Sit deinceps utercḡ arcuũ a b,b g maior quadrante. Dico cḡ arcus a
g erit minor quarta circumferentiæ . Abfcindam enim utruncḡ arcuũ g t & b h
quadrantem,per duocḡ puncta t & h producam arcum circuli magni occurrenḍ
tem arcui g a continuato in n,quoniam itacḡ angulus b eft rectus,& arcus b h
quarta circuli,erit h polus circuli b g,& angulus apud t rectus.cuncḡ arcus t g
fit quarta,erit g polus circuli t n,quare arcus g n eft quarta, & ideo arcus a g
minor quadrante.Poftremo fit arcus a b maior quarta,& arcus b g minor. Diḍ
co cḡ arcus a g maior erit quadrante.Fiat enim utercḡ arcuũ g e & b h quarta
circuli,prolongando quidem arcum g b ufcḡ ad e ,arcũ autem b h refecando ex
arcu a b,incedatcḡ arcus circuli magni per duo puncta h & e,occurfurus arcui a
g in puncto k.Cum igitur arcus b h eft quarta,& angulus b rectus ,erit h poḍ
lus circuli g b e,quare & angulus apud e rectus.eft autẽ & arcus e g quadrãs.
g' igitur eft polus circuli e k,quare arcus g k erit quarta circuli,unde & a g arḍ
cus quadrantem fuperare dinofcitur,quæ fuere concludenda.

v.

In omni triangulo rectangulo fi fuerit latus rectum fubtendens anḍ
gulum quarta circuli,erit alterum ex duobus rectũ ambientibus quar
ta cír

But if we take arc *AB* [to be] less than a quadrant, let that [arc *AB*] be extended all the way to point *D*, until *BD* is a quarter of a circle, and therefore *D* is the pole of circle *BG*. Then, when arc *DG* is drawn, ∠ *DGB* will be a right angle and ∠ *AGB* will be less than a right angle. Thus we have made the truth of our theorem apparent.

Theorem 4

In every right triangle, if one of the sides surrounding the right angle is a quarter of a circle, the side subtending the right angle will also be a quarter of a circle. But if the sides including the right angle are either both greater than or both less than a quadrant, the side subtending the right angle will be less than a quarter of a circle. But if one [of these surrounding sides] is greater than a quadrant and the other is less [than a quadrant], the side opposite the right angle will be declared greater than a quadrant.

If △ *ABG* has a right ∠ *B*, and side *AB* is a quadrant of the circumference, then side *AG* subtending the right angle will be a quarter of the circle. *A* will be the pole of circle *BG*; therefore, by Th. [1] above, arc *AG* will be a quarter of the circle.

However, if each of the arcs *AB* and *BG* is less than a quarter [of a circle], then arc *AG* will be less than a quadrant. Let arc *GE*, as well as arc *BD*, be a quarter of a circle, and let arc *DE* of a great circle pass through the two points *D* and *E*. Moreover, let arc *GA* be extended until it meets arc *DE* at point *Z*. Then, because the angles at *B* are right [angles], and [because] arc *BD* is a quarter of a circle, *D* will be the pole of circle *GE*. Therefore ∠ *E* will be a right angle. Furthermore *EG* is a quarter of the circle. Thus point *G* is the pole of circle *DE*, and hence

arc *GZ* is a quadrant of the circumference. Then the arc portion *AG* is less than a quadrant.

Again, if each of the arcs *AB* and *BG* is greater than a quadrant, then arc *AG* will be less than a quarter of the circumference. Let each of the arcs *GT* and *BH* be intercepted [as a] quadrant, and through the two points *T* and *H* let an arc of a great circle be extended, meeting arc *GA* extended at *N*. Then, because ∠ *B* is a right [angle] and arc *BH* is a quarter of the circle, *H* will be the pole of circle *BG* and the angle at *T* will be a right angle. And since arc *TG* is a quarter [of the circle], *G* will be the pole of circle *TN*. Therefore arc *GN* is a quarter [of the circle], and thus arc *AG* is less than a quadrant.

Finally, if arc *AB* is greater than a quarter [of a circle] and arc *BG* is less [than a quadrant], then arc *AG* will be greater than a quadrant. Let each of the arcs *GE* and *BH* be a quarter of a circle by extending arc *GB* all the way to *E* and by cutting arc *BH* from arc *AB*. Let the arc of a great circle pass through the two points *H* and *E* to meet arc *AG* at point *K*. Then, since arc *BH* is a quarter [of a circle] and [since] ∠ *B* is a right [angle], *H* will be the pole of circle *GBE*, and therefore the angle at *E* will be a right [angle]. Moreover, arc *EG* is a quadrant; therefore *G* is the pole of circle *EK*. Thus arc *GK* will be a quarter of a circle, and hence arc *AG* is determined to exceed a quadrant. Q.E.D.

Theorem 5

In every right triangle, if the side subtending the right angle is a quarter of a circle, one [side] of the two surrounding the right angle will be a quarter

ta circuli. Si uero fuerit minus quadrante, erit utruncp reliquoru aut
maius quadrante, aut minus eo, cp si ipsum fuerit maius quadrante,
erit alterum ex duobus rectum ambientibus quadrante maius, reliquum autem minus eo.

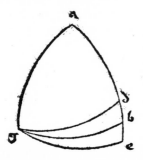

Hæc est conuersa præcedentis. Sit itacp in trian
gulo a b g angulum b rectum habente, latus a g
quadrans circumferentiæ. Dico cp altegz duoru late
rum a b, b g erit quarta circuli. Super puncto em
a facto polo, secundu quantitatem a g describat
circulus, quem oportebit esse magnu, cp arcus a g
sit quarta circumferentiæ magnæ. si igitur circumferentia eius transibit per punctu b, planum est ar
cum a b esse quadrantem. si uero prætereat ipsum,
aut ibit supra aut infra ipsum. si supra, secabit arcu
a b in puncto qui sit d, per itacp huius duo arcus
a d & d g orthogonaliter sibi inuicem insistent, est
autem & arcus g b orthogonalis ad arcum a b propter angulu b rectum ex hy
pothesi, quare per huius g est polus circuli a b, & ideo per huius arcus g b,
latus uidelicet trianguli propositi erit quarta circumferentiæ. Similem pristinis
concludes medijs, si circuferentia circuli super a polo descripti, transibit infra pun
ctum b, occurrens arcui a b prolongato in puncto e. Sit demum arcus a g mi

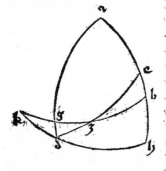

nor quadrante. Dico cp uterqp arcum a b, b g aut
erit maior quarta, aut minor ea. Crescat enim arcus a g in directum, donec habeat arcus a d quadrans, secundu cuius cordam circumducatur circulus magnus, qui nequacp offendet punctu b, alias
enim per huius & hypothesim sequeretur rectum
angulu esse minorem recto. Aut igitur secando arcum a b: si opus fuerit: prolongatum transibit supra b punctu aut infra. Transeat prius supra, secas
necessario arcum b g in puncto qui sit 3. erit itacp
per huius angulus 3 e b rectus, cucp sit etiam an
gulus e b 3 rectus ex hypothesi, erit per corollariu

2. huius 3 polus circuli a b, & ideo per primam huius arcus 3 b quarta circumfe
rentiæ. latus igitur b g quadrantem superabit, latus denicp a b quadrante exce-
dere nemini dubium erit: consideranti saltem arcum a e æqualem arcui a d esse
quartam circumferentiæ. Sed transeat circulus dictus infra punctu b, secando arcum a b protensum oportune in puncto h, arcum autem b g prolongatu in pu
cto k, erit itacp per media nunc commemorata, ne iusto crebrius primam atcp secundam huius repetam propositiones, angulus apud h rectus, sed & angulus apd
b rectus ex hypothesi, quare punctus k erit polus circuli a b, ideocp arcus k b
quadrans, & latus g b trianguli nostri quadrante minus, latus autem a b quadrante breuius esse nemo dubitabit. Tandem arcus a g rectu subtendens angulu
quartam superare ponatur. Dico cp altegz duoru laterum a b, b g maius erit qua
drante, reliquu uero minus. Resecto enim quadrante a d, cordam ipsius sup a po
lo circumferentes creabimus circulum magnu, cuius profecto circumferentia pun
ctum k præteribit, alias enim monstrum nasceretur mathematicu, uno uerius im
possibile

of the circle. However, if [the side subtending the right angle] is less than a quadrant, each of the other [sides] will be either greater than or less than a quadrant. But if [the side subtending the right angle] is greater than a quadrant, one of the two [sides] surrounding the right angle will be greater than a quadrant, and the other will be less than [a quadrant].

This is the converse of the preceding [theorem]. Thus, if in a △ *ABG*, having right ∠ *B*, side *AG* is a quadrant of the circumference, then one of the two sides *AB* and *BG* will be a quarter of the circle. Let a circle be described around point *A* as pole with the quantity *AG* [as polar line]. This [circle] must be a great circle because arc *AG* is a quarter of a great circumference. Therefore, if its circumference will pass through point *B*, it is evident that arc *AB* is a quadrant. But if [the circumference] by-passes [point *B*], it will go either above or below that [point]. If above, it will cut arc *AB* at point *D*. Then by Th. [1] above the two arcs *AD* and *DG* rest on each other orthogonally. Moreover, arc *GB* is orthogonal to arc *AB* because ∠ *B* is a right [angle] from the hypothesis. Therefore by Th. [2] above *G* is the pole of circle *AB*, and thus by Th. [1] above, arc *GB* — namely, the side of the given triangle — will be a quarter of the circumference. You may conclude a similar [situation] by the previous means if the circumference of the circle described around pole *A* passes beneath point *B*, meeting arc *AB* extended at point *E*.

Next, if arc *AG* is less than a quadrant, then each of the arcs *AB* and *BG* will be either greater than a quarter [of the circle] or less than one. Let arc *AG* increase in one direction until arc *AD* is a quadrant. Let a great circle be circumscribed with the chord of [*AD* as polar line]. This [circle] by no

means hits point *B*, for otherwise, by Th. [3] above and the hypothesis, it would follow that a right angle is less than a right angle. Therefore, in intersecting arc *AB*, prolonged if necessary, [the circle] will pass either above or below point *B*. First, let it pass above [point *B*], necessarily cutting arc *BG* at point *Z*. Thus by Th. [1] above ∠ *ZEB* will be a right angle. And since ∠ *EBZ* is also a right [angle] from the hypothesis, by the corollary of Th. 2 above *Z* is the pole of circle *AB*, and therefore, by Th. 1 above, arc *ZB* is a quarter of the circumference. Thus side *BG* will exceed a quadrant, and finally no one will doubt that side *AB* exceeds a quadrant — at least when one recognizes that arc *AE*, equal to arc *AD*, is a quarter of the circumference. But let the mentioned circle pass below point *B*, by intersecting arc *AB*, extended as necessary, at point *H* and arc *BG* extended at point *K*. Then by the method just mentioned, lest I reiterate the first and second theorems above more frequently than [is] reasonable, the angle at *H* will be a right [angle]. Furthermore the angle at *B* is a right [angle] from the hypothesis. Therefore point *K* will be the pole of circle *AB*. Thus arc *KB* is a quadrant, and side *GB* of our triangle is less than a quadrant. Moreover, no one will doubt that side *AB* is shorter than a quadrant.

At last, if arc *AG* subtending the right angle is taken to exceed a quarter [of a circle], then one of the two sides *AB* and *BG* will be greater than a quadrant, while the other [will be] less. For when a quadrant *AD* is cut [from *AG*], then, if we move the chord of [*AD*] around pole *A*, we will produce a great circle, whose circumference will indeed by-pass point *B*, for otherwise a mathematical anomaly — in fact, a veritable impossibility — would arise:

possibile,rectus scilicet angulus recto maior.Tran
seat itacɜ prius supra punctū b ,secando arcum a b
in puncto e ,arcum uero b g oportune extensum
in puncto 3 ,quem constabit esse polum circuli a b
angulis apud b & e rectis existentibus,unde & ar
cus 3 b circumferentiæ quarta pars habebitur . la=
tus igitur b g trianguli a b g quadrante breuius
relinquitur,latus autem reliquū a b quadrantē su=
perabit proculdubio. Qd̄ si circumferentia dicta
descenderit infra punctū b ,secando arcum a b sa
tis porrectū in puncto h ,arcum autem b g (nam
ita fieri necesse est)in puncto k ,pristino freti syllo=

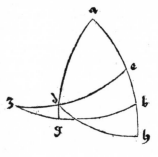

gismo,declarabimus k esse polum circuli a b ,& arcum b k quadrantem circum
ferentiæ,unde consequitur latus b g trianguli nostri quadrante longius esse,reli=
quū uero latus a b quadrantis a h longitudinem haud attingit. Tres igitur the=
orematis partes confirmauimus,quod libuit efficere.   Possumus præterea pro=
positionem nostram stabilire,aduersario opposi tū asserenti concludentes impossi
bile per   huius . Nam ponendo latus a g quadrantem
si dixerit neutrū reliquoɜ laterum esse quadrantē,erit ne=
cessario utrunɋ eoɜ aut maius uel minus quadrante , aut
alterū maius & reliquū minus.quocunɋ autem illoɜ exi=
stente,sequitur per   huius arcum a g non esse quadran=
tem,sic idem arcus quadrans & non quadrās aduersario
proteruienti habebitur,quod est impossibile.Si uero latus
a g fuerit minus quadrante,& non fuerit utrunɋ reliquo
rum laterum aut maius quadrante , aut minus sententia
quidem aduersarij,erit necessario alteɜ eorum quadrans,

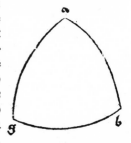

aut alterum eoɜ maius quadrante,& reliquū minus eo,quo ita existente , sequitur
per   huius arcum a g esse quadrantem aut maiorem eo,qui pridem ponebatur
minor eo.Qd̄ si latus a g maius supponatur quadrante,& credat aduersarius,al=
terum duorū laterum a b & b g esse quadrantem,aut utrunɋ eorum maius uel
minus quadrante,concludemus ei per   huius arcum a g aut esse quadrantē,aut
quadrante breuiorem,quem nuperrime concessit esse maiorē quadrante. Destru=
ctis igitur impossibilibus,ad quæ duximus aduersarium,theorematis nostri firi=
dabitur ueritas.

VI.

In omni triangulo rectangulo,si alter duorum angulorum,quos su
stinet latus,rectum respiciens angulum,rectus fuerit, erit latus ipsum
quarta circuli.si uterɋ eorum aut obtusus aut acutus extiterit, latus di
ctum quadrante breuius erit.si uero alter eorum obtusus,reliquus aūt
acutus occurrat,latus ipsum quadrante maius coniicietur .

Triangulus a b g angulum g rectum habeat. Dico ɋ si alter duorū angu=
loɜ a & g rectus fuerit, latus a g erit quarta circumferentiæ.si uero uterɋ eoɜ
aut obtusus aut acutus extiterit,latus a g quadrante breuius habebit.& si angu=
lus a fuerit obtusus,angulus aūt g acutus uel econtra,arcus a g quadrantem su=
perabit.Si enim alter anguloɜ a & g rectus fuerit,erit per   huius alterum duo=
　　　　　　　　　　　　　　　　　　　　　　　　　　　　N　　　　rum la=

namely, a right angle greater than a right angle. Therefore, first, let [the circle] pass above point *B*, by intersecting arc *AB* at point *E* and arc *BG*, extended by necessity, at point *Z*. This [point *Z*] will be established to be the pole of circle *AB* because the angles at *B* and *E* are right [angles]. Hence arc *ZB* will be a quarter part of the circumference. Then side *BG* of △ *ABG* is left shorter than a quadrant; furthermore, the other side, *AB*, will exceed a quadrant without doubt. But if the mentioned circumference descends below point *B* to intersect arc *AB*, sufficiently extended, at point *H* and arc *BG* (for it necessarily happens that way) at point *K*, then, relying upon the previous deduction, we will declare *K* to be the pole of circle *AB* and arc *BK* to be a quadrant of the circumference. Hence it follows that side *BG* of our triangle is longer than a quadrant, while the other side *AB* by no means approaches the length of quadrant *AH*. Therefore we have proven the three parts of the theorem, which it was desired to accomplish.

We could establish our theorem further, proving, by contrary claims, the opposite to be impossible by Th. [4] above. When side *AG* is taken to be a quadrant, then, if one were to say that neither of the other sides is a quadrant, each of them will necessarily be either larger or smaller than a quadrant, or one [will be] larger and the other smaller. Moreover, if any one of those [possibilities] exists, it follows from Th. [4] above that arc *AG* is not a quadrant. Thus the same arc will be a quadrant and will not be a quadrant when the contrary is introduced, which is impossible. But if side *AG* is less than a quadrant, and [if] neither of the other sides is larger or smaller than a quadrant by the

opinion of the contrary, then one of the [sides] will necessarily be a quadrant or one of them [will necessarily be] greater than a quadrant and the other less than one. If this is so, it follows by Th. [4] above that arc *AG* is a quadrant or is greater than [a quadrant]; [but] this [arc] was already given to be less than [a quadrant]. But if side *AG* is given greater than a quadrant and the contrary [of our theorem] is believed, then one of the two sides *AB* and *BG* is a quadrant or each of them is larger or smaller than a quadrant; to this we argue, by Th. [4] above, that arc *AG* is either a quadrant or less than a quadrant. [But] it was already agreed that this [arc *AG*] is greater than a quadrant. Thus with [these] impossibilities eliminated, about which we stated the contrary, the truth of our theorem will be made firm.

Theorem 6

In every right triangle, if one of the two angles which the side opposite the right angle bears is a right [angle], that side will be a quadrant of the circle. If each of those [two angles] is either obtuse or acute, the mentioned side will be less than a quadrant. But if one of them is obtuse while the other is acute, that side is concluded to be greater [than a quadrant].

Let △ *ABG* have right ∠ *G*.* Then if one of the two angles *A* and *G* is a right [angle], side *AG* will be a fourth of the circumference. But if each of them is either obtuse or acute, side *AG* will be shorter than a quadrant. And if ∠ *A* is obtuse while ∠ *G* is acute, or the converse, then arc *AG* will exceed a quadrant.

If one of the angles *A* and *G* is a right [angle], by Th. [3] above one of the two

*For *G* read *B*.

rum laterum a b & b g quarta circuli, quare per huius arcus a g erit quarta circuli. Si uero uterqʒ anguloȝ a & g aut obtuſum aut acutū ſeſe præbeat, erit per allegatā huius utrincʒ duoȝ laterum a b & b g aut maius quadrante, aut minus eo, unde per huius latus a g quadrante breuius arguetur. Qd' ſi alter angulorū a & g obtuſus, reliquus autem acutus extiterit, erit ex huius alter duoȝ arcuū a b & b g maior quadrante, reliquus autem mſuor eo, quare per huius latus a g quadrantem ſuperabit, quæ cenſuimus explananda.

VII.

Si latus rectum angulum trianguli ſphæralis reſpiciens quadrans circumferentiæ fuerit, alter angulorum ſibi inſidẽtium rectus indicabitur. ſi uero quadrante minus extiterit, erit uterqʒ dictorum angulorum aut obtuſus aut acutus. & ſi latus ipſum quadrante longius offeratur, alter dictorū angulorū obtuſus, reliquus aũt acutus prædicabit.

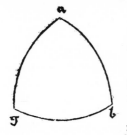

Hæc eſt cõuerſa prioris. Sit triangulus a b g rectangulus, angulū uidelicet b rectum habens. Dico ꝗ ſi latus a g fuerit quadrans circumferentiæ, alter angulorū a & g rectus erit. ſi uero arcus a g minor fuerit quadrãte, uterqʒ anguloȝ a & g aut obtuſus erit, aut uterqʒ acutus. & ſi arcus a g quadrantem excedat, erit alter angulorum a & g obtuſus, reliquus autem acutus. Nam arcu a g quadrante exiſtente, erit alter duorum arcuū a b & b g quarta circumferentiæ per huius, quare per huius alter anguloȝ a & g rectus concludetur. Si uero arcus a g quadrante minor extiterit, erit per eandem huius ũtercʒ arcuū a b & b g aut maior quadrante, aut minor eo, quare per huius utercʒ angulorū a & g aut obtuſus erit, aut acutus. Qd' ſi arcus a g quarta circumferentiæ maior occurrat, erit alter duorū arcuum a b & b g maior quadrante, & reliquus minor eo, quare per huius alter angulorum a & g obtuſus erit, & reliquus acutus, quæ fuere declaranda.

VIII.

Si quis triangulus ſphæralis duos acutos habeat angulos, aut duos obtuſos, arcus egrediens à uertice tertij anguli, lateri ſe reſpicienti perpendiculariter occurſurus, intra triangulū reperiet. ſi uero alter eorū acutus, & reliquus obtuſus extiterit, extra triangulū neceſſario cadet.

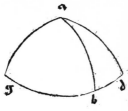

Sit triangulus huiuſmodi a b g, duos angulos b & g habens acutos, aut ambos obtuſos. Dico ꝗ arcus ab a puncto lateri b g : ꝗuis utrĩcʒ indefinito, perpendiculariter occurſurus, intra triangulū a b g conſiſtet. Non enim poterit egredi triangulū ipſum, quod ſi poſſibile arbitreris, ſit arcus ille a d coincidens arcui g b, quantū ſat eſt porrecto dextrorſum in puncto d, erit itacʒ arcus à d latus commune duobus triangulis a d b & a d g rectangulis

sides *AB* and *BG* will be a quarter of a circle. Therefore by Th. [4] above arc *AG* will be a quarter of a circle.

But if each of the angles *A* and *G* is either obtuse or acute, then, by the allegation of Th. [3] above, each of the two sides *AB* and *BG* is either greater than a quadrant or less than one. Hence by Th. [4] above side *AG* is proven shorter than a quadrant.

But if one of the angles *A* and *G* is obtuse while the other is acute, then, by Th. [3] above, one of the two arcs *AB* and *BG* will be greater than a quadrant and the other less than [a quadrant]. Therefore, by Th. [4] above, side *AG* will exceed a quadrant. Q.E.D.

circumference by Th. [5] above. Therefore, by Th. [3] above, one of the angles *A* and *G* is concluded [to be] a right [angle]. But if arc *AG* is less than a quadrant, then, by the same Th. [5] above, each of the arcs *AB* and *BG* will be either greater than a quadrant or less than [a quadrant]. Therefore by Th. [3] above each of the angles *A* and *G* will be either obtuse or acute. But if arc *AG* is greater than a quarter of the circumference, one of the two arcs *AB* and *BG* will be greater than a quadrant and the other less than a [quadrant]. Therefore, by Th. [3] above, one of the angles *A* and *G* will be obtuse and the other acute. Q.E.D.

Theorem 7

If the side opposite a right angle of a spherical triangle is a quadrant of the circumference, one of the angles resting on that [side] will be declared a right [angle]. But if that [side] is less than a quadrant, each of the mentioned angles will be either obtuse or acute. And if this side is given longer than a quadrant, one of the mentioned angles will be found obtuse while the other is acute.

This is the converse of the preceding [theorem]. Let △ *ABG* be a right triangle, that is, having right ∠ *B*. Then if side *AG* is a quadrant of the circumference, one of the angles *A* and *G* will be a right [angle]. But if arc *AG* is less than a quadrant, each of the angles *A* and *G* will be either obtuse or acute. And if arc *AG* exceeds a quadrant, one of the angles *A* and *G* will be obtuse while the other is acute.

When arc *AG* is a quadrant, one of the two arcs *AB* and *BG* will be a quarter of the

Theorem 8

If any spherical triangle has two acute angles or two obtuse [angles], the arc passing from the vertex of the third angle to meet the side opposite that [third angle] perpendicularly will be found within the triangle. But if one of these [angles] is acute and the other obtuse, [the arc] necessarily falls outside the triangle.

If *ABG* is such a triangle, having two angles *B* and *G* [both] acute or both obtuse, then the arc [that is drawn] from point *A* to meet side *BG* perpendicularly — [BG] being completely unbounded on both sides — is found within △ *ABG*.

[The perpendicular arc] cannot leave this triangle. But if you were to judge [it] possible, let that [hypothetical] arc be *AD*, meeting arc *GB*, extended to sufficient length on the right, at point *D*. Then arc *AD* would be a side common to the two right triangles *ADB* and *ADG*.

rectangulis,ſi igitur duo anguli b & g triãguli a b g fuerint acuti,erit angulus a b d obtuſus,quare per huius arcus a d minor erit quadrante propter angu= lum a g d trianguli a g d acutum,& per eandem maior quadrãte propter an= gulum a b d trianguli a b d obtuſum.idem itacɜ arcus minor erit quadrãte cir cumferentiæ eiuſdem,& maior eo,quod eſt impoſſibile,arcus ergo perpendicula= ris huiuſmodi non cadet extra triangulũ.Prohibet autem hypotheſis,ne arcus ille tranſeat per alterũ punctoɜ b & g ,ſic enim angulus acutus aut obtuſus haberet rectus,quod eſt impoſſibile.Deſtructis autem incõuenientibus illis relinquitur ꝙ non niſi intra triangulũ permaneat,quod pollicebatur prima pars theorematis.

Sit demũ alter prædictorũ anguloɜ,uerbi gratia , b ob tuſus,& reliquus g acutus.Dico ꝙ perpendicularis ca det extra triangulũ.Non enim poteſt coincidere alteri duorũ laterũ a b & a g ,ſic enim angulus obtuſus aut acutus eſſet rectus,ſed necɜ intra triangulum conſiſtere poterit.Si enim ita putaueris,ſit arcus a e perpendicu laris ad arcum b g,cui ſecundum punctũ e communi cat.erit itacɜ latus a e commune duobus triangulis a b e & a g e rectangulis,quoɜ unus angulũ a b e ha= 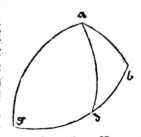 bet obtuſum,reliquus autem angulũ a g e acutũ,quare per huius bis aſſumptã arcus a e maior erit quarta circumferentiæ,& minor ea, quod eſt inconueniens. Cum igitur arcus perpendicularis huiuſmodi non poſſit coincidere alteri duorum laterũ a b & a g ,necɜ intra triangulũ cadere,reliquũ eſt neceſſario ut extra trian gulum reperiatur,quæ cupiebas addiſcere.

Cuiuslibet trianguli tres angulos acutos habentis , trium laterum unumquodcɜ minus quadrante pronunciabitur.

Trianguli a b g tres anguli ſint acuti. Dico ꝙ unũ quodcɜ latus eius quadrante minus erit.Fiant enim apud duo puncta b & g duo anguli recti,eductis duobus ar= cubus circuloɜ magnorũ b e & g e in puncto e cõcur= rentibus,quem per huius oportet eſſe polum circuli b g . â polo igitur e per punctũ a procedat circulus ma= 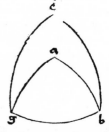 gnus,qui neceſſario ſecabit arcum b g ,ſi enim non ſeca= ret,neceſſario alterũ duoɜ arcuũ a b & a g ,fieretcɜ pars quadrantis ſemicircumferentia , quod eſt inconueniens, ſecet igitur in puncto d ,erit autem utercɜ anguloɜ apud d rectus per huius,& ideo utercɜ trianguloɜ a b d & a g d rectangulus,cuncɜ duo anguli a b d & b a d ſint acuti,totum enim b a g acutum tradidit hypotheſis,erit per huius ar= cus a b minor quadrante,ſimiliter probabimus arcũ a g minorem quadrante, reſtat igitur,ut arcum b g quadrante breuiorem oſtendamus,quod quidem habe bitur,ſi apud duo puncta a & b duos rectos,quemadmodũ nuperrime apud duo puncta b & g conſtituerimus.& hoc erat peragendum.

Si quis triangulus duos acutos habeat angulos æquales,utruncɜ la terum eos reſpicientiũ minus quadrante prædicabitur.

Sit triangulus a b g ,duos angulos b & g acutos habens æquales.Dico ꝙ

N 2 utruncɜ

Thus, if the two angles *B* and *G* of △ *ABG* were acute, ∠ *ABD* would be obtuse. Then by Th. [3] above arc *AD* would be less than a quadrant because ∠ *AGD* of △ *AGD* is acute, and, by the same theorem, [arc *AD*] would be greater than a quadrant because ∠ *ABD* of △ *ABD* is obtuse. Thus the same arc would be less than a quadrant of the same circumference and greater than [a quadrant], which is impossible. Therefore this perpendicular arc does not fall outside the triangle. Moreover, the hypothesis also prohibits the arc from passing through one of the points *B* and *G*, for then an acute or an obtuse angle would be a right angle, which is impossible. Furthermore, with these incongruities eliminated, what the first part of the theorem promised remains, because [the perpendicular arc can do] nothing except stay within the triangle.

Finally, if one of the aforementioned angles — *B*, for example — is obtuse and the other, *G*, is acute, then the perpendicular will fall outside the triangle. [The perpendicular] cannot coincide with one of the two sides *AB* and *AG*, since then an obtuse or an acute angle would be a right [angle]. Nor can [the perpendicular] be within the triangle. For if you were to consider it in that way, then let arc *AE** be perpendicular to arc *BG*, to which it is joined by point *E*. Thus side *AE* would be common to the two right triangles *ABE* and *AGE*, of which the one has obtuse ∠ *ABE* while the other has acute ∠ *AGE*. Therefore by Th. [3] above, applied twice, arc *AE* would be greater than a quarter of the circumference and less than one, which is contradictory. Then, since this perpendicular arc cannot coincide with one of the two sides *AB* and *AG* nor [can it] fall within the triangle, it necessarily remains that it will be found outside the triangle. Q.E.D.

Theorem 9

In any triangle that has three acute angles, every one of the three sides will be declared less than a quadrant.

If three angles of △ *ABG* are acute, then every single side of it will be less than a quadrant.

Let two right angles be formed at the two points *B* and *G* when two arcs, *BE* and *GE*, of great circles are drawn, meeting at point *E*. By Th. [2] above, [*E*] must be the pole of circle *BG*. Then from pole *E* through point *A*, let a great circle proceed, which will necessarily intersect arc *BG*; for if it does not intersect [arc *BG*], it would have to be one of the two arcs *AB* and *AG*, and part of a quadrant will be made a semicircumference, which is contradictory. Thus [the circle] will intersect [arc *BG*] at point *D*.† Moreover, each of the angles at *D* will be a right [angle] by Th. [1] above, and thus each of the triangles *ABD* and *AGD* will be a right [triangle]. And since the two angles *ABD* and *BAD* are acute, for the hypothesis gave the total [angle] *BAG* as acute, then by Th. [6] above arc *AB* will be less than a quadrant. We will prove similarly that arc *AG* is less than a quadrant. Thus it remains for us to show that arc *BG* is shorter than a quadrant; this would indeed be done if we were to set up two right [angles] at the two points *A* and *B*, as [we] recently [did] at the two points *B* and *G*. Q.E.D.

Theorem 10

If any triangle has two equal acute angles, each of the sides opposite those [angles] will be found less than a quadrant.

If *ABG* is a triangle that has two equal acute angles *B* and *G*, then

*In the figure, for arc *gdb* read *geb*.
†In the figure, for arc *gb* read *gdb*.

utrunqʒ laterūreius, a b & a g minus erit quarta circuli.
Creabimus enim apud duo pūcta b & g duos rectos an
gulos, arcui b g infeſſuros eductis duobus arcubus b d
& g d in puncto d confluentibus, quem huius non ſi
nitelle polum circuli b g, & ideo per huius uterqʒ ar
cuum b d & g d quadrans habebitur. Cū itaqʒ duo ar
cus b a & g a, ex punctis terminalibus lateris b g tri
angulí d b g egreſſi intra triangulum d b g concurrāt,
erunt per tertij huius duo arcus b a & a g coniuncti
breuiores duobus arcubus b d & d g. Vnde & medietas
horʒ arcus uidelicet b a breuior erit medietate iſtorʒ, quadrante ſcilicet b d . ſimi
liter arcū g a minorē eſſe cōstabit ipſo quadrāte g d . oportet em duos arcus b a
& a g æquales eſſe, tertij huius enunciante, quod libuit attingere.

X I.

Trianguli duos obtuſos habentis angulos æquales utruncʒ laterū
eos reſpicientiū, maius quadrante reperietur.

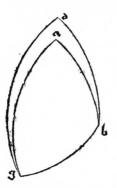

Sint duo anguli b & g, trianguli a b g obtuſi æqua
les. Dico cp utruncʒ laterum a b & a g quadrantem ſu
perabit. Factis em duobus rectis angulis apud duo pun
cta b & g, arcus b g, educendo duos arcus b d & g d,
quos conſtat intra triangulum a b g concurrere, quod ſi
at in puncto d, erunt queadmodum in præmiſſa argumē
tabar duo arcus b d & b g, quos oportet eſſe quadrātes,
minores duobus arcubus b a & a g lateribus trianguli
noſtri, & ideo ſicut duo arcus b a, a g æquales, ppter an
gulos a b g & a g b æquales, ſuperabūt duos quadran
tes b d & d g, ita utercʒ arcuum b a & a g quadrante
maior intelliget, quod uoluimus exponere.

X I I.

Si quis triangulus inæquales habeat duos angulos acutos, latus mī
nori eorum oppoſitum minus erit quadrante.

Triāguli a b g duo angulí b & g ſint acuti, ſitcʒ an
gulus g minor angulo b . Dico cp latus a b eſt minus
quadrante. Nā conſtituendo duos rectos apud duo pun
cta b & g , productis arcubus b d & g d in puncto d
concurrentibus, erūt duo arcus b a & a g minores duo
bus b d & d g , qui ſemicircūferentiam perficiunt, quo
niam utercʒ eorum eſt quadrans circumferentiæ, d polo
circuli b g exiſtente, duo igitur arcus b a & a g mino
res ſunt ſemicircumferentia, cūcʒ arcus b a minor ſit ar
cu a g per tertij huius, cp angulus g minor exiſtat an
gulo b, erit arcus b a minor quadrante circumferentiæ,
quod erat demonſtra ndum.

X I I I.

Trianguli duos obtuſos habentis angulos inæquales, latus maiori
eorū oppoſitū maius quadrante pronunciabitur.

Habeat

each of its sides *AB* and *AG* will be less than a quarter of a circle.

We will form two right angles, to rest on arc *BG*, at the two points *B* and *G* when two arcs, *BD* and *GD*, are drawn, meeting at point *D*. By Th. [2] above, [*D* can] be nothing other than the pole of circle *BG*. And therefore, by Th. [1] above, each of the arcs *BD* and *GD* will be a quadrant. Then, since the two arcs *BA* and *GA*, emerging from the end points of side *BG* of △ *DBG*, meet within △ *DBG*, then, by Th. III.[38] above, the two arcs *BA* and *AG* combined will be shorter than the two arcs *BD* and *DG*. Hence half of the former [two arcs] — namely, arc *BA* — will be shorter than half of the latter [two arcs] — namely, *BD*. Similarly, it will be established that arc *GA* is less than quadrant *GD*. Then the two arcs *BA* and *AG* must be equal, by the declaration of Th. III.[40] above. Q.E.D.

Theorem 11

In a triangle that has two equal obtuse angles, each of the sides opposite these [angles] will be found greater than a quadrant.

If in △ *ABG* there are two equal obtuse angles *B* and *G*, then each of the sides *AB* and *AG* will exceed a quadrant.

When two right angles are made at the two points *B* and *G* of arc *BG* by construction of the two arcs *BD* and *GD*, which one establishes to meet within △ *ABG* at point *D*, then, as was shown in the preceding [theorem], the two arcs *BD* and *BG*,* which must be quadrants, will be less than the two arcs *BA* and *AG*, the sides of our triangle. And therefore, as the two arcs *BA* and *AG*, [which are] equal because of equal angles *ABG* and *AGB*, exceed the two quadrants *BD* and *DG*, so each of the arcs *BA* and *AG* will be understood greater than [one] quadrant. Q.E.D.

Theorem 12

If any triangle has two unequal acute angles, the side opposite the smaller of these [angles] will be less than a quadrant.

If two angles *B* and *G* of △ *ABG* are acute and ∠ *G* is smaller than ∠ *B*, then side *AB* is less than a quadrant.

With the formation [of] two right angles at the two points *B* and *G* when arcs *BD* and *GD*, meeting at point *D*, are drawn, the two arcs *BA* and *AG* will be less than the two [arcs] *BD* and *DG*. [*BD* and *DG* together] make up a semicircumference because each of them is a quadrant of the circumference since *D* is the pole of circle *BG*. Therefore the two arcs *BA* and *AG* are less than a semicircumference. And since arc *BA* is less than arc *AG* by Th. III.[42] above because ∠ *G* is less than ∠ *B*, then arc *BA* will be less than a quadrant of the circumference. Q.E.D.

Theorem 13

In a triangle that has two unequal obtuse angles, the side opposite the larger of these [angles] will be declared greater than a quadrant.

*For *BG* read *DG*.

Habeat triangulus a b g duos angulos b & g ob
tusos inæquales, g scilicet maiorem angulo b. Dico φ
latus a b maius est quadrante. A punctis enim b & g
arcus erecti orthogonaliter concurrent intra triangulū
a b g, q̄d sit in puncto d polo scilicet circuli b g, erunt
φ duo arcus b a & a g maiores duobus arcubus b d,
d g, qui æquantur semicircumferentiæ, quare duo arcus b a, a g semicircumferentiam superabunt, unde &
maior eorum scilicet arcus a b quadrante maior habebitur, quod erat absoluendum.

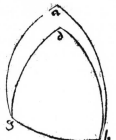
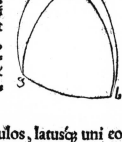

XIIII.

Si quis triangulus duos acutos habuerit angulos, latúsq́ uni eorū
oppositum non minus quadrante, erit reliquus eius angulus obtusus,
latúsq́ ei oppositum maius quadrante.

Sit triangulus a b g duos angulos b
& g acutos habens, cuius latus a b alterū
acutoꝝ respiciens, non sit minus quadrāte.
Dico φ angulus eius a erit obtusus, & latus
b g maius quadrāte. Fiāt em̄ apud duo pun
cta b & g duo anguli recti, eductis duobus
arcubus in puncto d extra triangulū a b g
concurrentibus, quem quidem oportet esse
polum circuli b g, demittaturq́ à polo d p
punctū a quadrans d a e, incidens arcui b
g in puncto e. Cum igitur arcus a g non
sit minor quadrante, erit ipse aut quadrans,
aut maior eo. Si quadrans, per huius alterum duoꝝ laterū a e, e b trianguli rectan-

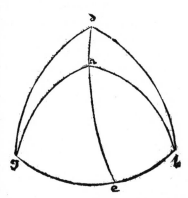

guli a e b erit quarta circumferentiæ. est autem arcus a e minor quadrante, reliquus ergo e b quadrans erit necessario, quare totus arcus b g maior erit quadrā
te. Similiter per huius probabimus angulū b a e esse rectū, cum angulus a b e
sit acutus ex hypothesi, totus igitur angulus b a g maior est recto. Qd̄ si latus a
b maius fuerit quadrante, erit per huius arcus b e maior quadrante, cum reliquus a e sit minor quadrante, arcus ergo b g multo maior erit quadrāte. Simili
ter per huius oportebit angulū b a e esse obtusum, cum angulus a b e sit acutus per hypothesim, angulus ergo b a g multo magis erit obtusus, patet igitur
propositio.

XV.

Si fuerint in sphæra duo circuli magni ad se inclinati, signentúrq́
in circumferentia unius eorum duo puncta, aut in utriusq́ circumferentia punctus unus, & producatur ex unoquoq́ punctorum ad circū
ferentiam circuli alterius perpendicularis arcus, proportio sinus arcus
qui est inter unum illorum punctorum, & punctum sectionis circulorum ad sinum arcus perpendicularis ex eo protracti ad circulum alterum est, ut proportio sinus arcus comprehensi inter punctum alium

N 3 &punctū

If △ *ABG* has two unequal obtuse angles *B* and *G* such that *G* is greater than ∠ *B*, then side *AB* is greater than a quadrant.

The arcs erected perpendicularly from points *B* and *G* will meet within △ *ABG* at point *D* — namely, the pole of circle *BG*. The two arcs *BA* and *AG* will be greater than the two arcs *BD* and *DG*, which [together] will equal a semicircumference. Therefore the two arcs *BA* and *AG* [together] will exceed a semicircumference. Hence the larger of them — namely, arc *AB* — will be greater than a quadrant. Q.E.D.

Theorem 14

If any triangle has two acute angles and the side opposite one of them is not less than a quadrant, the other [third] angle of [the triangle] will be obtuse and the side opposite it will be greater than a quadrant.

If *ABG* is a triangle which has two acute angles *B* and *G* and whose side *AB*, opposite one of the acute [angles], is not less than a quadrant, then ∠ *A* of [the triangle] will be obtuse and side *BG* will be greater than a quadrant.

Let two right angles be formed at the two points *B* and *G* when two arcs are drawn meeting outside △ *ABG* at point *D*. [Point *D*] is necessarily the pole of circle *BG*. And let a quadrant *DAE* be drawn from pole *D* through point *A*, falling upon arc *BG* at point *E*. Then, since arc *AG** is not less than a quadrant, it will be either a quadrant or greater than [a quadrant]. If it is a quadrant,

by Th. [5] above one of the two sides *AE* and *EB* of right △ *AEB* will be a quarter of the circumference. But arc *AE* is less than a quadrant; therefore the other [arc], *EB*, will necessarily be a quadrant. Therefore the total arc *BG* will be greater than a quadrant. Similarly, by Th. [7] above, we may prove ∠ *BAE* to be a right [angle] since ∠ *ABE* is acute by the hypothesis; thus the total ∠ *BAG* is greater than a right [angle]. But if side *AB* is greater than a quadrant, by Th. [5] above arc *BE* will be greater than a quadrant since the other [arc], *AE*, is less than a quadrant. Therefore arc *BG* will be much greater than a quadrant. Similarly, by Th. [7] above, ∠ *BAE* must be obtuse since ∠ *ABE* is acute by the hypothesis. Thus ∠ *BAG* will be considerably more obtuse. Q.E.D.

Theorem 15

If, in a sphere, two great circles are inclined toward each other, and if two points are marked on the circumference of one of them or one point [is marked] on the circumference of each, and if a perpendicular arc is drawn from either one of the points to the circumference of the other circle, then the ratio of the sine of the arc that is between one of those points and the point of intersection of the circles to the sine of the perpendicular arc extended from that [same marked point] to the other circle is as the ratio of the sine of the arc found between the other [marked] point

*For *AG* read *AB*.

& punctum ſectionis ad ſinum arcus producti ex illo puncto.

Non abſterreat obſecro te uerboſa præſens propoſitio,& primo aſpectu íntri
cata:rebus enim Mathematicis uix ſatis lucidum,ne dixerim uenuſtũ, accomoda
bis ſermonem;fructũ profecto dulciſſimũ hac arbore ʒʒuis rigida decerpes, quem
ubi perſenſeris,totũ fermē præſentē librum intelliges.Sint igiſ duo circuli magni

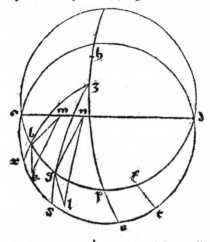

in ſphæra ad ſe inuicem inclinati a b g
d & a e d,quoʒ circumferentiæ ſecent
ſe in punctis a & d,ſignenſʒ duo pun
cta b g in circumferentia circuli a b g
d,à quibus deſcendant duo arcus ppen
diculares b r,g s ad circumferentiam
circuli a e d.Dico ꝙ proportio ſinus
arcus a b ad ſinum arcus b r eſt ut ſi
nus arcus a g ad ſinum arcus g s.A
punctis enim b & g binæ perpendicu
lares rectæ demittãtur,unæ quidem ad
ſectiõem communem circuloʒ ſcilicet
lineam a d,quæ ſint b m,g n,alteræ
uero ad ſuperficiem circuli a e d,quæ
ſint b k & g l ductis lineis k m & l n
quoniam itacʒ duæ lineæ m b,b k an

gulariter coniunctæ,æquediſtant duabus n g,g l angulariter coniunctis,b
m enim æquediſtat lineæ g n per primi,b k autem ipſi g l per undeci
mi,erit angulus m b k æqualis angulo n g l,uterʒ autem anguloʒ b k m &
g l n rectus eſt ex diffinitione lineæ perpendicularis ad ſuperficiem,quare per 32
primi duo trianguli m b k & n g l ſunt æquianguli,& ideo per 4.ſexti permuta
tim arguendo proportio m b ad b k eſt,ut proportio n g ad g l.eſt autem m
b ſinus rectus arcus a b per tertij & diffinitionem,n g uero ſinus arcus a g,
item b k ſinus arcus b r,g l autem ſinus arcus g s.proportio igitur ſinus arcus
a b ad ſinum arcus b r eſt ut ſinus arcus a g ad ſinum arcus g s.Habemus er
go partem propoſitionis ueram,quando duo puncta in una circũferentia ſignan
tur.Signemus ea demum in duabus circumferentijs dictorũ circuloʒ,ſtʒ unus b
in circumferentia circuli a b g d,reliquus autem t in circumferentia circuli a e
d,deſcendat à puncto t perpendicularis arcus t x ad circumferentiam circuli a
b g d.Dico ꝙ proportio ſinus arcus a b ad ſinum arcus b r eſt,ut proportio ſi
nus arcus d t ad ſinũ arcus t x.Sint enim poli duorũ circuloʒ h & 3,per quos
tranſeat circulus magnus h 3 p e ſecans circumferentias dictorũ circuloʒ in pũ
ctis p & e.In circumferentia itacʒ circuli a b g d ſignata ſunt duo puncta b &
p,à quibus deſcendunt duo ppendiculares b r & p e,quare proportio ſinus arcus
a b ad ſinum arcus b r eſt ut ſinus arcus a p ad ſinum arcus p e.Rurſus cum in
circumferentia circuli a e d ſignata ſint duo puncta e & t,à quibus duo perpen
diculares oriuntur e p & t x,erit proportio ſinus arcus d t ad ſinum arcus t x,
ſicut ſinus arcus d e ad ſinum arcus e p.eſt autem proportio ſinus arcus d e ad
ſinum arcus e p,ſicut ſinus arcus a p ad ſinum eiuſdem arcus p e,uterʒ enim ar
cuũ a p & d e eſt quarta circumferentiæ unius,proportio igitur ſinus arcus a b
ad ſinum arcus b r eſt ut ſinus arcus d e ad ſinũ arcus t x,quæ fuere peragenda.
Sed forſitan adhuc animus pendet,attento ꝙ à quolibet puncto in ſuperficie ſphæ
rica præ-

and the point of intersection to the sine of the arc drawn from that [second marked] point.

I pray that the present verbose and intricate theorem should not frighten you away at first sight, for in things mathematical you will scarcely make the language be clear enough, much less graceful. Truly you will pluck the sweetest fruit from this tree, however unyielding; and when you have savored this [fruit], you will understand almost the entire present book.

Thus, if two great circles *ABGD* and *AED* in a sphere are inclined toward each other [and] their circumferences intersect each other at points *A* and *D*, and if on the circumference of circle *ABGD* two points *B* and *G* are marked, from which two perpendicular arcs *BR* and *GS* descend to the circumference of circle *AED*, then the ratio of the sine of arc *AB* to the sine of arc *BR* is as that of the sine of arc *AG* to the sine of arc *GS*.

From points *B* and *G* let two rectilinear perpendiculars apiece be drawn, one [pair], *BM* and *GN*, to the line of intersection of the circles — namely, line *AD* — and the other [pair], *BK* and *GL*, to the surface of circle *AED*. When lines *KM* and *LN* are drawn, and since the two lines *MB* and *BK*, [which are] joined at an angle, are parallel to the two [lines] *NG* and *GL*, [which are] joined at an angle, because *BM* is parallel to line *GN* by [Euclid] I.[not given] and *BK* to *GL* by [Euclid] XI.[not given], then ∠ *MBK* will be equal to ∠ *NGL*. Moreover, each of the angles *BKM* and *GLN* is a right [angle] from the definition of a line perpendicular to a surface. Therefore by [Euclid] I.32 the two triangles *MBK* and *NGL* are similar, and thus, by [Euclid] VI.4 with [its] reasoning rearranged, the ratio of *MB* to *BK* is as the ratio of *NG* to *GL*. Moreover, *MB* is the right sine of arc *AB* by [Euclid] III.[not given] and [by] definition, while *NG* is the sine of

arc *AG*. Similarly, *BK* is the sine of arc *BR* and *GL* is the sine of arc *GS*. Thus the ratio of the sine of arc *AB* to the sine of arc *BR* is as that of the sine of arc *AG* to the sine of arc *GS*. Therefore we have part of the proposition true when the two points are marked on one circumference.

Finally, if we mark these [points] on the two circumferences of the mentioned circles, and if *B* is the one on the circumference of circle *ABGD* and *T* is the other on the circumference of circle *AED*, [and] if a perpendicular arc *TX* descends from point *T* to the circumference of circle *ABGD*, then the ratio of the sine of arc *AB* to the sine of arc *BR* is as the ratio of the sine of arc *DT* to the sine of arc *TX*.

Let the poles of the two circles be *H* and *Z*, through which a great circle *HZPE* passes, intersecting the circumferences of the mentioned circles at points *P* and *E*. Then on the circumference of circle *ABGD* two points *B* and *P* are marked, from which two perpendiculars *BR* and *PE* descend. Therefore the ratio of the sine of arc *AB* to the sine of arc *BR* is as that of the sine of arc *AP* to the sine of arc *PE*. Again, since, on the circumference of circle *AED*, the two points *E* and *T* are marked, from which the two perpendiculars *EP* and *TX* originate, then the ratio of the sine of arc *DT* to the sine of arc *TX* will be as that of the sine of arc *DE* to the sine of arc *EP*. Moreover, this ratio of the sine of arc *DE* to the sine of arc *EP* is as that of the sine of arc *AP* to the sine of this same arc *PE*, for each of the arcs *AP* and *DE* is a quarter of one circumference. Therefore the ratio of the sine of arc *AB* to the sine of arc *BR* is as that of the sine of arc *DE** to the sine of arc *TX*. Q.E.D.

But perhaps even now the mind is uncertain; then from any point marked on the surface of the sphere,

*For *DE* read *DT*.

rica præter polum circuli cuiuslibet,extra tamen circumferentiam eius signato,
geminos demittere liceat arcus perpendiculares ad circumferentiam ipsius circu
li.possibile est enim per punctum signatū & polum circuli transire circulū alium
magnum tertij huius docente.circuli igitur hoc pacto descripti,duo arcus inter
punctū signatum & circumferentiam circuli iacentis intercepti, perpendiculares
erunt ex huius ad circumferentiam circuli iacentis.In figuratione itaq; præsen
ti à puncto b prodibit perpendicularis,unus ad partem inclinatiōis,reliquus au=
tem ad partem oppositam.Idem quoq; accidet cæteris punctis in utraq; circumfe
rentiarū signatis,uerum hoc non interturbabit syllogismū nostrum, nam hij duo
perpendiculares cum sint æquales semicircumferentiæ per tertij huius, commu
nis animi conceptio eundem ipsis communem donabit sinum.quicquid igitur de
uno arcuū prædicabit theorema nostrum,& de reliquo demonstratū habebit.

XVI.

**In omni triangulo rectangulo omnium laterum sinus ad sinus an=
gulorum,quos subtendunt, eadem est proportio.**

Sit triangulus a b g angulum b rectum habēs.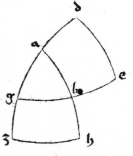
Dico ɋ, pportio sinus lateris a b ad sinum anguli a
g b eadem est proportioni sinus lateris b g ad sinū
anguli b a g ,& proportioni,quam habet sinus late=
ris a g ad sinum anguli a b g ,quod sic demonstra=
bimus.Necesse est, aut utrunq; angulog̃ a & g esse
rectum,aut alterū eog̃ duntaxat,aut nullum.Si uter=
q; eorum rectus est,erit per hypothesim & huius pū
ctus a polus circuli b g,b autem polus circuli a g
& g polus circuli a b,quare per diffinitionem unus=
quisq; trium arcuū dictog̃ determinabit quantitatem anguli se respicientis. idem
igitur erit sinus cuiuslibet trium later̃ & anguli sibi oppositi,& ideo sinus omniū
laterū ad sinus angulog̃ se respicientiū,proportionem eandem accipiunt uidelicet
æqualitatis.Si autem alter duntaxat angulog̃ a & g fuerit rectus , sit ille, uerbi
gratia,angulus g ,tradidit autem & hypothesis angulū b rectum,quare per hu
ius a est polus circuli b g ,& per huius uterq; arcuū b a,& a g quadrans cir=
cumferentiæ magnæ.per diffinitionem igitur unusquisq; arcuū a b ,b g & g a
quantitatem anguli se respicientis determinabit,eritq; idem sinus cuiuslibet late=
ris & anguli se respicientis conuertendo diffinitionem sinus anguli. unde postre=
mo constabit omnium laterum sinus ad sinus angulog̃ se respicientiū, eandem ha
bere proportionem scilicet æqualitatis.Qd' si neuter angulog̃ a & g rectus offe
ratur,non poterit aliquis trium arcuū lateralium esse quadrans circumferentiæ,
quemadmodū ex huius trahitur,uerum in triplici uarietate habebitur. Si enim
uterq; angulog̃ a & g fuerit acutus,erit per huius uterq; arcuū a b & b g mi
nor quadrante,unde & per huius arcus a g minor quarta circumferentiæ. Cre
scat igitur arcus g a uersus a donec fiet quadrans a d ,secundum cuius cordam
quæ est costa quadrati magni super puncto g facto polo describatur circulus ma
gnus,secans arcum g b prolongatū in puncto e .prolongetur demum arcus a g
ad partem puncti g ,donec quadrans habebitur a ʒ ,cuius corda super polo a cir
cumducta pariat circulū occurrentem arcui a b continuato in puncto h .Pinxi=
mus hoc pacto unam figurationem,Si uero uterq; angulorum a & g obtusum se
praebeat

besides the pole of a [selected] circle yet outside [its] circumference, one may drop a pair of perpendicular arcs to the circumference of this circle, because it is possible for another great circle to pass through the designated point and the pole of the circle by the teaching of Th. III.[15] above. Thus of the circle described in this way, the two arcs intercepted between the designated point and the circumference of the underlying circle will be perpendicular to the circumference of [that] underlying circle by Th. [1] above. Therefore, in the present construction, a perpendicular will proceed from point *B*, one [arc of this perpendicular proceeding] in the direction of the inclination and the other in the opposite direction. The same also happens with the other points marked on each of the circumferences. However, this will not confuse our deduction, for, since these two perpendiculars are equal to a semicircumference by Th. III.[19] above, an axiom will give them a common sine. Therefore whatever our theorem predicts concerning one of the arcs will be proven of the other also.

Theorem 16

In every right triangle the ratio of the sines of all the sides to the sines of the angles which [the sides] subtend is the same.

If *ABG* is a triangle with right ∠ *B*, then the ratio of the sine of side *AB* to the sine of ∠ *AGB* is the same as the ratio of the sine of side *BG* to the sine of ∠ *BAG* and as the ratio which the sine of side *AG* has to the sine of ∠ *ABG*, which we will prove thus.

It is necessary that each of the angles *A* and *G* be a right angle or that only one of them [be a right angle] or that none [of them be a right angle]. If each of them is a right angle, then, by the hypothesis and Th. [2] above, point *A* will be the pole of circle *BG*, *B* will be the pole of circle *AG*, and *G* will be the pole of circle *AB*. Therefore by definition each of the three mentioned arcs will determine the size of the angle opposite it.

Therefore the sine of any one of the three sides and [the sine] of the angle opposite that [side] will be the same, and thus the sines of all the sides to the sines of the angles opposite them will have the same ratio — namely, [that of] equality.

But if only one of the angles *A* and *G* is a right [angle], let that [angle] be ∠ *G*, for example. Moreover, the hypothesis gave ∠ *B* [to be] a right [angle]. Therefore, by Th. [2] above, *A* is the pole of circle *BG*, and, by Th. [1] above, each of the arcs *BA* and *AG* is a quadrant of the great circumference. Then by definition each of the arcs *AB*, *BG*, and *GA* will determine the size of the angle opposite itself, and, by the converse of the definition of the sine of an angle, the sine of any side and the angle opposite it will be the same. Hence, finally, it will be established that the sines of all the sides to the sines of the angles opposite them have the same ratio — namely, [that] of equality.

But if neither of the angles *A* and *G* is given [to be] a right [angle], some one of the three arcs [comprising] the sides cannot be a quadrant of the circumference, as was concluded from Th. [3] above; indeed, they may be in a triplicity [of combinations]. For if each of the angles *A* and *G* is acute, by Th. [3] above each of the arcs *AB* and *BG* will be less than a quadrant, and thus by Th. [6] above arc *AG* will be less than a quarter of the circumference. Therefore let arc *GA* extend beyond *A* until it becomes quadrant *AD*,* [and] with its chord [*GD*], which is the side of the great square, [as polar line] let a great circle be described around point *G* as pole, intersecting arc *GB* extended at point *E*. Finally, let arc *AG* be extended beyond point *G* until it becomes quadrant *AZ*,† whose chord [as polar line], rotated around pole *A*, produces a circle that meets arc *AB* extended at point *H*. In this way we have drawn one construction.

But if each of the angles *A* and *G* is obtuse,

*For *AD* read *GD*.
†In the figure, on p. 224, for arc *hh* read *zh*.

præbeat,per media nuperrime commemorata uterc̄p arcuū a b & b g quadrantem superabit,arcus autem a g quadrante minorem confitebimur . prolongato
igitur ut antea utrínc̄p arcu a g ,donec.& arcus g d quarta nascetur & arcus a 3,

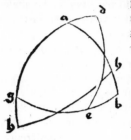

super polis g & a duo circuli magni describantur, quoc̄p
unius super g scilicet descripti circumferentia necessario
secabit arcum g b maiorem quadrante,quod fiat in pun
cto e ,reliqui uero super a descripti ,secabunt arcū a b,
quod contingat in puncto h .Sic altera surgit figuratio.
Qd̄ si alter angulo₃ a & g obtusus fuerit, reliquus autē
acutus, sit a obtusus & g acutus,erit itac̄p ex pallegatis
locis uterc̄p arcuū b g & g a quadrante maior,arcus au
tem a b minor eo.abscindant ergo ex arcu a g duo qua
drātes g d & a 3 ,in arcu d 3 participantes, descriptóc̄p
circulo ut prius super g polo,circumferentia eius secabit arcum b g maiorē qua
drante,quod fiat in puncto e ,circumferentia autem circuli super a descripti,non
secabit arcum a b,cum sit minor quadrante,sed occurret ei quantū sat est porrecto,quod fiat in puncto h .Neutro igitur angulo₃ a & g recto existente,tametsi

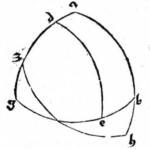

figuratióe triplici utamur,syllogismus tamē erit uni-
cus.Cum enim duo circuli g d & g e ad se inclinati
sint,& in circumferentia circuli g d signent duo pūcta,ex quibus duo perpendiculares a b & d e descen
dunt,erit per præcedentem proportio sinus arcus g a
ad sinū arcus a b sicut sinus arcus g d ad sinū arcus
d e,& permutatim sinus g a ad sinū g d ,sicut sinus
a b ad sinū d e.Item duo circuli a 3 & a h ad se inclinati sunt,signanturc̄p duo puncta g & 3 in circūferentia circuli a 3 ,ex quibus descendunt duo arcus
perpendiculares g b & 3 b ,quare per præmissam ,p
portio sinus a g ad sinum g b est tanc̄p sinus a 3 ad sinum 3 h ,& permutatim
sinus a g ad sinum a 3 sicut sinus g b ad sinum 3 h .est autem sinus a g ad sinum a 3 ,sicut sinus g a ad sinū g d ,c̄p uterc̄p arcuū a 3 & g d quadrans habea
tur.Sinus igitur lateris a b ad sinum d e, & sinus lateris b g ad sinum 3 h ;ean-
dem habent proportionē,quā uidelicet sinus lateris a g ad sinum quadrantis. Si
nus autē d e est sinus anguli a g b, arcus enim d e derterminat quantitatē anguli a g b, puncto g polo existente circuli d e.Similiter sinus 3 h est sinus angu
li b a g.sinus autē quadrantis est sinus anguli recti.quare sinus lateris a b ad sinum anguli a g b,& sinus lateris b g ad sinum anguli b a g ,sinus quoc̄p lateris a g ad sinum anguli recti a b g unam & eandem suscipiunt proportionem,
quod erat declarandum.

XVII.

In omni triangulo non rectangulo sinus laterum ad sinus angulo
rum eis oppositoru eandē habent proportionem.

Quam præcedens de triangulis rectangulis prædicabat passionem,præsens
quoc̄p de triangulis non rectangulis enunciat.Sit igitur triangulus a b g, nullū
habens rectum angulū.Dico c̄p sinus lateris a b ad sinum anguli g , sinusc̄p late
ris a b ad sinum anguli b,& sinus lateris b g ad sinum anguli unā & eandē pro
portionem accipiunt.Demittam em ex puncto a perpendicularem a d inciden
tem

then, by the method just mentioned, each of the arcs *AB* and *BG* will exceed a quadrant, while arc *AG* will be found less than a quadrant. Then when arc *AG* is prolonged on both sides, as before, until arc *GD*, as well as arc *AZ*, becomes a quarter [of a circle], let two great circles be drawn around poles *G* and *A*. The circumference of one of these [circles] — namely, [that] described around *G* — will necessarily intersect arc *GB*, [which is] greater than a quadrant, at point *E*; and [the circumference] of the other [circle] described around point *A* will intersect arc *AB* at point *H*. Thus the [second] construction arises.

But if one of the angles *A* and *G* is obtuse and the other acute, let *A* be obtuse and *G* acute; then, from the previously cited theorems, each of the arcs *BG* and *GA* will be greater than a quadrant and arc *AB* will be less than [a quadrant]. Then let two quadrants *GD* and *AZ*, sharing in arc *DZ*, be intercepted from arc *AG*. When a circle is described, as before, around *G* as pole, its circumference will intersect arc *BG*, greater than a quadrant, at point *E*; furthermore, the circumference of the circle described around *A* will not intersect arc *AB*, since [*AB*] is less than a quadrant, but [the circle] will meet [arc *AB*], extended as much as is sufficient, at point *H*.

Thus, when neither of the angles *A* and *G* is a right [angle], although we use a triplicity of constructions, the deductions will nonetheless be one and the same [for all three]. For since the two circles *GD* and *GE* are inclined toward each other, and since on the circumference of circle *GD* two points are marked from which two perpendiculars *AB* and *DE* descend, then, by the preceding [theorem], the ratio of the sine of arc *GA* to the sine of arc *AB* is as that of the sine of arc *GD* to the sine of arc *DE*, and by recombination, the sine of *GA* to the sine of *GD* is as the sine of *AB* to the sine of *DE*. Similarly, the two cir-

cles *AZ* and *AH* are inclined toward each other, and on the circumference of circle *AZ* two points *G* and *Z* are marked, from which two perpendicular arcs *GB* and *ZB* descend. Therefore, by the preceding [theorem], the ratio of the sine of *AG* to the sine of *GB* is as that of the sine of *AZ* to the sine of *ZH*, and by recombination, the sine of *AG* to the sine of *AZ* is as the sine of *GB* to the sine of *ZH*. Moreover, the sine of *AG* is to the sine of *AZ* as the sine of *GA* to the sine of *GD* because each of the arcs *AZ* and *GD* is a quadrant. Thus the sine of side *AB* to the sine of *DE* and the sine of side *BG* to the sine of *ZH* will [both] have the same ratio — namely, that of the sine of side *AG* to the sine of a quadrant. Furthermore, the sine of *DE* is the sine of ∠ *AGB*, for arc *DE* determines the size of ∠ *AGB* since point *G* is the pole of circle *DE*. Similarly the sine of *ZH* is the sine of ∠ *BAG*. Moreover, the sine of the quadrant is the sine of the right angle. Therefore the sine of side *AB* to the sine of ∠ *AGB*, and the sine of side *BG* to the sine of ∠ *BAG*, and also the sine of side *AG* to the sine of right ∠ *ABG* [all] have one and the same ratio. Q.E.D.

Theorem 17

In every nonright triangle the sines of the sides to the sines of the angles opposite those [sides] will have the same ratio.

What proof the preceding [theorem] indicated concerning right triangles, the present [theorem] states also concerning nonright triangles. Thus if *ABG* is a triangle having no right angle, then the sine of side *AB* to the sine of ∠ *G*, and the sine of side *AB** to the sine of ∠ *B*, and the sine of side *BG* to the sine of angle [*A*, all] will have one and the same ratio.

From point *A* let perpendicular *AD* be dropped, falling upon

*For *AB* read *AG*.

tem arcui b g si intra triangulū manserit, aut occurren=
tem arcui b g oportune prolongato si extra triangulum
ceciderit, quæ neutri arcuum a b & a g sibi contermi=
nalium coincidere poterit. sic em alter anguloꝶ b & g
rectus haberetur, quē hypothesis nostra non rectum tra
didit. Cadat itacꝗ prius intra triangulū, distinguens du=
os triangulos a b d & a g d rectangulos, erit ergo per
præcedentē terminis permutatis proportio sinus a b ad
sinum a d, sicut sinus anguli a d b rectī ad sinum angu

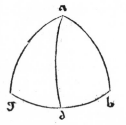

li a b d. sed & per eandem præmissam proportiō sinus a d ad sinum a g tancꝗ
sinus anguli a g d ad sinum anguli a d g recti. cūcꝗ sit idem sinus anguli a d g
& anguli a d b, cꝑ utercꝗ eoꝶ rectus est. erit per æquā proportionalitatē indirectā
sinus a b ad sinum a g ueluti sinus anguli a g b ad sinum anguli a b g, & per=
mutatim sinus lateris a b ad sinum anguli a g b tancꝗ sinus lateris a g ad sinū
anguli a b g. Eandem denicꝗ concludes proportionē sinus lateris b g ad sinum
anguli b a g, si prius ab altero punctoꝶ angularium b & g ad latus sibi opposi
tum demiseris arcum perpendicularē. Quod si perpendicularis a d extra trian=
gulum ceciderit mutata parumper figuratione pristinum repetemus syllogismū.
Erit em ex præmissa permutatim arguendo sinus a b ad
sinum a d tancꝗ sinus anguli a d b ad sinum anguli a b
d recti. itemꝗ sinus a d ad sinum a g, sicut sinus anguli
a g b ad sinum anguli a d g recti. quare per æquam indi
rectam erit sinus lateris a b ad sinum lateris a g, sicut si=
nus anguli a g b ad sinum anguli a b d. sinus autem an=
guli a b d est etiam sinus anguli a b g per communem
scientiam. sinus igitur a b ad sinum a g est ut sinus an=
guli a g b ad sinum anguli a g b, & ideo permutatis ter=

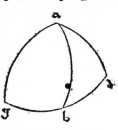

minis, erit sinus lateris a b ad sinum anguli a g b, tancꝗ sinus lateris a g ad sinū
anguli a b g. Hanc demū conuincemus esse proportionem sinus lateris b g ad
sinum anguli b a g, quemadmodū hactenus processimus. Passionem igitur quæ
de triangulis rectangulis & non rectangulis singulatim in hisce demonstrabatur
theorematibus, communiter tandem de omnibus triangulis qualescuncꝗ fuerint
concludere licebit, quæ res quantos & cꝗ iucundos allatura sit fructus pedetentim
intuebimur.

In omni triangulo unicum habente rectum angulum, proportio si=
nus anguli nō recti ad sinum anguli recti est, ut sinus complementi an
guli reliqui ad sinum complementi lateris eum subtendentis.

Sit triangulus a b g, cuius angulus b quidem sit rectus, neuter autem duoꝶ
a & g rectus habeatur. Dico cꝑ proportio sinus anguli b a g ad sinum anguli b
recti est, ut sinus complementi anguli a g b ad sinum complementi lateris a b,
similiter sinus anguli a g b ad sinum anguli b recti se habet, tancꝗ sinus comple=
menti anguli b a g ad sinum complementi arcus b g, quod sic demōstrabimus.
Quoniam neuter angulorum a & g est rectus, erit aut utercꝗ eorum acutus uel
obtusus, aut alter acutus, & reliquus obtusus. Sit primo utercꝗ acutus, quamobre
per huius utruncꝗ laterū a b & b g minus erit quadrante, ideocꝗ per huius
O   & arcus

arc *BG* if [*AD*] is found [to be] within the triangle, or meeting arc *BG*, necessarily extended, if [*AD*] falls outside the triangle. [*AD*] cannot coincide with either of the arcs *AB* and *AG* adjacent to it, for in that case one of the angles *B* and *G* would be a right [angle], which our hypothesis gives [to be] a nonright [angle].

Thus, first, let [it] fall within the triangle, determining two right triangles *ABD* and *AGD*. Then, by the preceding [theorem], with the terms changed around, the ratio of the sine of *AB* to the sine of *AD* is as that of the sine of right ∠ *ADB* to the sine of ∠ *ABD*. Furthermore, by the same previous [theorem], the ratio of the sine of *AD* to the sine of *AG* is as that of the sine of ∠ *AGD* to the sine of right ∠ *ADG*. And since the sine of ∠ *ADG* is the same [as the sine] of ∠ *ADB* because each of those [angles] is a right [angle], then, by indirect equal proportionality, the sine of *AB* to the sine of *AG* is as the sine of ∠ *AGB* to the sine of ∠ *ABG*; and, by a rearrangement, the sine of side *AB* to the sine of ∠ *AGB* is as the sine of side *AG* to the sine of ∠ *ABG*. Finally, you may infer the same ratio for the sine of side *BG* to the sine of ∠ *BAG* if first you would drop a perpendicular arc from one of the vertices *B* and *G* to the side opposite it.

But if perpendicular *AD* falls outside the triangle, we will return to the previous deduction, with a slight change in the construction. For, from the preceding proof by an interchange [of terms], the sine of *AB* to the sine of *AD* will be as the sine of right ∠ *ADB* to the sine of ∠ *ABD*.* And similarly the sine of *AD* to the sine of *AG* is as the sine of ∠ *AGB* to the sine of right ∠ *ADG*. Therefore, by indirect equal [proportionality], the sine of side *AB* to the sine of side *AG* will be as the sine of ∠ *AGB* to the sine of ∠ *ABD*. Moreover, the sine of ∠ *ABD* is also the sine of ∠ *ABG* by axiom. Therefore the sine of *AB* to the sine of *AG* is as the sine of ∠ *AGB*

to the sine of ∠ *AGB*,† and thus, by an interchange of the terms, the sine of side *AB* to the sine of ∠ *AGB* will be as the sine of side *AG* to the sine of ∠ *ABG*. Finally, we may prove that this is the ratio of the sine of side *BG* to the sine of ∠ *BAG*, just as we have proceeded up to this point.

Thus the proof concerning right triangles and nonright [triangles], which was shown individually in these theorems, may at last be inferred concerning all triangles, no matter what kind they are. This fact will gradually bring fruits of much size and succulence for us to consider.

Theorem 18

In every triangle having only one right angle, the ratio of the sine of a nonright angle to the sine of the right angle is as the sine of the complement of the remaining angle to the sine of the complement of the side subtending that [angle].

If *ABG* is a triangle whose ∠ *B* is a right [angle] but neither [of whose other] two, *A* and *G*, is a right [angle], then the ratio of the sine of ∠ *BAG* to the sine of right ∠ *B* is as that of the sine of the complement of ∠ *AGB* to the sine of the complement of side *AB*. Similarly, the sine of ∠ *AGB* to the sine of right ∠ *B* will be as the sine of the complement of ∠ *BAG* to the sine of the complement of arc *BG*, which we shall prove in this way.

Because neither of the angles *A* and *G* is a right [angle], each of them will be either acute or obtuse, or one of them will be acute and the other obtuse. First, let each be acute; whereupon by Th. [3] above each of the sides *AB* and *BG* will be less than a quadrant, and thus by Th. [6] above

*In the Latin text, for *anguli adb ad sinum anguli abd recti* read *anguli adb recti ad sinum anguli abd*.

†For *AGB* read *ABG*.

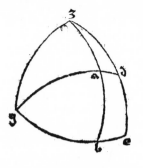

& arcus a g minor exiftet quadrante, protendat
ergo g a ad partem puncti a, donec fiet quadrans
g d, fecundum cuius cordā fuper g facto polo circumducatur circulus fecans arcum g b fatis prolongatum in puncto e. arcus deniq̃ b a ad parté
puncti a porrectus coincidat circumferentiæ circuli fuper g polo circumducti in pūcto 3, qué ex
huius conftabit effepolum circuli g b e. per autem huiuṣ uterq̃ arcuū 3 b & 3 e quadrans habe-
bĭtur, & angulus 3 e g rectus. quare per diffinitionem arcus a 3 erit complementum lateris a b, arcus autem d 3 cōplementum arcus d e. cūq̃ arcus
d e determinet quantitatem anguli a g b, puncto g polo exiftéte circulo 3 d e,
determinabit arcus 3 d quantitatem cōplementi anguli a g b, quod facilius ducto arcu g 3 cōfiteberis. Eft autem proportio finus anguli b a g ad finum anguli a b g recti per   huius conuerfim arguendo, ficut proportio finus b g ad finū
g a. finus demum b g ad finum g a, ficut finus d 3 ad finum 3 a per   huius.
quoniam in circumferentijs duoṛ circuloṛ g a d & 3 a b fignata funt duo pūcta
g & 3, à quibus alternatim defcendūt duo arcus perpendiculares ad circumferentias huiufmodi circuloṛ, g b quidé propter angulum a b g rectum ex hypothefi, 3 d auté propter angulum 3 d g rectum, g polo circuli 3 d exiftente. Sinus auté 3 d eft finus cōplementi anguli a g b, & arcus a 3 eft complementum lateris
a b. Proportio igitur finus anguli b a g ad finum anguli a b g recti, eft tanq̃
finus cōplementi anguli a g b ad finum complementi lateris eum fubtendentis.

Quòd fi fuerit uterq̃ anguloṛ a & g obtufus per
huius utrūq̃ lateṛ a b & b g quadranté fuperabit, unde & latus a g minus quarta enunciabitur. Crefcat igĭt arcus g a, donec arcus g d
per a punctū incedens fiet quadrans, fecūdū cuius cordā defcribatur circulus magnus, fecans ne
ceffario arcū g b, quod fiat in puncto e, fecabit
aūt circulus ille neceffario arcum a b in puncto
qui fit 3. per itaq̃ huius duo arcus g e & e 3
perpendiculariter fibi inuicem infiftent, cūq̃ per
hypothefim arcus a b perpendicularis fit ad arcum b g angulo b recto exiftente, erit per   huius 3 polus circuli b g, & per corollarium eiufdé

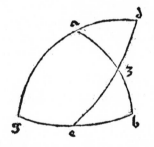

uterq̃ arcuum 3 b & 3 e quadrans circumferentiæ. quare per diffinitionem arcus a 3 erit complementum arcus a b, arcus etiam 3 d determinabit compleme
tum anguli a g b, quod apertum uidebitur fi arcū g 3 produxerimus. erit enim
angulus h g 3 rectus, arcu 3 b quadrante exiftente, finus ergo arcus 3 d erit finus complementi anguli a g b. Cum itaq̃ in circumferentijs duorum circuloṛ
a g & a b ad fe inclinatorum fignauerimus duo puncta g & 3, à quibus defcen-
dunt duo perpendiculares, g b quidem ad arcum a b propter angulum b rectū
ex hypothefi, 3 d autem ad arcum g d propter g polū circuli d 3, erit per   huius conuerfis terminis proportio finus arcus b g ad finum arcus g a, taq̃ finus d
3 ad finum arcus 3 a, uidelicet tanq̃ finus complementi anguli a g b ad finum
complementi lateris a b, finus autem arcus b g ad finum arcus g a fe habet per
huius

arc *AG* will also be less than a quadrant. Then let *GA* be extended beyond point *A* until it becomes quadrant *GD*. With this chord [*GD* as polar line] around *G* as the pole, let a circle be drawn, intersecting arc *GB*, sufficiently extended, at point *E*. Next let arc *BA*, extended beyond point *A*, meet the circumference of the circle that was drawn around pole *G* at point *Z*. From Th. [2] above, this [point *Z*] is established to be the pole of circle *GBE*. Moreover, by Th. [1] above, each of the arcs *ZB* and *ZE* will be a quadrant and ∠ *ZEG* will be a right [angle]. Therefore, by definition, arc *AZ* will be the complement of side *AB* and arc *DZ* will be the complement of arc *DE*. And since arc *DE* determines the size of ∠ *AGB* because point *G* is the pole of circle *ZDE*, arc *ZD* will determine the size of the complement of ∠ *AGB*, which you may easily show when arc *GZ* is drawn. Moreover, the ratio of the sine of ∠ *BAG* to the sine of right ∠ *ABG* is as the ratio of the sine of *BG* to the sine of *GA* by the converse reasoning of Th. [16] above. Next, the sine of *BG* to the sine of *GA* is as the sine of *DZ* to the sine of *ZA* by Th. [15] above, because, on the circumferences of the two circles *GAD* and *ZAB*, two points *G* and *Z* are marked, from which, in turn, two perpendicular arcs descend to the circumferences of these circles — *GB* [being perpendicular] because ∠ *ABG* is a right [angle] by hypothesis and *ZD* [being perpendicular] because ∠ *ZDG* is a right [angle] since *G* is the pole of circle *ZD*. Furthermore, the sine of *ZD* is the sine of the complement of ∠ *AGB* and arc *AZ* is the complement of side *AB*. Thus the ratio of the sine of ∠ *BAG* to the sine of right ∠ *ABG* is as that of the sine of the complement of ∠ *AGB* to the sine of the complement of the side subtending it.

But if each of the angles *A* and *G* is obtuse,

by Th. [3] above each of the sides *AB* and *BG* will exceed a quadrant. Hence side *AG* will be declared less than a quarter [of a circle]. Then let arc *GA* increase until arc *GD*, passing through point *A*, is a quadrant. With its chord [*GD* as polar line], let a great circle be described, necessarily intersecting arc *GB* at point *E*. Moreover, that circle will necessarily intersect arc *AB* [also] at point *Z*. Therefore, by Th. [1] above, the two arcs *GE* and *EZ* rest on each other perpendicularly. And since, by the hypothesis, arc *AB* is perpendicular to arc *BG* because ∠ *B* is a right [angle], then, by Th. [2] above, *Z* will be the pole of circle *BG*, and, by the corollary of the same [theorem], each of the arcs *ZB* and *ZE* will be a quadrant of the circumference. Therefore, by definition, arc *AZ* will be the complement of arc *AB*. Arc *ZD* also will determine the complement of ∠ *AGB*; this would appear evident if we were to draw arc *GZ*, for ∠ *HGZ** will be a right [angle] since arc *ZB* is a quadrant. Therefore the sine of arc *ZD* will be the sine of the complement of ∠ *AGB*. Then since, on the circumferences of the two circles *AG* and *AB* that are inclined toward each other, we have marked two points *G* and *Z*, from which two perpendiculars descend — *GB* [being perpendicular] to arc *AB* because ∠ *B* is a right [angle] by the hypothesis and *ZD* [being perpendicular] to arc *GD* because *G* is the pole of circle *DZ* — then, by Th. [15] above with the terms reversed, the ratio of the sine of arc *BG* to the sine of arc *GA* will be as that of the sine of *DZ* to the sine of arc *ZA* — namely, as the sine of the complement of ∠ *AGB* to the sine of the complement of side *AB*. Furthermore, the sine of arc *BG* to the sine of arc *GA* is, by Th. [16] above

*For *HGZ* read *BGZ*.

huius permutato terminorum situ, tanქ finus anguli b a g ad finum anguli a
b g recti,& ideo finus anguli b a g ad finum anguli a b g recti fe habet,ficut fi
nus complementi anguli a g b ad finum complementi lateris a b ipfum fubten
dentis.Poftremo alter anguloჰ a & g fit obtufus, uerbi
gratia,angulus a , & reliquus g acutus,erit itaქ per me
dia fupradicta arcus a b minor quadrante, uterქ uero
arcuum a g & g b quadrantem fuperabit. refecabo igi
tur ex arcu a g quadrantem g d ,defcriptoქ circulo fe
cundum quantiatem g d fuper g polo , circumferentia
eius neceffario concurrens cu arcu g b quadrante fupe
rante fecabit arcum a b,quod fiat in puncto 3,qუ opor
tet effe polუ circuli b g,per huius propter binos angu
los apდ b & e rectos ,unde & ex corollario eiufდ uterქ
arcuum 3 b & 3 e quadrans conuincetur. Cuქ arcus d e
quantitatე anguli a g e determinet,arcu 3 e quadrante

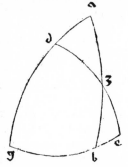

exiftente,arcus d 3 quantitatე cოplemეti anguli a g b determinabit,quod haud
incertum affirmabis,ubi arcუ g 3 produxeris.fimiliter a 3 erit cოplementუ late
ris a g.In circუferentijs autე duoჰ circuloჰ a g & a b ad fe inclinatoჰ (eifdem
enim characteribus nunc arcus nunc circulos fuos more noftro reprაfentamus)
fignata funt duo puncta g & 3,ა quibus alterni egrediuntur perpendiculares g b
& 3 d,quod nო nifi ex locis cომemoratis oftendemus, & tandem fyllogifimo freti
priftino concludemus finum anguli b a g fe habere ad finum anguli a b g recti
tanქ finუ cოplementi anguli a g b ad finum cოplementi arcus eum fubtenden
tis,quod iuuat attigiffe.Non aliter procedemus angulუ a g b uice anguli b a g
affumentes, cაteris ut res ipfa poftulat commutatis.

XIX.

In omni triangulo,cui unicus eft rectus angulus,finus cოplementi
lateris rectუ fubtendentis angulუ ad finუ cოplementi alterius rectum
ambientiუ,eam habet proportionem, quა finus complementi reliqui
lateris ad finum quadrantis.

Sit triangulus fphაralis a b g angulუ b rectუ ha
bens,& utrumქ reliqnorუ non rectum.Dico ფ propor
tio finus complementi lateris a g ad finum complemen
ti lateris a b, eft ut finus complementi lateris b g ad to
tum fcilicet finum quadrantis. itemქ finus complemen
ti eiufdem lateris a g ad finum cოplementi lateris b g
tanქ finus cოplementi lateris a b ad finum quadrātis.
Huius demonftrationem afferemus fimplicე,tametfi tri
uaria foleat effe figuratio.Si enim uterქ anguloჰ a & g acutus fuerit,erit p me
dia in prაmiffis fაpenumero adducta,unუquodქ laterუ trianguli noftri minus
quadrante.Prolongetur ergo latus g a ad partem puncti a , donec fiet arcus g
d quarta circუferentiა,fecundum cuius cordა fuper g polo circulus defcriptus fe
cabit arcum g b fatis continuatum,quod fiat in puncto e . fecabit autem & arcუ
b a prolongatუ,quod fiat in puncto 3.Si deniქ uterქ anguloჰ a & g obtufum
fe prაbeat,erit uterქ arcუ a b & b g maior quadrante,arcus autem a g qua
drante minor,quo crefcente donec fiet quadrans a d, fecundum cordა ipfius de

O 2 fcribatun

with the position of the terms changed, as the sine of ∠ *BAG* to the sine of right ∠ *ABG*. Therefore the sine of ∠ *BAG* to the sine of right ∠ *ABG* is as the sine of the complement of ∠ *AGB* to the sine of the complement of side *AB* subtending [∠ *AGB*].

Finally, let one of the angles *A* and *G* be obtuse — for example, ∠ *A* — and the other, *G*, be acute. Then by the above-mentioned method, arc *AB* will be less than a quadrant while each of the arcs *AG* and *GB* will exceed a quadrant. Then, from arc *AG*, quadrant *GD* is intercepted; and when a circle is described around *G* as pole with length *GD* [as polar line], the circumference of this [circle], necessarily meeting with arc *GB* that exceeds a quadrant, will intersect arc *AB* at point *Z*. [Point *Z*] must be the pole of circle *BG* by Th. [2] above, because the two angles apiece at *B* and *E* are right [angles]. Hence, from the corollary of the same [theorem], each of the arcs *ZB* and *ZE* is proven [to be] a quadrant. Now, since arc *DE* determines the size of ∠ *AGE*, and since arc *ZE* is a quadrant, arc *DZ* will determine the size of the complement of ∠ *AGB*. You may prove this [to be] a complete certainty if you would draw arc *GZ*. Similarly, *AZ* will be the complement of side *AG*. Moreover, on the circumference of the two circles *AG* and *AB* that are inclined toward each other (for, according to our custom, we represent now the arc and now its circle by the same characteristics), two points *G* and *Z* are marked, from which, in turn, perpendiculars *GB* and *ZD* emerge. We may show that only through the theorems [already] mentioned. At last, by relying upon the earlier deduction, we will conclude that the sine of ∠ *BAG* is to the sine of right ∠ *ABG* as the sine of the complement of ∠ *AGB* is to the sine of the complement of the arc subtending it. This [conclusion] was desired to be attained. Taking ∠ *AGB* in place

of ∠ *BAG*, we will not proceed any differently if the rest [of the sides and angles] are changed as this situation demands.

Theorem 19

In every triangle in which only one angle is a right [angle], the sine of the complement of the side subtending the right angle to the sine of the complement of one [of the sides] including the right [angle] has that ratio which the sine of the complement of the [third] side has to the sine of a quadrant.

If *ABG* is a spherical triangle having right ∠ *B* and having each of the other angles a nonright [angle], then the ratio of the sine of the complement of side *AG* to the sine of the complement of side *AB* is as that of the sine of the complement of side *BG* to the whole sine — namely, the sine of a quadrant. Similarly, the sine of the complement of the same side *AG* to the sine of the complement of side *BG* is as the sine of the complement of side *AB* to the sine of a quadrant. We may bring to bear a straightforward proof of this although the construction usually is trivariant.

If each of the angles *A* and *G* is acute, then, by the method repeatedly invoked in the preceding [theorems], every one of the sides of our triangle is less than a quadrant. So let side *GA* be extended beyond point *A* until arc *GD* becomes a quarter of the circumference. The circle that is described with the chord of [*GD* as polar line], around *G* as the pole, will intersect arc *GB*, sufficiently extended, at point *E*. Moreover, it will intersect arc *BA*, extended, at point *Z*.

Next, if each of the angles *A* and *G* is obtuse, each of the arcs *AB* and *BG* will be greater than a quadrant while arc *AG* will be less than a quadrant. When [arc *AG*] is increased until it becomes a quadrant *AD*, then with the chord of this [quadrant] let

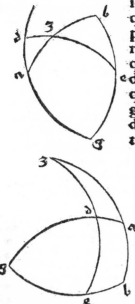

scribatur circulus magnus secans necessario arcum g b
quadrante maiore in puncto qui sit e, arcū autē a g in
puncto quem uocabimus 3. Et si postremo alter angulo
rum a & g obtusus fuerit, uerbi gratia angulus a, relí
quus autem acutus, erit uterᶜᵽ arcuum a g & g b qua
drante maior ex allegatis locis, arcus autem a b minor
quadrante. Abscindam igitur ex arcu g a quadrātem
g d, secandū cuius cordā super polo g circulū magnū
describo, qui secet arcū b g in puncto e, & arcū b a cō
tinuatum in puncto 3. Quibus ita ordinatis constabit p
huius 3 esse polum circuli b g propter binos angu
los apud b & e rectos, & per corollarium eiusdē utrū
ᶜᵽ arcuum 3 b & 3 e esse quadrantem, quare per diffi
nitionem arcus a 3 est complementū lateris a b, ar
cus aūt a d complementū lateris a g. & arcus b e cō
plementū lateris b g. Cunᶜᵽ duo circuli 3 b & 3 e ad se
inclinati sint, in quoᵹ unius scilicet 3 b circumferētia
signantur duo puncta a & b, ex quibus perpendicu
lares arcus ad circumferentiam alterius descendunt,
qui sunt a d & b e, propter binos angulos apud d &
e rectos g polo circuli 3 e existente, erit per huius
proportio sinus 3 a ad sinum a d, sicut sinus 3 b quadrantis ad sinum b e,
& permutatim sinus 3 a complementi, scilicet arcus a b ad sinum quadrantis 3
b tanᶜᵽ sinus a d complementi lateris a g rectum subtendentis angulū ad sinū
b e complemēti scilicet lateris b g rectū ambiētis, quod ᵽponebaᵗ cōfirmandū.

XX.

In omni triangulo non rectangulo perpendicularis à uertice angu=
li cuiuslibet ad latus sibi oppositū demissus cū duobus lateribus sibi
conterminalibus duos complectetur angulos, eritᶜᵽ proportio sinus
dextri ad sinum siniſtri eorum anguloru tanᶜᵽ sinus cōplementi angu
li dextri ipsius trianguli ad sinū cōplementi anguli siniſtri.

Angulum dextrū appello, quem facit perpendicularis cū latere trianguli dex
tro, Siniſtrum autem quē cum siniſtro. Sit triangulus a b g, nullū habens rectū

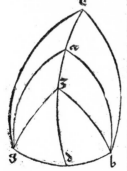

angulū, à cuius anguli uertice descendat arcus a d per
pendicularis ad arcum b g, cōplectens cum duobus late
ribus trianguli sibi cōterminalibus duos angulos, cum la
tere quidē a b dextro angulū b a d, cum latere autem si
niſtro a g angulū a g d. Dico ᵽ proportio sinus angu
li b a d ad sinū anguli g a d, est sicut ᵽportio sinus cōple
menti anguli a b g ad sinum cōplementi anguli a g b.
Perpendicularis em arcus aut cadit intra triangulū, aut
extra, cum neutri arcuum sibi conterminaliū coincidere
possit. sic naᶜᵽ alter angulorum b & g rectus fieret quē
hypothesis non rectum tradidit. unde etiam palam ᵽ per
pendicularis cum duobus lateribus duos continebit an
gulos. Cadat primum intra, & sit minor quadrante, quod quidē accidit dū uterᶜᵽ
angu

a great circle be described, necessarily intersecting arc *GB*, [which is] greater than a quadrant, at point *E* and [intersecting] arc *AG* at the point which we will call *Z*.

And finally, if one of the angles *A* and *G* were obtuse — ∠ *A*, for example — and the other acute, then each of the arcs *AG* and *GB* will be greater than a quadrant, from the [same] theorems cited [before], while arc *AB* will be less than a quadrant. Then, from arc *GA*, let quadrant *GD* be intercepted, with whose chord [as polar line] a great circle is described around pole *G*. This [circle] intersects arc *BG* at point *E* and arc *BA*, extended, at point *Z*.

Thus, with these [constructions] arranged, it will be established by Th. [2] above that *Z* is the pole of circle *BG* because the two angles apiece at *B* and *E* are right [angles], and by the corollary of the same [theorem] each of the arcs *ZB* and *ZE* is a quadrant. Therefore, by definition, arc *AZ* is the complement of side *AB*, arc *AD* is the complement of side *AG*, and arc *BE* is the complement of side *BG*. Since the two circles *ZB* and *ZE* are inclined toward each other, [and since] on the circumference of one of [them] — namely, *ZB* — two points *A* and *B* are marked, from which [two] perpendicular arcs *AD* and *BE* — [being perpendicular] because the two angles apiece at *D* and *E* are right [angles] since *G* is the pole of circle *ZE* — descend to the circumference of the other [circle *ZE*], then by Th. [15] above the ratio of the sine of *ZA* to the sine of *AD* is as that of the sine of quadrant *ZB* to the sine of *BE*. By a recombination, the sine of the complement of *ZA* — namely, [the sine] of arc *AB* — to the sine of quadrant *ZB* is as the sine of the complement of *AD*, [which is the sine] of side *AG* subtending the right angle, to the sine of the complement of *BE* — namely, [the sine] of side *BG* that includes the right [angle]. Q.E.D.

Theorem 20

In every nonright triangle, the perpendicular drawn from the vertex of any angle to the side opposite that [angle] will form two angles with the two sides adjacent to itself, and the ratio of the sine of the right-hand [one] of these angles to the sine of the left-hand [one] will be as the sine of the complement of the right-hand angle of this triangle to the sine of the complement of the left-hand angle.

I call the right-hand angle that [angle] which the perpendicular makes with the right-hand side of the triangle; the left-hand [angle is] that [which it makes] with the left-hand side. Let *ABG* be a triangle, having no right angle, from whose vertex [*A*] to arc *BG* a perpendicular arc *AD* descends, forming, with the two sides of the triangle adjacent to [*AD*], two angles — the one [formed] with the right-hand side *AB* [being] ∠ *BAD* while that [formed] with the left-hand side *AG* [is] ∠ *AGD*. Then the ratio of the sine of ∠ *BAD* to the sine of ∠ *GAD* is as the ratio of the sine of the complement of ∠ *ABG* to the sine of the complement of ∠ *AGB*.

The perpendicular arc will fall either inside or outside the triangle since it can coincide with neither of the arcs adjacent to itself, for, in that case, one of the angles *B* and *G*, which the hypothesis gave to be nonright, would be a right [angle]. Hence I may repeat that the perpendicular will form two angles with the two sides.

First, let [the perpendicular] fall within and be less than a quadrant, which happens when each of the

angulorum b & g acutus fuerit, quemadmodum ex huius trahitur, non poteſt
autem eſſe quadrans angulo à b g non exiſtente recto . Extendatur ergo d a uſcp ad e donec arcus d e quadrans habeatur, eritcp per huius e
polus circuli b g,â quo duobus punctis b & g occurrant duo arcus circuloꝝ ma
gnoꝝ e b & e g,quos oportet eſſe quadrantes perpendiculariter quidẽ arcui b g
inſiſtentes.eritcp per diffinitionem angulus a b e cõplementum anguli a b g,
& angulus a g e complementũ anguli a g b. Per autẽ huius conuerſis terminis eſt proportio ſinus anguli a b e ad ſinum arcus a e tan̄cp ſinus anguli e a b
ad ſinum quadrantis e b. & per eandem ſinus arcus a e ad ſinum anguli a g e,ſi
cut ſinus quadrantis e g ad ſinum anguli e a g . Sinus autem quadrantis e g eſt
etiam ſinus quadrantis e b,per æquã igiť ſinus anguli a b e ad ſinũ anguli a g e
proportionem habet,quã ſinus anguli e a b ad ſinum anguli e a g. Cuncp ſinus
anguli e a b ſit etiã anguli b a d, φ ſimiliter ſinus anguli e a g ſinus habeá anguli d a g, φ bini binis rectis æquipolleant, erit proportio ſinus anguli b a d
ad ſinũ anguli g a d,tan̄cp ſinus anguli a b e, ſcilicet cõplementi anguli a b g
ad ſinum anguli a g e,ſcilicet complementi anguli a g b,quod libuit efficere. Si
autem arcus a d quadrante maior extiterit,quod quidem euenit utrocp anguloru
b & g obtuſo exiſtente,abſcindatur ex eo quadrans z d,& à puncto z quẽ hu
ius polũ circuli b g demõſtrat,duo arcus procedãt duobus punctis b & g occur
ſuri,quos liquet eſſe quadrantes per huius orthogonaliter inſidentes arcui b g,
angulus igitur a b z eſt complementum anguli a b g per diffinitionem,itemcp
angulus a g z complementum anguli a g b diffinietur. Eſt autem per huius
conuerſim arguendo proportio ſinus anguli b a z ſiue b a d ad ſinum quadrantis z b,ueluti ſinus anguli a b z ad ſinum arcus a z.& per eandem ſinus quadrã-
tis z g ad ſinum anguli g a z ſiue g a d,tãcp ſinus arcus a z ad ſinũ anguli a
g z.per æquam igitur ſinus anguli b a d ad ſinum anguli g a d erit ut ſinus an
guli a b z,ſcilicet complementi anguli a b g ad ſinum anguli a g z,ſcilicet cõplementi anguli a g b.Cadat demum perpendicularis extra triangulum, altero
angulorum b & g obtuſo exiſtente,& reliquo acuto,quemadmodum ex huius
trahitur.Obuiabit autem perpendicularis arcui g b prolongato ad partem anguli obtuſi,qui uerbi gratia ſit b,eritcp minor quadrante
φ angulus a b d acutus habeatur . extẽdamus igitur eũ
donec quadrans fiet ad punctum quidem h terminatus,
quẽ oportet eſſe polum circuli b g huius arguente. ab
eo itacp polo duo arcus egrediantur h b & h g , qui e-
runt quadrantes orthogonaliter arcui b g incidentes p

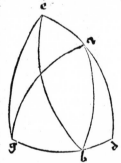

huius, quo fit ut angulus a b h complementũ anguli
a b g habeatur, anguluscp a g h cõplementum anguli
a g b. eſt autẽ per huius conuerſis terminis proportio
ſinus anguli b a h ad ſinum quadrantis h b ,tancp ſinus
anguli a b h ad ſinum arcus a h. & per eandem ſinus
quadrantis h g ad ſinum anguli h a g. ueluti ſinus ar-
cus a h ad ſinum anguli a g h. per æquã igitur proportioni ſinus anguli b a h
ad ſinum anguli h a g, æqualis eſt proportio ſinus anguli a b h ad ſinum angu
li a g h. ſinus autem anguli b a h eſt et ſinus anguli b a d . itemcp ſinus anguli g a h eſt ſinus anguli g a d. quare proportio ſinus anguli b a d ad ſinum an
guli g a d, tancp ſinus anguli a b h complementi uidelicet anguli a b g ad ſinũ anguli a g h cõplementi ſcilicet anguli a g b habebitur . Ratũ igitur exegi
mus quod proponebatur. O z Si quis

angles B and G is acute, just as was concluded from Th. [8] above. Moreover, it cannot be a quadrant because ∠ ABG is not a right [angle]. Then let DA be extended all the way to E until arc DE is a quadrant. By Th. [2] above E will be the pole of circle BG. From this [pole] to the two points B and G two arcs, EB and EG, of great circles run out, which [arcs] must be quadrants that rest perpendicularly on arc BG. By definition, ∠ ABE will be the complement of ∠ ABG and ∠ AGE will be the complement of ∠ AGB. Moreover, by Th. [17] above, when the terms are reversed, the ratio of the sine of ∠ ABE to the sine of arc AE is as that of the sine of ∠ EAB to the sine of quadrant EB. And by the same [theorem], the sine of arc AE to the sine of ∠ AGE is as the sine of quadrant EG to the sine of ∠ EAG. Furthermore, the sine of quadrant EG is also the sine of quadrant EB. Then by simultaneous solution, the sine of ∠ ABE to the sine of ∠ AGE has that proportion which the sine of ∠ EAB [has] to the sine of ∠ EAG. Now since the sine of ∠ EAB is also that of ∠ BAD, and, similarly, since the sine of ∠ EAG is the sine of ∠ DAG, and since the two pairs are each equal to two right angles, then the ratio of the sine of ∠ BAD to the sine of ∠ GAD will be as the sine of ∠ ABE — namely, [the sine] of the complement of ∠ ABG — to the sine of ∠ AGE — namely, [the sine] of the complement of ∠ AGB. Q.E.D.

If arc AD is greater than a quadrant, which happens when each of the angles B and G is obtuse, then let quadrant ZD be cut from it, and from point Z, which Th. [2] shows to be the pole of circle BG, let two arcs proceed to meet the two points B and G. These [arcs] are clearly quadrants by Th. [1] above, resting perpendicularly on arc BG. Therefore ∠ ABZ is the complement of ∠ ABG by definition, and similarly ∠ AGZ is defined [to be] the complement of ∠ AGB. Moreover, by Th. [17] above, with the reasoning reversed, the ratio of the sine of ∠ BAZ, or BAD, to the sine of quadrant ZB is as that of the sine of

∠ ABZ to the sine of arc AZ. And by the same [theorem], the sine of quadrant ZG to the sine of ∠ GAZ, or GAD, is as the sine of arc AZ to the sine of ∠ AGZ. By simultaneous solution, then, the sine of ∠ BAD to the sine of ∠ GAD will be as the sine of ∠ ABZ — namely, [the sine] of the complement of ∠ ABG — to the sine of ∠ AGZ — namely, [the sine] of the complement of ∠ AGB.

Next, let the perpendicular fall outside the triangle, with one of the angles B and G being obtuse and the other acute, as was concluded from Th. [8] above. Furthermore, the perpendicular will meet arc GB extended on the side of the obtuse angle, which may be B, for example. [The perpendicular] will be less than a quadrant because ∠ ABD is acute. Then we may extend this [arc AD] until it becomes a quadrant that is terminated at point H,* which must be the pole of circle BG by the reasoning of Th. [2] above. From that pole let two arcs HB and HG emerge, which will be quadrants falling orthogonally on arc BG by Th. [1] above. From this it happens that ∠ ABH is the complement of ∠ ABG and ∠ AGH is the complement of ∠ AGB. Moreover, by Th. [17] above with the terms reversed, the ratio of the sine of ∠ BAH to the sine of quadrant HB is as that of the sine of ∠ ABH to the sine of arc AH, and by the same [theorem] the sine of quadrant HG to the sine of ∠ HAG is as the sine of arc AH to the sine of ∠ AGH. Therefore, by simultaneous solution, the ratio of the sine of ∠ ABH to the sine of ∠ AGH is equal to the ratio of the sine of ∠ BAH to the sine of ∠ HAG. Furthermore, the sine of ∠ BAH is also the sine of ∠ BAD. And similarly the sine of ∠ GAH is the sine of ∠ GAD. Therefore the ratio of the sine of ∠ BAD to the sine of ∠ GAD is as that of the sine of ∠ ABH — namely, the complement of ∠ ABG — to the sine of ∠ AGH — namely, the complement of ∠ AGB. Thus we have completed the determination which was proposed.

*In the figure, for △ *ged* read *ghd*.

XXI.

Si quis arcus notus minor semicircumferentia in duos diuidatur, quorum sinus proportionem habeant datam, uterҩ eorū notus erit.

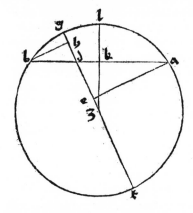

Sit arcus a g b datus minor semicircūfe
rentia diuisus in duos arcus a g & g b, sitҩ
proportio sinus arcus a g ad sinum arcus g
b data. Dico ɋ uterҩ arcuū partialium a g
& g b datus habebitur. Subtendatur enim
arcui a g b corda sua a b, ducaturҩ per
punctum g & centrum circuli 3 diameter
circuli secans cordam a b in puncto d. ex
punctis autem a & b arcum a b terminan
tibus duæ rectæ descendant perpendiculares
ad diametrum, quæ sint a e & b h, quarum
utranҩ constat esse sinū rectum arcus sibi cō
terminalis, a e quidem arcus a g, & b har
cus b g. educatur etiam semidiameter 3 l

orthogonaliter secans cordam a b in puncto k. Si igitur proportio sinuū data
fuerit proportio æqualitatis, erunt duo arcus a g & g b æquales per commūne
scientiam sinibus suis æqualibus existentibus. cunҩ totus arcus a g b sit notus,
erit & uterҩ arcuū a g & g b notus ex primi huius. Si uero proportio dictoҩ
sinuū nō fuerit proportio æqualitatis, erit alter eoҩ altero maior, sit itaҩ a e ma
ior sinu bꝁ, unde & arcus a g maior erit arcu g b. cum autem proportio a e ad
b h sit nota, oportet eam in terminis notis reperiri per diffinitionem proportiōis
datæ, & ideo per quintā primi huius in numeris notis, qui sint r & s, r quidē ma
ior, & s minor, ita ut sit proportio sinus a e ad sinum b h sicut r ad s. cum autē
duo trianguli a e d & b h d rectanguli duos angulos apud contrapositos habe
ant æquales, erunt ipsi per 3 2. primi æquianguli, & ideo per 4.sexti proportio a e
ad b h sicut a d ad d b,, pportio autem a e ad b h erat tanҩ r ad s, quare & p
portio a d ad d b est ueluti r ad s, & coniunctim a b ad b d sicut r s ad s ,
tres autem harum quantitatū notæ sunt, r s quidem propter duos numeros r &
s notos, numerus autem s ex eis quæ supradicta sunt, corda deniҩ a b ppter ar
cum a g b notum, intercedente tabula sinuū aut cordaҩ, quarta igitur scilicet li
nea b d nota ueniet per primi huius, est autem b k nota medietas cordæ a b p
tertij, reliqua igitur d k nota erit. educta insuper semidiametro 3 a, erit tri
angulus 3 a k rectangulus, cuius duo latera 3 a & a k nota sunt, unde & per
primi huius linea 3 k nota prodibit. triangulus itaҩ 3 d k duo latera 3 k & k d
habens cognita, angulū d 3 k cognitum afferet per primi huius, qui quidem ad
quatuor rectos eam habet proportionem, quā arcus g l ad totam circumferentiā,
quemadmodū ex ultima sexti trahitur, arcus igitur g l notus habebitur, quem si
arcui a l dimidio scilicet arcus a g b addideris, resultabit arcus a g notus, ipse
demum ex eodem arcu dimidio, siue ex arcu b l ablatus relinquet arcum g b co
gnitum, uterҩ igitur arcuū partialium notus habebitur, quod pollicebatur nostrū
theorema. Possumus autem & ea quæ demonstrauimus applicare ad arcum semi
circumferentia maiorem, ut si arcus a t b notus diuidatur in duos arcus a t & t
b, quorum sinus proportionem habeant notam, dum tamen uterҩ arcuum par
tialium fuerit minor semicircumferentia, sic enim necesse est diametrū circuli per

punctum

Theorem 21

If any known arc less than a semicircumference is divided into two [arcs], whose sines have a given ratio, each of them will be known.

If arc *AGB*, given less than a semicircumference, is divided into two arcs *AG* and *GB*, and if the ratio of the sine of arc *AG* to the sine of arc *GB* is given, then each of the arc portions *AG* and *GB* may be found.

Let chord *AB* subtend [its] arc *AGB*, and let a diameter of the circle be drawn through point *G* and the center, *Z*, of the circle, cutting chord *AB* at point *D*. Furthermore, from points *A* and *B* that terminate arc *AB*, let two rectilinear perpendiculars *AE* and *BH* descend to the diameter. It is established that each of these [perpendiculars] is the right sine of the arc adjacent to itself — *AE* of arc *AG* and *BH* of arc *BG*. Also let radius *ZL* be drawn intersecting chord *AB* perpendicularly at point *K*.

Then, if the given ratio of the sines is the ratio of equality, the two arcs *AG* and *GB* will be equal by axiom because their sines are equal. And since the total arc *AGB* is known, each of the arcs *AG* and *GB* will be known by Th. I.[19] above.

However, if the ratio of the given sines is not the ratio of equality, one of them will be larger than the other. Then let [the sine of] *AE* be greater than the sine of *BH*, and hence arc *AG* will be larger than arc *GB*. Furthermore, since the ratio of *AE* to *BH* is known, then by the definition of a given ratio, it must be found in known terms and therefore, by Th. I.5 above, in known numbers, which may be *R* and *S* — *R* the larger and *S* the smaller — so that the ratio of the sine of *AE* to the sine of *BH* is as that of *R* to *S*. Moreover, since the two right triangles *AED* and *BHD* have two equal [vertical] angles at opposite

places, they will be equal-angled by [Euclid] I.32, and therefore by [Euclid] VI.4 the ratio of *AE* to *BH* is as that of *AD* to *DB*. Furthermore, the ratio of *AE* to *BH* was as *R* to *S*, and therefore the ratio of *AD* to *DB* is as that of *R* to *S*, and by addition *AB* to *BD* is as *R* [+] *S* to *S*. Now three of these quantities are known — *R* [+] *S* because the two numbers *R* and *S* are known, number *S* from those [things] which are mentioned above, and finally chord *AB* with the assistance of the table of sines or chords because arc *AGB* is known. Therefore the fourth [quantity] — namely, line *BD* — becomes known by Th. I.[19] above. Moreover, *BK*, [which is] half of chord *AB*, is known by [Euclid] III.[not given]; thus the remainder, *DK*, will be known. In addition, if radius *ZA* is drawn, △ *ZAK* will be a right [triangle] whose two sides *ZA* and *AK* are known. Hence, by Th. I.[26] above, line *ZK* will also become known. Thus △ *ZDK*, having two known sides *ZK* and *KD*, will yield ∠ *DZK* known by Th. I.[27] above. [Angle *DZK*] has the [same] ratio to four right [angles] as arc *GL* has to the total circumference, as was concluded from the last theorem of [Euclid] VI. Therefore arc *GL* will be known. If you add [*GL*] to arc *AL* — namely, half of arc *AGB* — it will cause arc *AG* [to be] known. Finally [*GL*] itself, subtracted from the same half-arc [*AL*], or from arc *BL*, leaves arc *GB* known. Then each of the arc portions will be known, which the theorem promised us.

Furthermore, we can apply what we have demonstrated to an arc greater than a semicircumference — for example, suppose known arc *ATB* is divided into two arcs *AT* and *TB*, whose sines have a known ratio — provided, however, that each of the arc portions is less than a semicircumference. Then the circle's diameter *TG*

punctum t transeuntem,quæ sit t g secare cordam a b arcus a t b,quæ & arcui a g b communis,unde & secabit arcum a g b minorem semicircumferētia,
& distinguet ex eo duos particulares arcus scilicet a g & b,quorum sinus propor
tionem habebunt datam,quoniam & huiusmodi sinus communes sunt duobus ar
cubus a t & t b,utrunq; igitur arcuum a g & g b ex supradictis cognitum sub
trahemus à semicircumferentia,& relinquetur socius suus,arcus uidelicet in sinu
secum participans.Quòd si arcus a g b fuerit semicircumferentia ,& diuidatur
in duos arcus a g & g b ut cōtingit,tametsi fuerit data proportio sinus illius ad
sinum istius,quā oportet esse proportionem æqualitatis per communem scientiā,
non tamen alter eoꝝ necessario dabitur,infinitis enim modis potest diuidi arcus
ille,qui est semicircumferentia proportione sinuum,quos habent arcus particula
res,nō mutata. ▽Operationē hoc pacto perficies.Si ꝓportio sinuum data fuerit
æqualitatis,arcum datū dimidiabis,& habebis duos particulares arcus cognitos.
Si uero fuerit inæqualitatis,duos terminos eius congregabis,collectumq; pro pri
mo statuas numero,minorem autem terminū ꝓportionis datæ pro secundo,& nu
merum cordæ arcus dati pro tertio.multiplica igitur secundū per tertiū,& produ
ctum diuide per primum,quodq; exibit à dimidia corda arcus dati auferas,& resi
duum custodias,deinde semidiametro circuli in se multiplicata,aufer quadratum
dimidiæ cordæ arcus dati,quod atitem relinquetur quadrato eius,quod custodiri
præcepimus,coniunge,& collecti radicem elice quadratam,custoditū deniq; per
sinū totum extende,& productū in radicem elicitam distribuas,exibit enim sinus differentiæ,quæ est inter dimidiū arcum datum,& utrunq; arcuū quæsitoris
quā ex tabula sinus inuentā minue ex dimidio arcu dato,& relinquetur arcus minor quæsitus,aut eidem adde,ut resultet arcus maior,In exemplo. Ponatur arcus
40.graduum,& proportio sinus arcus a g maioris ad sinum arcus g b minoris,
sicut 7.ad 4.colligo 7.& 4.fiunt 11.pro primo numero;4 autem accipiam pro se
cundo,& 41042.scilicet cordam arcus dati pro tertio.multiplico 41042.per 4. ꝓ
ducuntur 164168.quæ diuido per 11.exeunt 14924.linea uidelicet d b ,quā subtrahe à medietate cordæ manenti 5597.custodienda pro linea d k .Item semidia
meter siue sinus totus 60000.quæ duco in se,producuntur 3600000000.à quo au
fero quartū dimidiæ cordæ,quod est 421111441,manebunt 3178888559.hoc ad
do quadrato lineæ d k scilicet 31326409.resultāt 3210214968.huius radix quadrata ferè est 56659.quam seruo,deinde multiplico numerum lineæ d k per sinum totum,producuntur 335820000.quæ diuido per radicem seruatam,exeunt
5927.huius arcus 5̄ .40.ferè,quèm deme ex dimidio arcu dato scilicet 2̄0.manēt
1̄4.20.arcus scilicet g b ,item eidem ipsum addo,ueniunt 2̄5.40.& totus ferè ha
bebitur arcus a g .

XXII.

Si data fuerit differentia duorum arcuum cum proportione sinuū suorum,uterq; eorum cognitus habebitur.

Duos arcus a g & g b cōterminales intelligantur,minorꝗ qui est g b pars
maioris a g ,quorum differentia sit data arcus uidelicet a b ,eorumꝗ sinus habe
ant datā proportionem.Dico ꝙ uterq; eorum notus reddetur.Incedat enim per g
terminum communem arcuū dictoꝝ & centrum circuli ꝫ linea recta utrinq; inde
finita,diametrum tamen circuli g t complectens,educaturꝗ semidiameter ꝫ m
secans cordam a b orthogonaliter in puncto l,à punctis autem a & b cordā a
b termi

tnat passes through point T necessarily intersects chord AB of arc ATB, [this chord] also being common to arc AGB. Hence it will intersect arc AGB, [which is] less than the semicircumference, and [will divide AGB into] two arc portions — namely, AG and B,* whose sines will have a given ratio. Because these sines are common to the two arcs AT and TB, then we subtract each of the arcs AG and GB, known from the above, from the semicircumference, and [each one's] associated [arc] is left — namely, the arc that shares the [same] sine.

But if arc AGB is a semicircumference and is divided into two arcs AG and GB, then, as it happens, even if the ratio of the sine of the former to the sine of the latter is given, [a ratio] which must be the ratio of equality by axiom, still not one of these [arcs] will necessarily be found; for this arc that is a semicircumference can be divided in countless ways without the ratio of the sines, which the arc portions have, being changed.

You may accomplish the mechanics in this way. If the ratio of the sines is given [to be that] of equality, you will take half the given arc, and you will have the two arc portions known. However, if [the ratio] is [that] of inequality, you will add their two terms and take the sum for the first number, the smaller term of the given ratio for the second, and the number of the chord of the given arc for the third. Then multiply the second by the third and divide the product by the first. Subtract what results from the half-chord of the given arc and set the remainder aside. Next, after the radius of the circle is squared, subtract [from it] the square of half the chord of the given arc. Now add what is left to the square of that which we [just] admonished should be set aside, and of this sum take the square root. Next multiply what was to be set aside by the whole sine and divide the product by the [previously] elicited square root, for the sine of the difference between half the given arc and each of the desired arcs will result. Subtract [the arc] that is found [for this sine] in the table of sines from half the given arc, and the desired smaller arc will be left; or add the same [two arc portions] so that the greater arc results.

For example, let an arc of 40° be given, and let the ratio of the sine of the greater arc AG to the sine of the smaller arc GB be as 7 to 4. When 7 and 4 are added, they make 11, [which is taken] for the first number. Moreover, 4 is taken for the second number and 41042 — namely, the chord of the given arc — for the third. Then 41042 multiplied by 4 gives 164168. This, divided by 11, leaves 14924 — namely, line DB. Subtract this from half the chord, leaving 5597, to be set aside as line DK. Now, the radius, or the whole sine, is 60000, which is squared, making 3600000000. From this subtract the square† of the half-chord, which is 421111441, leaving 3178888559. To this add the square of line DK — namely, 31326409 — making 3210214968. The square root of this is approximately 56659, which should be set aside. Next, the value of line DK is multiplied by the whole sine, giving 335820000. This is divided by the square root that was set aside, to leave 5927. Of this the arc is about 5° 40, which is subtracted from half the given arc — namely 20° — leaving 14° 20 — namely, arc GB. Similarly, the same [arcs] added give 25° 40, and this approximate total is arc AG.

Theorem 22

If the difference of two arcs is given, together with the ratio of their sines, then each of them may be found.

Let two adjacent arcs AG and GB be understood [such that] the smaller, GB, is a portion of the larger AG, [and] let their difference be given — namely, arc AB; also let their sines have a given ratio. Then each of them can be found.

Let a straight line, unbounded on either end but including the diameter GT of the circle, pass through G, the common end of the given arcs, and the center Z of the circle. And let radius ZM be drawn cutting chord AB perpendicularly at point L. Moreover, from points A and B terminating chord AB,

*For B read GB.

†In the Latin text, for *quartum* read *quadratum*.

142

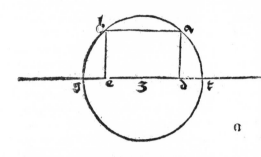

b terminantibus duæ perpendicu
lares a d & b e ad diametrum de
scendant,quas constat esse duos si
nus arcuũ a g & g b. St itaq̃ ipsi
fuerint æquales,hoc est, proportio
sinuũ data fuerit proportio æquali
tatis,erit per communem scientiã
arcus g b æqualis arcui a t.dem=
pto igitur a b noto per hypothe=
sim ex semicircumferentia nota,re
sidui medietas, arcus scilicet b g

minor cognitus erit,cui si arcum a b notum adieceris,prodibit arcus a g maior
cognitus. Si uero alter sinuum maior reliquo extiterit,sit uerbi gratia arcus maio
ris a g sinus,maior sinu arcus minoris b g, abscindaturq̃ ex sinu a d linea re=
cta k d æqualis ipsi b e,ducta linea b k,quæ per primi æquedistabit lineæ e d,
unde & per primi angulus e b k rectus habebitur angulo e d k recto existẽte
& ideo angulus a b e rectum superabit.Producta autem linea ab a per b inde=
finita ex parte puncti b, erit reliquus angulus apud b acutus. cumq̃ sit angulus
g b e g rectus,linea dicta a b satis porrecta concurret cũ linea t g oportune pro
longata,quod fiat in puncto h. Quoniam igitur proportio a d ad b e data est,
in numeris eã reperiemus per corollariũ quitæ primi huius,qui sint r & s,r quidẽ

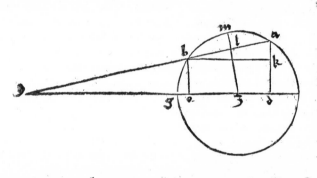

maior numero s,
q̃rũ differentia sit
x, Est autẽ per
primi & quartam
sexti proportio a
d ad b e,& ideo r
ad s tanq̃ a h ad
h b, quare disiun=
ctim a b ad b h si
cut differentia nu
mero ꝗ r & s,ui=
delicet x ad ipsum

numeꝗ s, cumq̃ tres harum quantitatum proportionalium sint datæ,a b quidẽ
cordam arcus sui notificat per tabulã sinuum aut cordarum,erit & linea b h no
ta,& ideo tota a h cognita ueniet.item b l medietas lineæ a b notæ non erit in
cognita,unde & linea h l data comparabitur. Quod igitur sub a h & h b con=
tinetur per primi huius notum erit,ipsum autem æquatur ei quod sub t h & h
g per tertij,quamobrem quod sub t h & h g continetur notum erit,cui si qua=
dratum semidiametri ȝ g notæ adiecerimus, resultabit per sextã 2. quadratũ li=
neæ h ȝ notum,unde & ipsa h ȝ linea cognoscetur.Trianguli ergo ȝ h l rectan
guli duo latera ȝ h & h l nota sunt, quare per primi huius, angulus eius h ȝ l
notus erit,cuius deniq̃ numerus arcum g m notum faciet,ad quem si arcũ m a,
medietatem scilicet arcus a b dati adiunxeris, arcum a g notum habebis, itẽq̃
si ex eodem arcu g m dimidium arcum datum reieceris,arcus b g notus relin=
quetur,quod hactenus expectauimus.Quod si maior duorum sinuum minoris fu
erit arcus, ut si quis offerat proportionem sinus arcus b t ab sinum arcus a t, cũ
arcu a b noto,non aliter q̃ nuperrime peragendũ erit, donec elicietur arcus b g
cognitus

let two perpendiculars *AD* and *BE* descend to the diameter. It is established that these [perpendiculars] are the two sines of arcs *AG* and *GB*.

Then, if these [sines] are equal — that is, the given ratio of the sines is the ratio of equality — then, by axiom, arc *GB* will be equal to arc *AT*. Then when *AB*, known by the hypothesis, is subtracted from the known semicircumference, half of the remainder — namely, the smaller arc *BG* — will become known. If you add known arc *AB* to this, the larger arc *AG* will be made known.

However, if one of the sines is greater than the other, then let the sine of the larger arc *AG*, for example, be larger than the sine of smaller arc *BG*. And from sine *AD* let straight line *KD* be taken equal to *BE*. When line *BK* is drawn, which by [Euclid] I.[not given] will be parallel to line *ED*, then by [Euclid] I.[not given] ∠ *EBK* will be a right [angle] because ∠ *EDK* is a right [angle]. Therefore ∠ *ABE* will exceed a right [angle]. Moreover, when an unbounded line from *A* through *B* is extended beyond point *B*, the other angle at *B* will be acute. And since ∠ *GBEG** is a right [angle], then when the mentioned line *AB* is extended sufficiently, it will meet with line *TG*,† necessarily extended, at point *H*. Then, because the ratio of *AD* to *BE* is given, we may represent it in numbers by the corollary of Th. I.5 above. These [numbers] are *R* and *S*, *R* being larger than number *S*, [and] their difference is *X*. Moreover, by [Euclid] I.[not given] and [Euclid] VI.4 the ratio of *AD* to *BE*, and therefore of *R* to *S*, is as *AH* to *HB*. Therefore, by subtraction, *AB* to *BH* is as the difference of the numbers *R* and *S* — namely, *X* — to the number *S*. Now, since three of these proportional quantities are given — [*AB*'s] arc identifying chord *AB* from the table of sines or chords — line *BH* will be found, and therefore the total [line] *AH* becomes known. Similarly *BL*, half of the known line *AB*, will not be unknown; and hence line *HL* will be found. Therefore the product of *AH* and *HB* will be known by Th. I.[16] above. Moreover, this [product] will be equal to the product of *TH* times *HG* by [Euclid] III.[not given], whereupon the product of *TH* times *HG* will be known. If to this [product] we add the square of known radius *ZG*, by [Euclid] II.6 the square of line *HZ* will become known. Hence line *HZ* itself will be found. Therefore two sides *ZH* and *HL* of right △ *ZHL* are known. Thus by Th. I.[27] above its ∠ *HZL* may be found. The value of this [angle] makes arc *GM* known; if to this you would add arc *MA* — namely, half of given arc *AB* — you will have arc *AG* known. And similarly if you subtract half the given arc from the same arc *GM*, arc *BG* is left known, which we had expected up to this point.

But if the larger of the two sines is [that] of the smaller arc — for example, if it is given as the ratio of the sine of arc *BT* to the sine of arc *AT*, with arc *AB* known — then the procedure will be no different from that just above until arc *BG* is found.

*For *GBEG* read *BEG*.
†In the figure, for line *hgezd* read *hgezdt*.

cognitus, quo dempto ex semicircumferentia, relinquetur arcus b t notus, cui si arcum a b ex hypothesi notum subtraxeris, arcum minorem a ●notũ relinques. ⊽Operatio. Si data proportio sinuũ fuerit æqualitatis, subtrahe arcũ datum ex semicircumferentia, & residui dimidium erit arcus minor quæsitus, cui si arcũ da= tum adieceris, resultabit arcus maior. Si uero fuerit proportio inæqualitatis, & si= nus maioris arcus maior sinu arcus minoris, differentiã terminorum proportiõis datæ constitue primum numerum, terminum autem minorẽ pro secundo, & cor= dam arcus dati, qui est differentia arcuum quæsitoris, pro tertio. Multiplica igitur secundum per tertium, & productum diuide per primũ, & quod exibit addas cor= dæ dimidiæ arcus dati, collectumᵍᵃ serua. Idem quoᵍᵃ adde toti cordæ arcus dati, & collectum multiplica per id quod cordæ toti & eius medietati addidisti, eiᵍᵃ ᵭᵈ producitur, quadratum semidiametri, scilicet sinus totius adijcias, huius demum aggregati radicem elice quadratã. Deinde quod supra seruatum iri iussimus, per sinum totum multiplica, & productum diuide in radicem iam elicitam, ab exeun= tisᵍᵃ arcu dimidium arcũ datum minue, & relinquetur arcus minor quæsitus, quẽ & eidem si addideris, maiorem arcum quæsitum numerabis. In exemplo. Ponat̃ proportio sinus arcus a g, ad sinum arcus b g, sicut 20. ad 13. sitᵍᵃ differentia ar= cuum 40. huius corda est 41042. differentia 20. 4. 13. est 7. Multiplico igit̃ 41042. per 13. producuntur 533546. quæ diuido per 7. exeunt 76221. linea scilicet b h. huic addo medietatem cordæ a b, scilicet 20521. resultant 96742. linea scilicet l h. item colligo 76221. & 41042. ueniunt 117263. quæ multiplico per 76221. pro creantur 8937903123. quibus addo quartum sinus totius, quod est 3600000000. colligunt̃ 12537903123. huius radix quadrata est 111973. quã seruo, deinde mul̃ tiplico 96742. p 60000. pducunt̃ 5804520000. ᵍ diuido p 111973. exeũt 51839 sinus scilicet arcus m g, qui erit 5̄9̄ 45. à quo aufero 20. manebit arcus b g 39.46 item adde 2̄0̄. ueniunt 79.46. & tantus computabitur arcus a g.

XXIII.

Si angulum notum in duos particulares secueris angulos, quorum sinus proportionem habeant datam, uterᵍᵃ eorum notus erit.

Hæc ad angulos superficiales & planos rectilineos & sphærales accõmodabit̃. Sit angulus huiusmodi notus a 3 g diuisus in duos a 3 b & b 3 g, sitᵍᵃ propor= tio sinus anguli a|3 b ad sinum anguli b 3 g data. Dico ᵠ utrumᵍᵃ eorum co= gnoscet Geometra. Super uertice enim anguli, scilicet puncto 3 facto cẽtro, si pla nus & rectilineus fuerit, secundũ quantãlibet distantiam describatur circulus a b g, aut super puncto 3 facto polo secundum cordã quadrantis magni circulus a b g circumducatur, prolongenturᵍᵃ tria latera angulos particulares complecten= tia, donec obuiabunt circumferentiæ descripti circuli in punctis a, b, & g. Vnus̃ quisᵍᵃ igitur trium arcuum, a g quidem totalis, & duorum partialiũ a b & b g quantitatem anguli se respicientis determinat, unde & sinus illorum arcuum tãᵍᵃ sinus angulorum suorũ censebimus. Oportet autem arcum a g esse notum per 19 primi huius, sed & sinus duorum arcuum a b & b g, qui sunt etiam angulorum su orum, proportionem habebunt notam. quare per   huius uterᵍᵃ illorum arcuum notus erit, & ideo per corollarium primi huius uterᵍᵃ angulorum a 3 b & b 3 g notus habebitur, quod erat deducendum. Operatio huius ab opere   huius in nul lo discrepat, nisi ᵠ uice angulorum arcus se determinantes accipias.

XXIIII.

Data differentia duorum angulorum, & proportione sinuum suo=
P rum

When this [arc *BG*] is subtracted from the semicircumference, arc *BT* is left known. If, from this [arc], you would subtract arc *AB*, known from the hypothesis, you will leave the smaller arc *A*.*

The mechanics: If the given ratio of the sines were [that] of equality, subtract the given arc from the semicircumference, and half of the remainder will be the desired smaller arc. If you add the given arc to this, the greater arc will result.

But if the ratio were [that] of inequality and the sine of the larger arc were greater than the sine of the smaller arc, then establish the difference of the terms of the given ratio as the first number, the smaller term for the second, and the chord of the given arc, which is the difference of the desired arcs, for the third. Then multiply the second by the third and divide the product by the first. Add what results [*BH*] to half the chord of the given arc, and set aside the sum [*HL*]. Add the same [fourth term, *BH*,] to the total chord of the given arc, and multiply the sum by that [same term *BH*] to which you [formerly] added the total chord and [prior to that] its half [hence, *AH* times *BH*]. To that [product] add the square of the radius — namely, [the square] of the whole sine. Next, of this sum [*HZ* squared] take the square root. Finally, multiply what we ordered above to be set aside [*HL*] by the whole sine, and divide the product by the square root just taken [*HZ*]. [This is the chord of the arc from the midpoint of the difference to the unshared end of the larger arc; hence the arc may be obtained from the table of sines or chords.] From the resulting arc subtract half the given arc, and the desired smaller arc is left. And if you add this to the same [half-arc], you will compute the desired larger arc.

For example, let the ratio of the sine of arc *AG* to the sine of arc *BG* be given as 20 to 13. And let the difference of the arcs be 40. The chord of that is 41042. The difference of 20 [and] 13 is 7. Then 41042 multiplied by 13 gives 533546, which is divided by 7 leaving 76221 — namely, line *BH*. To this is added half of chord *AB*, 20521, yielding 96742 — namely, line *LH*. Similarly add 76221 and 41042, obtaining 117263, which is multiplied by 76221, making 8937903123. To this is added the square† of the whole sine, which

is 3600000000, totaling 12537903123. The square root of this is 111973, which is set aside. Next multiply 96742 by 60000, making 5804520000, which is divided by 111973, leaving 51839 — namely, the sine of arc *MG*, which will be 59° 4[6]. From this, 20 is subtracted [and] arc *BG* is left as 39[°] 46. Similarly, add 20° to obtain 79° 46, and this much will be considered arc *AG*.

Theorem 23

If you were to cut a known angle into two angle portions whose sines have a given ratio, each of [the angles] may be found.

This will apply to both rectilinear planar and superficial spherical angles. If this known ∠ *AZG* is divided into two [angles] *AZB* and *BZG*, and if the ratio of the sine of ∠ *AZB* to the sine of ∠ *BZG* is given, then geometry will find each of them.

Around the vertex of the angle — namely, around point *Z* — as center, if [the angle] is planar and rectilinear, with any length [as radius] let a circle *ABG* be described; or around point *Z* as pole [if the angle is spherical] with the chord of the great quadrant [as polar line], let circle *ABG* be drawn. And let the three sides that enclose the angle portions be extended until they meet the circumference of the described circle at points *A*, *B*, and *G*. Then every one of the three arcs will determine the size of the angle opposite it — *AG* [will determine the size] of the total [angle] and *AB* and *BG* [the size] of the two portions. Hence we may consider the sines of these arcs as the sines of their angles. Moreover, arc *AG* must be known by Th. I.19[20] above. Furthermore, the sines of the two arcs *AB* and *BG*, which are also [the sines] of their angles, will have a known ratio. Therefore by Th. [21] above, each of these arcs will be known, and thus, by the corollary of Th. I.[21] above, each of the angles *AZB* and *BZG* will be known. Q.E.D. The mechanics of this differ in no way from the mechanics in Th. 21 above except that, instead of the angles, take the arcs determining those [angles].

Theorem 24

When the difference of two angles is given and the ratio of their sines [is also given],

*For *A* read *AT*.
† In Latin, for *quartum* read *quadratum*.

rum, utriuſcꝫ eorum quantitatem addiſcemus.

Hanc ex huius non aliter ꝗ̃ præmiſſam ex emanare conſtat. nihil enim no
ui admiſcetur, quod non colligantia angulorum & arcuum ſe determinantium ad
miniſtret. Quæ hactenus lucubrauimus, fuere præambula arti triangulorū ſphæ
ralium, nunc rem ipſam ingrediamur. Habet autem triangulus ſphæralis tria la
tera & tres angulos, quorum tribus quibuſcūꝗ cognitis, reliqua tria cognoſcenda
uia parata eſt. Demonſtrabimus enim ꝗ in omni triangulo ſphærali ex arcubus
circulorum magnorum conſtante, cui non ſunt duo recti anguli, ſiue duo latera
cū uno angulo, ſiue duos angulos cum uno latere, aut tria latera, aut tres angulos
notos habuerimus, reliqua tria non latebunt. Cūcꝫ non poſſint eſſe plures cōbina-
tiones huiuſmodi, omnem hoc pacto triangulorum ſphæralium arte abſoluemus,
& quidem planiſſime, quæ res ꝗ utilis quamcꝫ iucunda ueniat Aſtronomo, non ſa
tis dici poteſt. Vt autem res ipſa cognita facilior exiſtat, libuit intermiſcere numeros exemplares, ſæpe etenim obſcuritatem propoſitionis, numerorū ſoluit ac-
cōmodatio. Huius operationem facile comparabis, ſi pro angulis uniuerſis arcus
ſe determinantes acceperis, quemadmodum in præcedenti.

X X V.

Omnis triangulus unicum habens rectum angulum, cum duobus lateribus cognitis, reliquum latus reliquosꝗ angulos latere non ſinet.

Trianguli a b g angulum b rectum habentis, duo latera quæcūꝗ ſint cognita. Dico ꝗ reliquum eius latus notum erit cū reliquis duobus angulis. Erit enim
proportio ſinus complementi arcus a g, ad ſinum complemēti a b, ſicut ſinus cōplementi b g ad ſinum quadrantis. Ex quatuor itaꝗ quantitatibus proportiona-

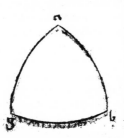

libus tres notæ ſunt propter duo latera nota per hypothe
ſim cum quadrante notos. Vnde & per 19. primi huius
quarta nota erit, uidelicet ſinus complementi a g tertij
lateris, quare ipſum complementum quod nuncꝫ quadrā
te maius exiſtit cognitum habebitur, quod quidem dem
ptum ex quadrāte, relinquet latus tertium notū, ſi ipſum
quadrante minor extiterit, aut complementum illud ad-
ditum quadranti, latus illud notum efficiet, ſi quadrantē
ſuperauerit. Vtrum autem minus quadrante aut maius
fuerit docebunt quantitates laterum datorum, huius
dirigente. Tria igitur trianguli latera ſunt cognita, cū
ꝗ ſinus ſingulorū ad ſinus angulorum ſibi oppoſitorum proportionē habeant notam, eam uidelicet quā habet ſinus a g ad ſinum anguli a b g recti, quemadmo
dū enunciauit huius, erunt & ſinus reliquoꝛ anguloꝛ noti, anguli aūt ipſi qua
les ſint, uolo dicere. maiores recto an minores duo latera ſe reſpicientiū quātita
tes edocebūt, huius intercedēte, quare & o͞es anguli no ti erūt. ꝗ Opatio. Si duo
latera rectū ambientia fuerint data, multiplica ſinū cōplementi alterius eoꝛ per
ſinum complementi lateris reliqui, quodcꝫ producetur in ſinum totū, ſcilicet qua
drantis partiare. exibit enim ſinus complementi lateris rectum ſubtendentis, cuius arcū, ſcilicet ipſum complementum demes ex quadrante, ſi utrūcꝫ datorum la
terum aut maius quadrante fuerit aut minus eo, & relinquetur quātitas lateris rectū reſpicientis angulum. ſi uero alterū ex datis lateribus quadrante maius, reliquā
uero minus extiterit, complementum ipſum quadranti adijcias, & reſultabit latus
quæſitū. Quod ſi alterū datorum laterū recto opponat, multiplicabis ſinū comple
menti

we will ascertain the size of each of them.

This is established to derive from Th. [22] above in the [same] way as that in which the preceding [theorem] derived from Th. [21] above. For nothing new is added which would hinder the summings of the angles and of the arcs determining those [angles].

Let what we have done up to this point be an introduction to the art of spherical triangles; now we will enter upon the art itself. Moreover, a spherical triangle has three sides and three angles; when any three of these are known, the way is prepared for the other three to be found. For we will demonstrate that, in every spherical triangle which is established from the arcs of great circles [and] in which there are not two right angles, whether we were to have two sides with one angle or two angles with one side, or three sides, or three angles, the other three will not be hidden. And since there cannot be any more combinations of this sort, we may prove every art of spherical triangles in this way — indeed, in the clearest way. This art is so useful and agreeable to astronomy that enough cannot be said. Moreover, so that this art may be known more easily, it will be helpful to intermix numerical examples, for, indeed, an application of numbers will often remove an obscurity of the proposition.

You may provide the mechanics of this [present theorem] easily if, in place of all the angles, you take the arcs determining those [angles], as [was done] in the preceding [theorem].

Theorem 25

Every triangle having one and only one right angle together with two known sides permits the other side and the other angles to be [found].

If in △ *ABG* having right ∠ *B*, any two sides are known, then its other [third] side will be known along with the other two angles.

The ratio of the sine of the complement of arc *AG* to the sine of the complement of *AB* is as the sine of the complement of *BG* to the sine of the quadrant. Then, of the four proportional quantities, three are known because two sides are known by hypothesis [and] because a quadrant is known. Hence and by Th. I.19[20] above, the fourth may be found — namely, the sine of the complement of the third side *AG*. Therefore this complement, which is never greater than a quadrant, will be known. Indeed, this complement, subtracted from a quadrant, leaves the third side known if [the third side] itself is less than a quadrant; or this complement, added to a quadrant, makes that [third] side known if [that side] exceeds a quadrant. Moreover, by the directions of Th. [4] above, the sizes of the given sides will tell whether [*AG*] is less than or greater than a quadrant. Therefore three sides of the triangle are known, and since the sines of the individual [sides] to the sines of the angles opposite them have a known ratio — namely, that which the sine of *AG* has to the sine of right ∠ *ABG* — as Th. [16] above declared, the sines of the other [two] angles will also be known. Moreover, I should like to comment about what type these angles are: Whether [they are] greater than a right [angle] or less than [a right angle] the sizes of the two sides opposite them will tell, by the help of Th. [3] above. Hence all the angles will be known.

The mechanics: If the two sides including the right [angle] are given, multiply the sine of the complement of one of these [sides] by the sine of the complement of the other side, and divide what is obtained by the whole sine — namely, [the sine] of the quadrant. Thus the sine of the complement of the side subtending the right [angle] results. Its arc — namely, the complement itself — you subtract from the quadrant if each of the given sides is either greater or less than a quadrant, and the length of the side opposite the right angle is left. But if one of the given sides is greater than a quadrant while the other is less, you add the complement itself to the quadrant and the desired side will result. However, if one of the given sides is opposite the right [angle], you will multiply the sine of the complement menti lateris rectum subtendentis per sinū quadrantis, producto enim diuiso per sinum reliqui lateris dati exibit sinus complementi lateris quæsiti, cuius arcū, scili cet complementum ipsum ex quadrante minuas, si utrumcp datorum laterū qua drante aut maius aut minus extiterit, si uero alterum eorum maius, & reliquū mi nus quadrante occurrat, complementum ipsum quadranti adiectum, latus tertiū manifestabit. Hæc pro latere tertio reperiendo. Duos autem angulos non rectos (rectus enim quilibet notus est)hoc pacto metieris. Sinum lateris oppositi angu lo quæsito per sinum quadrantis extende, productumcp per sinum lateris rectum subtendētis partiaris, exibit em sinus anguli quæsiti, cuius arcū in tabula sinus ac cipias; maiorē quidem si arcus ipsum respiciēs angulū maior quadrāte fuerit, mi norem uero si minor. omnem quippe sinum duobus respōdere arcubus perspicuū est. In exemplo. Offeratur arcus a b 20.graduum, & b g 36.libet inuenire arcū a g. Complemētum arcus a b est 70. cuius sinus est 56382. Complementū arcus b g est 54. cuius sinus est 48541, dū sinus totus, scilicet quadrantis est 60000. quē admodū in tabula nostra constituimus. Multiplico igitur 56382.per 48543.pro= ducuntur 2736838662.quæ diuido per sinum totum 60000. exeunt 45614. sinus scilicet complementi arcus a g, huius arcus est 49. 29.complementum scilicet ar= cus a g, quod minuo ex 90.relinquuntur 40. 31.& tantus erit arcus a g. Sed po natur latus a b $\overline{160}$.& latus b g $\overline{144}$. sinus complementorū sunt, quibus nunc usi sumus, uenietcp latus a g quantum antehac elicitum est, oportet enim arcum a g minorem esse quadrante. Quòd si latus a b fuerit $\overline{20}$.& b g 144.licet sinus prio= res redeant & complementum arcus a g idem quod prius, ipsum tamen nunc ad dendum est quadranti, ut habeatur arcus a g, cp alter arcuum a b & b g minor quadrante sit re, & reliquus maior eo. Sit demum latus a b $\overline{20}$.graduū, a g autē 50.complementū a b est 70.cuius sinus 56382. complementū a g 40.cuius sinus 38567.multiplico 38567.per 60000.producuntur 2314020000.quæ diuido per 56382.exeunt 41042.(quod enim propinquū est uero, ueritatis utimur uice) hic est sinus cōplemēti arcus b g, cuius arcus 43.10.ferè, quē demo ex 90 . relinquunt $\overline{46}$. 50.pro arcu b g.tantus denicp haberētur arcus b g, si arcus a b fuisset 160.& arcus a g 130.quoniā utrobicp arcū b g minorem quadrāte oporteret esse. Si ue ro arcus a b fuerit 20.& arcus a g 130.erit arcus b g maior quadrante, repeti tocp opere pristino, quoniam eidem erunt numeri, ueniet complementū arcus b g iterum 43.10.addendum quidem quadranti, quo facto congregabuntur $\overline{133}$. 10. tantuscp numerabitur arcus a g quæsitus. Hæc de lateribus. Postremo libeat in uenire angulum a g b arcu a b existente 20.& b g 50. Sinus 20.graduum est 20521. Sinus 50.est 45963.Multiplico 20521.per sinum totum, producuntur 1231260000. quæ diuido per 45963.exeunt 26788.sinus uidelicet anguli a g b, cuius arcus minor est $\overline{26}$. 31.determinans quantitatem anguli a g b, quoniā ar= cus a b minor est quadrāte.& tantus censebitur etiā angulus a g b. Si aūt arcus a b superaret, angulus a g b recto maior haberetur. Vnde accipiendus esset ar= cus maior respondens sinui prædicto, qui est $\overline{153}$. 29.& tantum pronunciaremus angulum a g b. Non aliter ad notitiam anguli b a g perducemur.

XXVI.

Tribus angulis trianguli rectanguli cognitis, omnia eius latera pa= tefient.

Sit triangulus a b g cuius tres anguli noti habeantur. Dico cp omnia eius la tera fient cognita. Aut enim duo eorum sunt recti, aut unus tantū. Si duo, sint ipsi

P   2   uerbi

of the side subtending the right [angle] by the sine of the quadrant; for, when the product is divided by the sine of the other given side, the sine of the complement of the desired side emerges. Its arc — namely, the complement itself — you subtract from the quadrant if each of the given sides is either larger than or smaller than a quadrant. But if one of them is greater and the other less than a quadrant, the complement itself, added to the quadrant, will reveal the third side. This is [enough] concerning ascertainment of the third side.

You will measure the two nonright angles (for any right [angle] is known) in this way. Multiply the sine of the side opposite the desired angle by the sine of the quadrant and divide the product by the sine of the side subtending the right [angle]. Thus the sine of the desired angle will result, whose arc you take from the table of sines — the greater [of the two complementary arcs given in the table] if the arc opposite the angle itself is larger than a quadrant, but the smaller if [that arc is] less [than a quadrant]. Certainly it is clear that every sine corresponds to two arcs.

For example, let arc AB be given as 20° and BG as 36; it is desired to find arc AG. The complement of arc AB is 70, whose sine is 56382. The complement of arc BG is 54, whose sine is 48541, while the whole sine — namely, [the sine] of the quadrant — is 60000, as we established in our table. Then multiply 56382 by 48543, producing 2736838662. This, divided by the whole sine 60000, gives 45614 — namely, the sine of the complement of arc AG. The arc of this is 49° 29 — namely, the complement of arc AG. This, subtracted from 90°, will leave 40° 31, and this value will be arc AG. But [now] let side AB be given as 160° and side BG as 144°. The sines of the complements are those which we have just used, and side AG is of that size determined above, for arc AG must be less than a quadrant. However, if side AB is 20° and BG 144°, then, although the previous sines return and the complement of arc AG is the same as before, that [complement] is now to be added to the quadrant in order that arc AG may be obtained, because one of the arcs AB and BG is smaller than a quadrant in this situation

and the other [of those two arcs] is greater than [a quadrant]. Next, let side AB be 20° while AG is 50°. The complement of AB is 70, whose sine is 56382. The complement of AG is 40, whose sine is 38567. Then 38567 is multiplied by 60000, giving 2314020000, which, divided by 56382, leaves 41042 (for what is close to true [i.e., rounded off] we use in place of the [exact] truth). That is the sine of the complement of arc BG, whose arc is about 43° 10. This [arc] is subtracted from 90° leaving 46° 50 for arc BG. Then this is the size [that] arc BG would be if arc AB were 160 and arc AG 130°, because with both [AB and AG being larger than a quadrant] arc BG must be less than a quadrant. But if arc AB were 20 and arc AG 130, arc BG would be greater than a quadrant; and when the earlier work is consulted again because the values will be the same, the complement of arc BG again becomes 43° 10, [but this time it is] to be added to the quadrant. When this is done, the sum will be 133° 10, and that amount will be considered [to be] the desired arc AG.* This has concerned the sides.

Finally, let it be desired to find $\angle AGB$ when arc AB is 20 and BG 50. The sine of 20° is 20521. The sine of 50° is 45963. Then 20521 is multiplied by the whole sine, giving 1231260000, which, divided by 45963, leaves 26788 — namely, the sine of $\angle AGB$. The smaller arc of this [sine] is 26° 31, which determines the size of $\angle AGB$ because arc AB is less than a quadrant. Then that value will also be considered [the size of] $\angle AGB$. But if arc AB were to exceed [a quadrant], $\angle AGB$ would be greater than a right [angle]. Hence the greater arc corresponding to the aforementioned sine would be taken, which is 153° 29, and we would declare $\angle AGB$ to be that size. We are brought to the identity of $\angle BAG$ in the same way.

Theorem 26

When three angles of a right triangle are known, all its sides may be disclosed.

If ABG is a triangle whose three angles are known, then all its sides may be found.

Either two of these [angles] are right [angles] or one is that size. If two are, let them be

*For AG read BG.

uerbi gratia, b & g, erit igitur per   huius pūctus a polus circuli b g, & per
huius utercz arcuum a b & a g quadrans cognitus, sed & arcus b g determinat
quantitatem anguli b a g noti, unde & ipse notus habetur. Tria itacz trianguli
latera nota reddidimus. Si uero unus duntaxat angulus, uerbi gratia, b sit rectus,

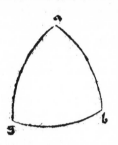

huius   consulemus. erit enim proportio sinus anguli b
a g ad sinum anguli a b g recti, tancz sinus complemē
ti anguli a g b ad sinum complementi lateris a b, tres
autem horum sinuū notos faciūt anguli per hypothesim
dati, quare & sinus complementi arcus a b cognitus ue
niet, cuius arcus, uidelicet ipsum complementū ex quarta circumferentiæ demptus relinquet arcum a b notum
si arcus a b minor quadrante extiterit, aut additus quadranti ipsum arcū a b notum cōstituet, si arcus a b qua
drantē supauerit. Arcus autē a b qualis fuerit respectu
quadrantis, angulus a g b huius dirigente indicabit. Similiter per omnia notū
reddemus latus b g, & tandē per præcedentē latus tertiū a g innotescet. Verū in
uento arcu a b hāc uia ingredi licebit. Proportio sinus anguli a g b per huius
ad sinum arcus a b est ut sinus anguli b a g ad sinum lateris b g. Tres autem hu
iusmodi sinus noti sunt, quare & quartus, & ideo arcus b g cognitus habebitur. si
militer reperiemus arcū a g mediante angulo a b g recto. Cautum profecto te
uelim esse in accipiendis arcubus per sinus datos, ne centies idem repetendo mem
brana cōtaminētur. unumquecz eim sinū minorē sinu toto duobus respondere arcubus sæpenumero dictū est, quocz alter quadrante maior, alter eo minor existit.
Volenti ergo sinui dato arcū suum reddere, cōsiderandum est, sit ne arcus suus ma
ior qnadrante aut minor eo, quòd nimirū superiores conclusiones satis apertū tra-
didere. Idem præterea de cōplementis arcuum & anguloz observandū est, queaσ
modum enim unumquodcz cōplementū arcuale duobus seruit arcubus, quocz alter quadrante maior, alter autem eo minor est, ita & omne cōplementum angula-
re duos respicit angulos, hunc quidē maiorē, illū autē recto minorem. Si igitur cō
plemento arcuali reddere conaris arcū suū, prius exploratū habeas, sit ne arcus ille maior quadrante aut minor eo . si maior, cōplementum suum additū quadranti arcum cōstituet quæsitū. si uero minor, cōplementū ex quadrante reiectum a-
cus quæsiti relinquet quantitatem. Non aliter circa angulos procedemus, nisi σ
ubi pridem arcus erat, nunc angulum intelligamus.   ⊮ Operatio huius. Si trian
gulus habuerit duos rectos, iam conclamatū est, utruncz eim laterū eos subtenden
tium erit quadrans notus. tertiū autem latus eū, quē angulus se respiciens sortitur
numerum. Si uero unus duntaxat rectus fuerit, multiplica sinum complemēti an
guli non recti, quem subtendit latus quæsitum per sinum totum, & productum diuide per sinum reliqui anguli non recti, exibit enim sinus complementi arcus quæ
siti, cum quo, ut supra monuimus, operabere. Ad reliqua demum cognoscenda latera, multiplicabis sinū arcus iam inuenti per sinū anguli respicientis aliud latus
quæsitum, siue rectus siue non rectus extiterit, & productum partieris per sinum
anguli quē subtendit latus nnper inuentū, exibit enim sinus lateris quæsiti, cū quo
ut antea præcepimus, latus ipsum elicies. In exemplo. Sit utercz anguloru b & g,
90. graduū, & angulus a 40, erit utercz arcuum a b & b g 90 graduū, & arcus b
g 40 . Sed ponatur angulus b rectus, angulus uero a 50 gr. & angulus g 70 . uolo inuenire arcum b g, sinus 50 gr. est 45963 . Sinus complementi 70 . gr. est
70521, quem duco in sinum totum, producuntur 1231260000 . hæc diuido
                                                                                          per

B and G, for example. Then, by Th. [2] above, point A will be the pole of circle BG, and by Th. [1] above, each of the arcs AB and AG will be known [to be] a quadrant. Furthermore, arc BG determines the size of known ∠ BAG, and hence [BG] itself will be known. Thus we have found the three sides of the triangle.

But if only one angle, for example B, is a right [angle], we consult Th. [18] above. For the ratio of the sine of ∠ BAG to the sine of right ∠ ABG will be as the sine of the complement of ∠ AGB to the sine of the complement of side AB. Furthermore, three of these sines are known, the angles having been given by the hypothesis. Therefore the sine of the complement of arc AB will become known. Its arc — namely, the complement itself — subtracted from a quarter of the circumference, will leave arc AB known if arc AB is smaller than a quadrant; or [the arc], added to the quadrant, will make arc AB known if arc AB exceeds a quadrant. Moreover, with the directions of Th. [3] above, ∠ AGB will indicate of what type arc AB is with respect to a quadrant. Similarly, by all [the same processes as above] we may find side BG, and finally, by the preceding [theorem], the third side AG will become known.

However, when arc AB is found, one might continue in this way. The ratio of the sine of ∠ AGB to the sine of arc AB is as that of the sine of ∠ BAG to the sine of side BG by Th. [16] above. Moreover, three of these sines are known and therefore the fourth, and thus arc BG, will be known. Similarly, we may find arc AG by the help [of] right ∠ ABG. I would like you truly to be cautious in selecting the arcs from the given sines, lest [your] parchment be spoiled by [your] repeating the same [thing] a hundred times. For it has often been mentioned that every sine that is less than the whole sine corresponds to two arcs, of which one is greater than a quadrant and the other is smaller than [a quadrant]. Therefore, when, from a given sine, one wishes to find its arc, it is to be decided whether its arc is greater than a quadrant or less than one; but it is clear that the previous conclusions certainly have taught enough [about this]. Furthermore, the same [caution] is to be observed concerning the complements of arcs and angles, for just as every arcual com-

plement serves two arcs, of which one is greater than a quadrant while the other is smaller than one, so every angular complement corresponds to two angles — the one greater than [a right angle] and the other less than a right [angle]. Therefore, if, from an arcual complement, you endeavor to find its arc, you would determine, [as] before, whether that arc is greater than a quadrant or less than one. If greater, its complement added to a quadrant will constitute the desired arc. However, if less [than a quadrant], the complement subtracted from a quadrant will leave the size of the sought arc. We may proceed in the same way concerning angles, except that where the arc was previously, we now understand the angle [to be].

The mechanics: If the triangle has two right [angles], everything is already declared, for each of the sides subtending those [angles] will be known [to be] a quadrant. Moreover, the third side, which the angle opposite that [side] determines, will be the value [of that angle itself]. But if only one [angle] is a right [angle], multiply the sine of the complement of the nonright angle which the desired side subtends [times] the whole sine and divide the product by the sine of the other nonright angle. Thus the sine of the complement of the desired arc will result. You will work with this [sine] as we advised above. Finally, for the other sides to be found, you will multiply the sine of the arc just found by the sine of the angle opposite the other desired side, whether [that angle] is right or nonright. Then you will divide the product by the sine of the angle which the side, just found, subtends. Thus the sine of the desired side will emerge, [and] with this, as we have instructed before, you will elicit the side itself.

For example, let each of the angles B and G be 90° and ∠ A 40°. Each of the arcs AB and BG* will be 90° and arc BG will be 40°. But [now] let ∠ B be a right [angle] while ∠ A is 50° and ∠ G is 70°. It is desired to find arc BG.† The sine of 50° is 45963. The sine of the complement of 70° is 20521, which is multipled by the whole sine, giving 1231260000. This is divided

*For BG read AG.
†For BG read AB.

per 45963.exeunt 26788. sinus scilicet complementi arcus a b, cuius arcus
scilicet ipsum complementum est 26.31.quem demo ex quadrante, φ arcum
a b minorem esse quadrante oporteat angulo g acuto existente,& relinquuntur
63.29.tantusφ computabitur arcus b g.Rursus pro latere b g metiendo, sinus
anguli g,qui ponebatur 70 est 56382.sinus lateris b g, quod iam nunc reperi-
mus 63.29.est 53688.quem multiplico per sinū anguli a,qui erat 45963, producuntur 2467661544.quæ diuido per 56382.exeunt 43767.sinus scilicet lateris b,
g,cuius arcus minor est 46,50.& tantus est arcus b g, quoniam angulus b a g
ponebatur acutus.Similiter reperiemus arcum a g,nihil prorsus uariando,nisi φ
loco anguli b a g angulum a b g rectum constituamus.

XXVII.

Vno latere trianguli rectanguli cum altero duorum non rectorum cognito,& angulum reliquū cum lateribus reliquis inuenire.

Sit triangulus a b g, angulum b rectum habes,
duosφ a & g non rectos,quorum alter, uerbi gratia, g sit datus cum uno latere quocunφ.Dico φ angulus a cognitus erit,& reliqua duo latera . Si enim
latus datum angulo dato opponatur, ut in figura est
latus a b   huius consulemus,erit enim conuersis terminis proportio sinus anguli a g b ad sinum lateris
a b tanφ sinus anguli a b g recti ad sinum lateris a

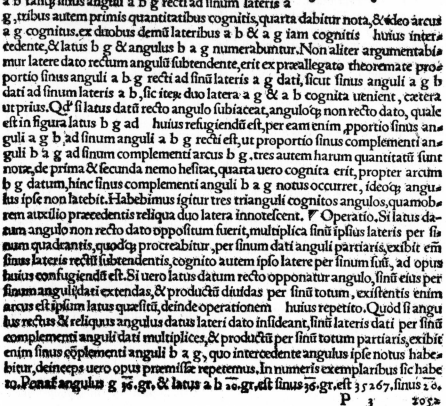

g ,tribus autem primis quantitatibus cognitis,quarta dabitur nota,& ideo arcus
a g cognitus.ex duobus demū lateribus a b & a g iam cognitis huius intercedente,& latus b g & angulus b a g numerabuntur.Non aliter argumentabi-
mur latere dato rectum angulū subtendente,erit ex præallegato theoremate proportio sinus anguli a b g recti ad sinū lateris a g dati,sicut sinus anguli a g b
dati ad sinum lateris a b,sic item duo latera a g & a b cognita uenient, cætera
ut prius.Qd̄ si latus datū recto angulo subiaceat,anguloφ non recto dato, quale
est in figura latus b g ad   huius refugiendū est,per eam enim ,pportio sinus anguli a g b ad sinum anguli a b g recti est,ut proportio sinus complementi anguli b a g ad sinum complementi arcus b g ,tres autem harum quantitatū sunt
notæ,de prima & secunda nemo hesitat,quarta uero cognita erit, propter arcum
b g datum,hinc sinus complementi anguli b a g notus occurret , ideoφ angulus ipse non latebit.Habebimus igitur tres trianguli cognitos angulos,quamob-
rem auxilio præcedentis reliqua duo latera innotescent. ▼ Operatio.Si latus datum angulo non recto dato oppositum fuerit,multiplica sinū ipsius lateris per sinum quadrantis,quodφ procreabitur ,per sinum dati anguli partiaris,exibit em̄
sinus lateris rectū subtendentis,cognito autem ipso latere per sinum suū, ad opus
huius confugiendū est.Si uero latus datum recto opponatur angulo,sinū eius per
sinum anguli dati extendas,& productū diuidas per sinū totum, existentis enim
arcus est ipsum latus quæsitū,deinde operationem huius repetito.Quòd si angulus rectus & reliquus angulus datus lateri dato insideant,sinū lateris dati per sinū
complementi anguli dati multiplices,& productū per sinū totum partiaris,exibit
enim sinus cōplementi anguli b a g, quo intercedente angulus ipse notus habebitur,deinceps uero opus præmissæ repetemus.In numeris exemplaribus sic habe-
to.Ponat angulus g 36.gr.& latus a b 20.gr.est sinus 36.gr.est 35267.sinus 20.

P 3

by 45963, leaving 26788 — namely, the sine of the complement of arc *AB*. Its arc — namely, the complement itself — is 26° 31. This is subtracted from a quadrant, because arc *AB* must be less than a quadrant since ∠ *G* is acute, and 63° 29 are left. Arc *BG** will be considered [to be] this value. On the other hand, for side *BG* to be measured, the sine of ∠ *G*, which is 70°, is 56382. The sine of side *BG*,† which we just now found [to be] 63° 29, is 53688. This [sine] is multiplied by the sine of ∠ *A*, which was 45963, giving 2467661544. That [product], divided by 56382, leaves 43767 — namely, the sine of side *BG*. The smaller arc of this [sine] is 46° 50, and this value is arc *BG* because ∠ *BAG* was given acute. Similarly, we may find arc *AG*, with no variation at all except that we take right ∠ *ABG* in place of ∠ *BAG*.

Theorem 27

When one side of a right triangle is known together with one of the two nonright [angles], to find the other angle as well as the other sides.

If *ABG* is a triangle having right ∠ *B* and two nonright [angles] *A* and *G*, of which one — *G*, for example — is given together with any one side, then ∠ *A* as well as the other two sides will be known.

If the given side is opposite the given angle, as side *AB* is in the figure, then we may consult Th. [16] above, because, by a reversal of the terms, the ratio of the sine of ∠ *AGB* to the sine of side *AB* will be as that of the sine of right ∠ *ABG* to the sine of side *AG*. Furthermore, since three of these quantities are already known, the fourth will become known, and thus arc *AG* may be found. Next, since the two sides *AB* and *AG* are now known, then, with the intervention of Th. [25] above, both side *BG* and ∠ *BAG* will be computed.

We will reason in the same way when the given side subtends the right angle. By the already-cited theorem [16] the ratio of the sine of right ∠ *ABG* to the sine of the given side *AG* will be as that of the sine of the

given ∠ *AGB* to the sine of side *AB*. Thus, again, the two sides *AG* and *AB* become known, and the rest follows as before.

But if the given side, which is *BG* in the figure, underlies the right angle, and if the nonright angle [*G*] is given [as before], one should refer to Th. [18] above, for by that [theorem] the ratio of the sine of ∠ *AGB* to the sine of right ∠ *ABG* is as the ratio of the sine of the complement of ∠ *BAG* to the sine of the complement of arc *BG*. Moreover, three of these quantities are known — no one will doubt about the first and second while the fourth will be known because arc *BG* was given. Hence the sine of the complement of ∠ *BAG* will be known and thus the angle itself will not be hidden. Therefore we will have three known angles of the triangle, whereupon, with the help of the preceding [theorem], the other two sides may be found.

The mechanics: If the given side is opposite a given nonright angle, multiply the sine of this side by the sine of the quadrant, and divide what will be produced by the sine of the given angle, for the sine of the side subtending the right [angle] will emerge. Moreover, when this side is known by its sine, one should refer to the work of [Th. 25] above. However, if the given side is opposite the right angle, multiply its sine by the sine of the given angle and divide the product by the whole sine, for the arc of [the sine that] results is the sought side itself. Then one may return to the mechanics of Th. [25] above.

But if the right angle and the other given angle rest on the given side, multiply the sine of the given side by the sine of the complement of the given angle and divide the product by the whole sine, for the sine of the complement of ∠ *BAG* will result. By the assistance of this [complement], the angle itself will be known, and finally we may refer to the work of the preceding [theorem].

In numerical examples one may [work] thus. Let ∠ *G* be 36° and side *AB* 20°. The sine of 36° is 35267. The sine of 20°

*For *BG* read *AB*.
†For *BG* read *AB*.

20521.quem multiplico per finum totum,producuntur 123 1260000.hunc diui
do per 35267.exeunt 34912.finus fcilicet lateris a g .inuenio igitur ex tabula la
tus a g 3 5.35.deinde pro latere & angulo reliquis ad numeros   huius refugio.
Sed maneat angulus g quantus erat,& fit latus a g 20.gr. multiplico finum 36.
gr.qui eſt 35267.per finum 20.gr.fcilicet 20521.procreantur 723714107. quæ di
uido per finum totū,exeunt 12062.finus fcilicet arcus a b, quare ipſe arcus a b
erit 11.36.reliqua per   huius numerabimus.Ponatur demum angulus a g b,ut
prius 36.& arcus b g 20.Sinus 36 eſt 35267.Sinus complementi 20.eſt 56382.
quem duco in 35267 . producuntur 1988423994.hoc diuido per finum totū,exe
unt 33140.fcilicet finus complementi anguli b a g,cuius arcus eſt 33.32. hic dem
ptus ex 90.relinquet 56.28.& tantus habebit angulus b a g . ipſum em minorē
eſſe quarta circumferentiæ,arguit arcus b g datus minor quadrāte.Reliqua tan
dem per operationē præcedentis abſoluemus.Non egreſeras,fi ſolito prolixiores
in his tribus propoſitionibus uideamur,id enim poſtulat tenor operationis, nō ni
hil moræ attulit exemplaris numeroᴚ manuductio,in qua ſi te ſatis exercueris,to
tā fermè artem trianguloᴚ ſphæralium facilem arbitraberis.

<h3 style="text-align:center">XXVIII.</h3>

Cognitis duobus lateribus trianguli nō rectanguli datū angulum
cōtinentibus,reliquū latus reliquosǽ angulos cognitum iri.

Trianguli a b g nō rectanguli duo latera a b & b g ſint data cū angulo b.
Dico ǿ & latus a g,& duo anguli a & g innoteſcent. Deſcendat enim ex termi
no alteriᴚ datoᴚ laterū ad reliquum,perpendicularis,quæ neceſſario intra trian
gulum conſiſtet aut extra eum cadet,neutri enim laterū ſibi conterminaliū pote
rit coincidere,ſic em idem angulus & rectus haberetur & non rectus. Vtrum autē
hoᴚ fiat nondū ſciendi allata eſt facultas,id enim non immediate pendet ex hypo

theſi,ſed paulo poſt exploratum dabimus . Cum autem à
quolibet puncto ſublimi extra circūferentiam circuli ſig
nato poſſimus demittere duas perpendiculares,hoc in p.
poſito eam duntaxat eligemus,quæ ſubtendit angulū da
tum.Sit itaǽ hæc perpendicularis a d, habebit ergo tri
angulus a b d angulū b non rectum notum cum latere
a b,quare per præmiſſam uterǽ arcū a d & d b cogni
tus ueniet.ſi itaǽ arcus b d nunc inuentus minor fuerit
arcu b g dato,conſtabit perpendicularem cecidiſſe intra
trianguluͤ a b g,ſi uero maior extra,oportet autem differentiam duoᴚ arcuū b g
& b d notam eſſe,triangulus igitur a g d rectangulus duo latera a d & d g ha
bens nota per   huius latus ſuū a g, quod & triangulo a b g commune eſt, du
osǽ angulos d a g & d g a notificabit. triāgulus itaǽ propoſitus a b g tria la
tera nota ſunt,cū duobus angulis a b g quidē per hypotheſim, a g b autē per ar
gumentationem,erat autem & uterǽ anguloᴚ b a d & d a g notus, quibus col
lectis,ſi perpendicularis intra trianguluͤ ceciderit,aut minori eoᴚ à maiori ſubtra
cto,angulū b a g addiſcemus,quæ fuere declaranda.Diceres forſitan,proportio
ſinus arcus a g per argumentationem cogniti,ad ſinū anguli a b g, quem de
dit hypotheſis,ſit ut ſinus arcus b g noti per hypotheſim ad ſinū anguli b a g
huius demonſtrante.cuncǽ tres haᴚ quantitatū ſint notæ,& ideo oporteat ſinum
anguli b a g fieri notum,nonne facilius hoc pacto per unicam operationem in
ueniemus angulū b a g,ǿ ingeminando opus duobus angulis b a d & d a g di
uiſim

is 20521, which is multiplied by the whole sine, giving 1231260000. This is divided by 35267, leaving 34912 — namely, the sine of side *AG*. Therefore side *AG* is found from the table [to be] 35° 35. Finally, for the other side and angle, one has recourse to the calculations of Th. [25] above. But let ∠ *G* stay [the same] size, and let side *AG* be 20°. Multiply the sine of 36°, which is 35267, by the sine of 20° — namely, 20521 — giving 723714107. This is divided by the whole sine, leaving 12062 — namely, the sine of arc *AB*. Therefore arc *AB* itself will be 11° 36. The other [unknowns] we may calculate by Th. [25] above. Finally let ∠ *AGB* be given, as before, [to be] 36° and arc *BG* [to be] 20. The sine of 36 is 35267. The sine of the complement of 20° is 56382, which, multiplied by 35267, gives 1988423994. This, divided by the whole sine, leaves 33140 — namely, the sine of the complement of ∠ *BAG*, whose arc is 33° 32. This, subtracted from 90°, will leave 56° 28, and this value will be ∠ *BAG*. For arc *BG*, given as less than a quadrant, makes it clear that [the angle] itself is less than a quarter of the circumference. At last we may find the other [unknowns] by the mechanics of the preceding [theorem].

Do not be put out if we may seem more extensive than usual in these three theorems, for the course of the procedure demands it. Guidance in numerical examples has occasioned something of a delay. If you were to use this [guidance] enough, you would easily perceive almost the entire art of spherical triangles.

Theorem 28

When the two sides enclosing a given angle of a nonright triangle are known, the remaining side and the other angles may be found.

If, of nonright △ *ABG*, two sides *AB* and *BG* are given together with ∠ *B*, then side *AG* and the two angles *A* and *G* will become known.

Let a perpendicular drop from the end of one of the given sides to the other [side]. This [perpendicular] necessarily stays within the triangle or falls outside it. It would not coincide with either of the sides adjacent to it, for then one angle would be both right and nonright. Moreover, the means of knowing which of these it does has not yet been brought out, for it does not come immediately from the hypothesis. But we will find [it] after a little investigation. Now since, from any upper point marked outside the circumference of the circle, we can drop two perpendiculars, in this premise we will choose only that [one] which subtends the given angle. Then let this perpendicular be *AD*. Thus △ *ABD* will have nonright ∠ *B* known, together with side *AB*. Therefore, by the preceding [theorem], each of the arcs *AD* and *DB* will become known. Then if arc *BD*, now found, is less than the given arc *BG*, it will be established that the perpendicular has fallen within △ *ABG*. If [arc *BD* is] greater [than arc *BG*], [the perpendicular has fallen] outside. Moreover, the difference [*DG*] of the two arcs *BG* and *BD* is necessarily known. Then by Th. [25] above, right △ *AGD*, having two sides *AD* and *DG* known, will identify its side *AG*, which is common to △ *ABG*, and the two angles *DAG* and *DGA*. Thus [in] the given △ *ABG* three sides are known together with two angles — *ABG* by hypothesis and *AGB* by reasoning. Furthermore, each of the angles *BAD* and *DAG* was known. When these are added if the perpendicular fell within the triangle, or when the smaller of them is subtracted from the larger [if the perpendicular fell outside the triangle], we will also obtain ∠ *BAG*, which was to be declared.

Perhaps you might say that the ratio of the sine of arc *AG*, known by reasoning, to the sine of ∠ *ABG*, which the hypothesis gave, is as the sine of arc *BG*, known by the hypothesis, to the sine of ∠ *BAG* by the proof of Th. [17] above. And since three of these quantities are known, then the sine of ∠ *BAG* must also be known. Would we not as easily find ∠ *BAG* in this way by a single operation as we may determine ∠ *BAG* by increasing the work through a division [of the triangle] when the two angles *BAD* and *DAG*

uiſtm cognitis, ipſum angulum b a g eliciemus. Reſpondeo tibi ſinum quidem anguli b a g hac uia reperiri ҩ latiſſime, quæ tñ duobus reſpondens angulis incertum eſt uter eoҩ eligendus ſit, id autem minime turbabit uiam noſtram, quoniam uterҩ anguloҩ b a d & d a g qualis ſit reſpectu anguli recti certum tradidimus.

XXIX.

Cognitionem duorum laterum trianguli non rectanguli, & anguli uni eorum oppoſiti, inuentioni reliqui lateris & reliquorum angulorum minime ſufficere.

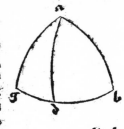

Similem paſſionem de triangulis planis rectilineis demonſtrauimus in primo huius angulo dato exiſtente acuto, quam nunc de ſphæralibus prædicabimus, ſiue angulus datus fuerit acutus, ſiue obtuſus. Sit enim angulus ſphæralis b a g, duobus arcubus æqualibus b a & a g contentus, quorū uterҩ minor ſit quadrante, copulenturҩ duo puncta b & g per arcum circuli magni b g, qui ſecetur per mediū in puncto d deſcendente arcu a d, in arcu autem g h prolongato, ſignetur punctus h ubilibet, ſic tamē, ҩ arcus g h ſit ſemicircuferentia ducto arcu a h. Cum igitur duo trianguli a b d & a g d ſint æquilateri, erunt & æquianguli per tertij huius, & ideo duo anguli apud d ſunt recti, & angulus b a d eſt acutus, erit per huius arcus b d minor quarta circumferentiæ, eſt autem & a b minor quarta, quare per huius & arcus a d minor quadrante declarabitur, unde & per huius angulus a h g acutus erit. Offerenti ergo nobis duo latera g a & a h cognita, aut duo b a & a h ipſis æqualia, cum angulo a h g, neҩ tertium latus, neҩ reliquos angulos reddere poterimus, nam duo trianguli g a h g & b a h, etſi in omnibus quantitatibus datis participent, latera tamē tertia ſortiuntur uaria, quemadmodū in figura claret. Idem declarabitur angulo h obtuſo exiſtente. Repetitæ enim priſtinæ figurationi iterum dabimus locū hoc uno uariato, ҩ uterҩ arcuū a b & a g æqualiū quadrantē excedat, erit enim ut nuper angulus b a d acutus, & ideo per huius arcus b d minor quarta, cunҩ ſit arcus a b maior quadrante, erit & arcus a d quadrante maior, & ideo per huius angulus a h g obtuſus, cætera ut antehac proſequemur.

XXX.

Duobus lateribus trianguli non rectanguli cognitis cum angulo alteri eorum oppoſito, ſi qua lege datum angulum reſpiciens perpendicularis cadat exploratum fuerit, reliquum latus reliquiҩ anguli non latebunt.

Sit triangulus a b g non rectangulus, duo latera a b & a g nota habēs, cū angulo b uni ſcilicet eoҩ oppoſito, ſitҩ certū quonam pacto cadat perpendicularis à comuni termino datoҩ laterū ad baſim, uidelicet an intra an extra triangulū, quæ ſit a d. Dico ҩ & latus b & duo anguli a & g noti uenter. Certi enim principio ſimus perpendicularis

are known? I respond to you that for the sine of ∠ *BAG* to be found in this way is so very broad that from this [method], since [the sine] corresponds to two angles, it is still uncertain which of those [two angles] should be chosen. Moreover, this [problem] will not disturb our method at all because we determined of what type each of the angles *BAD* and *DAG* is with respect to a right angle.

Theorem 29

Knowledge of two sides and of one angle opposite them [in] a nonright triangle does not suffice at all for determination of the other side and of the other angles.

We demonstrated a similar proof concerning rectilinear planar triangles in the First Book when the given angle was acute [see Th. I.51]. We shall now discuss this [proof] concerning spherical [triangles], whether the given angle be acute or obtuse.

Let *BAG* be a spherical angle contained by two equal arcs *BA* and *AG*, of which each is less than a quadrant, and let the two points *B* and *G* be joined by an arc of great circle *BG*, which is bisected at point *D* by a descending arc *AD*. Moreover, anywhere on arc *GB* extended, let point *H* be marked such that arc *GH*, when arc *AH* is drawn, is a semicircumference. Then since the two triangles *ABD* and *AGD* are equal-sided, they will be equal-angled by Th. III.[35] above, and thus the two angles at *D* are right [angles] and ∠ *BAD* is acute. By Th. [3] above arc *BD* will be less than a quarter of the circumference. Moreover *AB* is less than a quarter [of the circumference]; therefore by Th. [5] above arc *AD* will be declared less than a quadrant, and hence by Th. [3] above ∠ *AHG* will be acute. Thus, while this gives us two known sides *GA* and *AH*, or two equal [sides] *BA* and *AH*, as well as ∠ *AHG*, we can find neither the third side nor the other angles, for the two triangles *GAHG** and *BAH*,

although they share in every given quantity, have nonetheless been allotted different third sides, as is obvious in the figure.

The same may be said if ∠ *H* is obtuse. For we may assign the [same] situation to the first construction, which is used again, with one variation — that each of the equal arcs *AB* and *AG* exceeds a quadrant. For, as just [above], ∠ *BAD* will be acute, and thus by Th. [3] above arc *BD* will be less than a quarter of a circumference. And since arc *AB* is greater than a quadrant, arc *AD* will be greater than a quadrant, and thus by Th. [3] above ∠ *AHG* will be obtuse. The rest follows as before.

Theorem 30

When two sides of a nonright triangle are known together with the angle opposite one of them, [and] if it were established by what law the perpendicular opposite the given angle falls, then the remaining side and the other angles may be found.

If *ABG* is a nonright triangle having two sides *AB* and *AG*, as well as ∠ *B* opposite one of those [sides], known, and if it were certain in what way the perpendicular *AD*, from the point of intersection of the given sides, falls to the base — namely, whether inside or outside the triangle — then *B*† and the two angles *A* and *G* may be found.‡

To begin with, let us be certain that perpendicular

*For *GAHG* read *GAH*.
†For *B* read *BG*.
‡Regiomontanus must have worked his proofs of Ths. 30, 31, and 32 against figures with their letters reversed from those made by the engraver for this book. It is important to the sense of the following three theorems that *b* and *g* be reversed: *b* should be the lower left vertex of the given triangle, and *g* should be the lower right vertex of the given triangle. Thus line *ad* and the letters given in the statements and the proofs may remain as printed.

diculatem a d cadere intra triangulum. habebit itacp triangulus a d rectangu=
lus latus a b cognitum cum angulo a b d, quare per　huius duo eius latera a d
& d b nota uenient cum angulo b a d. Triangulus demū a g d duo latera a g

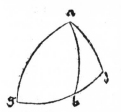

quidem per hypothefim, a d autem per argumentatione
habebit cognita, & ideo per　huius & latus eius g d & u=
tercp duog angulog a g d & g a d innotefcet. Quemad
modū autem ex duobus arcubus b d & d g feorfum no=
tis arcus b g notus refultat, ita & duo anguli b a d & d
a g collecti angulū totū b a g reddent cognitum. Quod
fi perpendicularis extra triangulum ceciderit, fyllogifmū
repetemus, nihil prorfus immutando, nifi cp arcū d g ex
arcu d b minuamus, ut notum relinquatur latus b g trianguli propofiti, angulū
denicp g a d ex angulo b a d, & angulum a g d ex duobus rectis demamus, re
linquentur em̄ duo anguli b a g & a g b noti, quod uolebamus attingere.

XXXI.

Si quis triangulus nõ rectangulus duos habuerit angulos cū latere
eos fuftinente datos, reliquū angulū & reliqua latera cognitū iri.

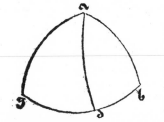

Habeat triangulus a b g nõ rectangulus no=
tos duos angulos a & b, latuscp ipfis fubiacens
a b cognitum. Dico cp & angulus a g b, & duo
latera a g & g b innotefcent. Defcendat em̄ à
uertice alterius datog angulog perpendicularis
uerfus latus nõ datum, fubtendens reliquum an=
gulum cognitū, & fit uerbi gratia a d, quæ cadat
ne intra an extra triangulum nondum fciendi eft
poteftas. id em̄ nõ ftatim noftrā cõfequitur hypothefim. Verum paulo longius p
fecti hoc explorabimus. Ex angulo igitur a b d cognito, & latere a b trianguli
rectanguli a b d per　huius & angulus b a d & utruncp laterum a d & d b co
gnitiõe fortientur. Si itacp angulus b a d fyllogifmo cognitus minorem fe offe
rat angulo b a g per hypothefim dato, certum eft perpendiculare intra triangu
lum cecidiffe. angulusq b a d ex angulo b a g fublatus relinquet angulū g a d
notum. Triangulus ergo g a d rectangulus, cui & arcum a d & angulum g a d
notos declarauimus, ductu　huius, angulum a g d, qui & triangulo a b g termi
nis habetur, cum utrocp latere a g & g d in luce depromet. Duo autem arcus b
d & d g fyllogifmo cogniti fi coadunentur totum arcū b g datum accipiemus.
Si uero angulus b a d ex argumentatione repertus maior occurrat angulo b a g
quē hypothefis tradidit, perfpicuum erit perpendiculare a d extra triangulū ceci
diffe. Proceffucp fuperiori cupitā attingemus metam, nihil immutando nifi cp ar
cum g d, quē nuperrime arcui d b adiecimus, nunc ex eo reijciamus, arcus reli
qui g d cognofcendi gratia. Sed & angulum a g d duobus rectis fubtrahendo
relictum metiemur arcum b g, quæ cenfebā explananda. Nõ poterat autem per
pendicularis a d alteri duog fibi conterminalium laterum coincidere, hypothefi
id prohibenti. fic enim alter duorum angulorum b g & g rectus eueniffet, quem
tamen hypothefis non rectum adminiftrabat.

XXXII.

Duobus angulis trianguli nõ rectanguli cognitis cum latere alterū
eorū fubtendēte, reliqū angulū reliquacp latera inueftigare.

Dati

AD falls within the triangle. Then right △ *AD** will have side *AB* known together with ∠ *ABD*. Therefore, by Th. [27] above, its two sides *AD* and *DB*, as well as ∠ *BAD*, become known. Finally, △ *AGD* will have two sides known, *AG* by the hypothesis and *AD* by reasoning; thus by Th. [25] above its side *GD* and each of the two angles *AGD* and *GAD* will become known. Moreover, just as arc *BG* becomes known from the two arcs *BD* and *DG*, known individually, so also the two angles *BAD* and *DAG* combined make the total ∠ *BAG* known.

But if the perpendicular were to fall outside the triangle, we use the [same] deduction with absolutely nothing changed except that we subtract arc *DG* from arc *DB* so that side *BG* of the given triangle is left known. Finally, we subtract ∠ *GAD* from ∠ *BAD* and ∠ *AGD* from two right [angles], for the two angles *BAG* and *AGB* are left known. Q.E.D.

Theorem 31

If any nonright triangle were to have two angles given together with the side bearing them, the remaining angle and the other sides would be found.

If nonright △ *ABG* has two known angles *A* and *B* and if side *AB*, underlying those [angles], is known, then ∠ *AGB* and the two sides *AG* and *GB* may be found.†

Let a perpendicular — *AD*, for example — descend from the vertex of one of the given angles toward a side that is not given, [the perpendicular] subtending the other known angle. Whether [the perpendicular] falls within or outside the triangle cannot yet be known, for it does not immediately follow from the hypothesis. But we will indeed investigate this a little longer. Thus, since ∠ *ABD* is known as well as side *AB* of right △ *ABD*, by Th. [27] above ∠ *BAD* and each of the sides *AD* and *DB* may be recognized.

Then if ∠ *BAD*, known by deduction, is less than ∠ *BAG*, given by the hypothesis, it is certain that the perpendicular has fallen within the triangle. And ∠ *BAD*, subtracted from ∠ *BAG*, leaves ∠ *GAD* known. Therefore, by the guidance of Th. [27] above, right △ *GAD*, in which we have declared arc *AD* and ∠ *GAD* known, will reveal [its] ∠ *AGD*, which is a term in △ *ABG* also, together with each of the sides *AG* and *GD*. Moreover, if the two arcs *BD* and *DG*, known by deduction, are combined, we will obtain the total arc *BG*.

However, if ∠ *BAD*, found by reasoning, is greater than ∠ *BAG*, which the hypothesis gave, it will be evident that perpendicular *AD* has fallen outside the triangle. By the above process, we will attain the desired goal with nothing changed except that arc *GD*, which we previously added to arc *DB*, we now subtract from it, so that the other arc *GD*‡ may be known. Furthermore, when ∠ *AGD* is subtracted from two right [angles], we may measure the other arc *BG*.§ Q.E.D.

Furthermore, perpendicular *AD* cannot coincide with one of the two sides adjacent to it since the hypothesis forbids that. For in that case one of the two angles *BG*** and *G* would have been a right [angle]; however, the hypothesis gave [each to be] nonright.

Theorem 32

When two angles of a nonright triangle are known together with the side subtending one of them, the remaining angle and the other sides may be investigated.

*For *AD* read *ABD*.
†Again recall that *b* and *g* on the figures must be reversed.
‡For *GD* read *GB*.
§For "arc *BG*" (*arcum bg*) read "∠ *AGB*" (*angulum agb*).
**For *BG* read *B*.

Sint duo anguli a b g & a g b trianguli a b g non
rectanguli cum latere, uerbi gratia, a b, alterum eo℞ sci=
licet angulū g subtendente. Dico ꝙ & anguli a & utri=
uſꝗ later℞ a g & g b noticiam conſequemur. Deſcendat
em ex uertice anguli a non dati, perpendicularis a d uer
ſus latus, quod duos ſuſtentat datos angulos, quæ cadat
ne intra an extra triangulū a b g duo anguli b & g co=
gniti huius manuducente declarabunt. Cadat prius in=
tra. Triangulus ergo a b d rectangulus, cum & latus a
b datum habeat, & angulū a b d non rectum, per huius & anguli ſui b a d &
duo℞ arcū a d & d b noticiam afferet, per eandem deniꝗ huius latere a d &
angulo a g d notis exiſtentibus, & angulus g a d, & duo latera a g & g d in=
noteſcent. eſt autem & a g commune triangulo propoſito, latus demū b g ex du
obus arcubus b d & d g ſingulatim notis, quemadmodū angulus b a g ex duo
bus angulis b a d & d a g inuentis conflabitur. Quòd ſi perpendicularis extra
triangulū ceciderit, non aliter ratiocinabimur, uerū an
gulum d a g, quem prius addidimus angulo b a d,
nunc ex eo minuemus, ut relinquatur angulus b a g co
gnitus. ſimiliter arcus g d ex arcu d b demptus, relin=
quet latus b g trianguli noſtri cognitum, anguloꝗ
tandem a g d ex duobus rectis ſublato, manebit angu
li propoſiti cognitio, planum ergo reddidimus quic=
quid præſens pollicebatur theorema.

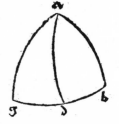

XXXIII.

Datis tribus angulis trianguli ſphæralis non rectanguli, tria eius latera menſurare.

Sit triangulus huiuſmodi a b g tres notos habens angu
los non rectos. Dico ꝙ tria eius latera fient cognita. Ex uer=
tice enim anguli cuiuſuis uerbi gratia a uerſus arcū ſibi op=
poſitum procedat perpendicularis a d, quæ cadat ne intra
an extra triangulū huius manuducente callebimus, utroꝗ
angulo℞ b & g noto per hypotheſim exiſtente, neutri enim
duo℞ later℞ a b & a g coincidet, ſic em alter angulo℞ b & g re
ctus eſſet, quod interdixit hypotheſis. Cadat ergo prius intra
triangulū, erit itaꝗ per huius proportio ſinus anguli b a d ad ſinum anguli d
a g, ſicut ſinus complementi anguli a b g ad ſinum complementi anguli a g b.
proportio antem ſinus complementi anguli a b g ad ſinum complementi angu
li a g b nota eſt, ꝓpter utrūꝗ angulo℞ b & g cognitū, quare & proportionem ſi=
nus anguli b a d ad ſinū anguli d a g datam non inficiaberis. cunꝗ totum an=
gulum b a g notum tradiderit hypotheſis, erit & per huius uterꝗ angulorum
apud a particulariū non ignotus. Triangulus igitur b a d rectangulus omnes
angulos ſuos habens cognitos, argumento huius duo latera ſua a b & b d co=
gnitioni noſtræ ſubijciet, non aliter trianguli a g d rectanguli, tres angulos ha=
bentis datos, duo latera a g & g d metiemur. ſic duo latera a b & a g trianguli
propoſiti gemino didicimus proceſſu, duobus autem arcubus b d & d g congre
Q gatis

If two angles *ABG* and *AGB* of nonright △ *ABG* are given together with side *AB*, for example, which subtends one of those [given angles] — namely, ∠ *G* — then we may determine the identity of ∠ *A* and of both of the sides *AG* and *GB*.*

Let perpendicular *AD* drop from the vertex of ∠ *A*, not given, toward the side which bears the two given angles. Whether this [perpendicular] falls within or outside △ *ABG* the two known angles *B* and *G* will reveal with the guidance of Th. [8] above.

First let it fall within. Then right △ *ABD*, since side *AB* is given as well as nonright ∠ *ABD*, will provide the identity of its ∠ *BAD* and the two arcs *AD* and *DB* by Th. [27] above. Next, by the same [Th. 27] above, since side *AD* and ∠ *AGD* are known, ∠ *GAD* and the two sides *AG* and *GD* will become known. Moreover, *AG* is also common to the given triangle. Finally side *BG* will be determined since the two arcs *BD* and *DG* are known individually, just as ∠ *BAG* will be determined since the two angles *BAD* and *DAG* [have been] found.

But if the perpendicular were to fall outside the triangle, we would reason in the same way, but we will now subtract ∠ *DAG*, which we previously added to ∠ *BAD*, from [*BAD*] so that ∠ *BAG* is left known. Similarly, arc *GD*, subtracted from arc *DB*, will leave side *BG* of our triangle known. Finally, when ∠ *AGD* from two right [angles], for the two knowledge of the intended angle will remain. Q.E.D.

Theorem 33

When three angles of a nonright spherical triangle are given, its three sides may be measured.

If *ABG* is a triangle of this sort, having three known nonright angles, then its three sides may be found.

From the vertex of any angle — *A*, for example — toward the arc opposite it, let a perpendicular *AD* proceed. Whether it falls inside or outside the triangle we will know with the guidance of Th. [8] above. Since each of the angles *B* and *G* is known by the hypothesis, [the perpendicular] may not coincide with either of the two sides *AB* and *AG*, for then one of the angles *B* and *G* would be a right [angle], which the hypothesis forbids. Thus, first, let [the perpendicular] fall within the triangle. Then, by Th. [20] above, the ratio of the sine of ∠ *BAD* to the sine of ∠ *DAG* is as that of the sine of the complement of ∠ *ABG* to the sine of the complement of ∠ *AGB*. Moreover, the ratio of the sine of the complement of ∠ *ABG* to the sine of the complement of ∠ *AGB* is known because each of the angles *B* and *G* is known. Hence you would not deny that the ratio of the sine of ∠ *BAD* to the sine of ∠ *DAG* is given. And since the hypothesis gave the total ∠ *BAG*, then, by Th. [23] above, each of the angle portions at *A* will not be unknown. Therefore right △ *BAD*, having all its angles known, will yield its two sides *AB* and *BD* to our knowledge by the reasoning of Th. [26] above. We may measure the two sides *AG* and *GD* of right △ *AGD*, having three given angles, in the same way. Thus we have found two sides *AB* and *AG* of the given triangle by [this] double process. Furthermore, when the two arcs *BD* and *DG* are added

*The letters *b* and *g* must again be reversed in both figures of this theorem.

gatis(nam eos fingulatim dimenfi fumus)refultabit tertium latus b g trianguli a b g cognitum.Si uero perpendicularis trianguli egreffa fuerit,erit ex præalle= gata   huius proportio finus anguli b a d ad finū anguli g a d nota,quoniā pro portio finus complementi anguli a b g ad finum complementi anguli a g b da ta eft.cunq; nota fit differentia duorum angulop b a d & g a d,uidelicet angu= lus b a g,erit per   huius uterq; angulop b a d & g a d cognitus.Triangulus ergo a b d rectangulus omnes angulos fuos habet notos,angulū enim a b d no tum relinquit angulus a b g,quem dedit hypothefis,posteaq; ex duobus rectis auferetur,quare per   huius latus fuū a b,quod & triangulo a b g commune eft notum habebitur cum arcu b d. Non aliter ad noticiam duop laterū a g & g d trianguli a g d rectanguli tres cognitos habentis angulos perueniemus,duo ita q; trianguli propofiti latera a b & a g nota reddidimus,dempto autem arcu b d ex arcu d g quos geminus elicuit fyllogifmus tertij lateris b g noticiam confe= quemur,quæ decreuimus promulgare. His autem poftremis theorematibus teno rem operationis numerosq; exemplares fubiungere non erat confilium, fatis eni res huiufmodi apud fuperiores conclufiones lucubrauimus, quas fi memoria tua perdiderit cenfeo repetendas. Tri anguli a b g tria latera fint data, fiant a & g poli circulop e k & d 3 , intelligaturq; paralellus h k fuper polo g defcriptus fecundū quantitatem arcus g b,quo facto erit opus omnino fimile ei , ubi ex altitudine folis data quærit diftan tia eius à meridie,fic habebis qua tuor modos demonftrandi proble ma de tribus datis lateribus trian= guli fphæralis.

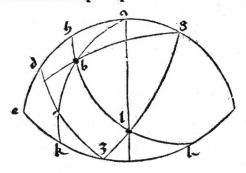

XXXIIII.

Cuiuslibet trianguli tria latera nota habentis, tres angulos redde= recognitos.

Sit triangulus a b g,cuius tria latera nota fint a b , a g & b g.Dico q; tres eius anguli noti habebūtur . Libeat primo inuenire angulū b a g.Si igit uterq; arcuū a b & a g quadrante minor extiterit,protendatur uterq; ad partem late= ris b g,donec fiant duo quadrantes a d & a e,& fuper polum a fecundū cordam quadratis a d de fcribatur circulus,quem conftat transire per pun= ctum e,qui fit d e,arcus itaq; d e quantitatem an guli b a g quem quærimus determinabit,qui ut notus emergat,procedant à centro fphæræ 3 tres femidiametri ad tria puncta a, d,& e ,quæ fint 3 a,3 d & 3 e. à punctis autem b & d cordam b g terminantibus,binæ ducantur perpendiculares g h quidem & b k ad duas femidiametros 3 e & 3 d,duæ uero reliquæ g l & b m ad femidiame= trū 3 a,quas quidem perpendiculares conftat effe finus arcuū,à quop terminis egrediūtur,duoq; tan dem pun

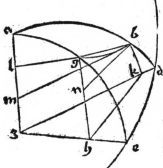

(for we measured them individually), the third side *BG* of △ *ABG* will result.

However, if the perpendicular had left the triangle, then, by the previously cited Th. [20] above, the ratio of the sine of ∠ *BAD* to the sine of ∠ *GAD* will be known because the ratio of the sine of the complement of ∠ *ABG* to the sine of the complement of ∠ *AGB* is given. And since the difference of the two angles *BAD* and *GAD* is known — namely, ∠ *BAG* — then by Th. [24] above each of the angles *BAD* and *GAD* will be known. Therefore right △ *ABD* has all its angles known, for ∠ *ABG*, which the hypothesis gave, leaves ∠ *ABD* known after [*ABG*] is subtracted from two right [angles]. Therefore, by Th. [26] above, its side *AB*, which is also common to △ *ABG*, may be found together with arc *BD*. In the same way we will arrive at the identity of the two sides *AG* and *GD* of right △ *AGD* that has three known angles. Thus we have found two sides *AB* and *AG* of the proposed triangle. Furthermore, when arc *BD* is subtracted from arc *DG*, which [two arcs] the twofold deduction has derived, we will obtain the identity of the third side *BG*. Q.E.D.

Now in these last theorems it was not the plan to attach a course of operation and numerical examples, for we have worked enough things of this sort in the above conclusions. I recommend that they be recalled if your memory has grown stale.

Of △ *ABG* let three sides be given; let *A* and *G* be poles of the circles *EK* and *DZ*; and let a [circle] *HK*, parallel [to *DZ*], be understood to be described around pole *G*

with the length of arc *GB* [as polar line]. When this is done, the mechanics will be entirely similar to those where, from a given altitude of the sun, its distance from the meridian is sought. Thus you will have four methods of solving a problem when three sides of a spherical triangle are given.

Theorem 34

In any triangle that has three known sides, the three angles may be found.

If *ABG* is a triangle whose three sides, *AB*, *AG*, and *BG*, are known, then its three angles will be known.

It is desired first to find ∠ *BAG*. Therefore, if each of the arcs *AB* and *AG* is less than a quadrant, let each be extended beyond side *BG* until they become two quadrants *AD* and *AE*. And around pole *A* with the chord of quadrant *AD* [as polar line], let a circle *DE* be described, which is established to pass through point *E*. Then arc *DE* will determine the size of ∠ *BAG*, which we seek. In order that this [angle] may be found, let three radii, which are *ZA*, *ZD*, and *ZE*, proceed from the center, *Z*, of the sphere to the three points *A*, *D*, and *E*. Moreover, from points *B* and *D** terminating chord *BG*, let two perpendiculars apiece be drawn — *GH* and *BK* — to the two radii *ZE* and *ZD*, and the two others *GL* and *BM* to the radius *ZA*. Indeed, it is established that these perpendiculars are the sines of the arcs from whose ends they emerge. And finally

*For *D* read *G*.

dem puncta h & k copulētur per líneam h k.Si ígítur duo arcus a b & a g. æ=
quales fuerint,erunt duæ líneæ g l & b m æquales & contermi[n]ales, itemq̃ li=
nea g h líneæ b k æqualis quidem propter arcus b d & g e æquales, æquedi=
stans autem per diffinitionem superficiei ad superficiem erectæ & quartā undeci=
mi,unde & per   primi duæ líneæ rectæ b g & k h æquales habentur. est autem
g b nota scilicet corda arcus g b dati,quare & linea h k nota fiet. Item duæ su=
perficies l h & m k sunt æquedistantiū laterū,quare per   primi g l nota æqua=
lis erit h 3 & b m cognita ipsi k 3 . tríangulus itaq̃ h 3 k planus rectilíneus
tria latera habens cognita per   primi huius angulū h 3 k manifestabit, cuius
quantitatem determinat arcus d e,& ideo arcus d e diffiniens quantitatem an=
guli b a g notus concluditur.habet igitur tríangulus sphæralis a b g duo latera
b a & a g cognita cum angulo b a g,unde & per   huius reliqui sui anguli non
latebunt. Aliter tamen & facilius procedere poterimus, si duo arcus a b & a g
æquales fuerint,hoc pacto,ex a puncto descendet perpendicularis arcus a d, qui
necessario satis continuatus secabit latus b g ipsius trianguli, quod fiat in pun=
cto d. eritq̃ b d æqualis g d per   .quare cū totus arcus b g sit notus,erit b d
datus,per   autē huius sinus cōplementi b a ad sinum cōplementi a d se habet,
sicut sinus cōplementi b d ad sinū totū.unde a d notus erit.
Hinc quoq̃ angulus a b d,& ei æqualis a g b. item tandē

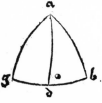

angulus b a g &c.Si autē alter duog̃ arcuū b a & a g reli=
quo maiorem se offerat, uterq̃ tñ minor quadrante. sit uerbi
gratia arcus a b maior arcu a g,quāobrem alternatim erit
arcus g e maior arcu b d,repetitaq̃ figuratione pristina e=
rit linea g h longior linea b k, abscindatur ergo ex ea por=
tio h n æqualis líneæ b k,ducta línea b n,quā ut supra opor
tet esse æqualē líneæ h k,similiter tam g l líneæ h 3 æqua=
lis erit, q̃ b m ipsi k 3 .erit autē angulus b n g rectus, cunq̃ duo latera b g &
g n trianguli g b n rectanguli sint cognita. est em b g corda arcus b g per hy
pothesim noti. g n autē differentia duog̃ sinuum b k & g h notog̃, erit & latus
eius b n per   primi huius notum.et ideo h k nota declarabitur . Duas autē li=
neas h 3 & 3 k quēadmodum ante hac notas declarabimus. tria igitur latera tri=
anguli h 3 k nota sunt.unde angulū h 3 k & reliqua ut nuperrime dimetiemur.
Quòd si uterq̃ arcuum a b & a g quadrante ma=
ior extiterit,protendantur ipsi donec in puncto s
coincidēt,quo fit ut duo arcus b s & s g noti red
dantur.Est enim per   primi huius uterq̃ arcuum
a b s,& a g s semicircumferentia nota. Triangu
lus ergo b s g trium notog̃ laterum ex iā dictis
angulum b s g æqualē angulo b a g sortietur cognitum. Ad uiā itaq̃ pristinā
perducti,ex duobus lateribus a b & a g cū angulo b a g cognitis, reliquos du=
os angulos   huius dirigente inuestigabimus. Postremo sit alter duog̃ arcuum a
b & a g maior quarta circumferentiæ,reliquus uero minor. & sit uerbi gratia a
b quadrante maior, a g autē minor eo,resumptaq̃ figuratione nihil in ea uaria=
bimus,nisi q̃ lineam g h continuemus ad partē puncti h,donec h n fiet æqua=
lis sinui b k arcus b d. similiter a 3 semidiameter prolongetur, ita ut b m ipsi
perpendiculariter insidere possit.quibus ita dispositis argumētabimur hoc pacto
Corda b g nota est propter arcum suum notū. linea g cōplectitur sinum arcus
g e noti,& lineā h n æqualē b j k sinui arcus b d noti.tota ergo g n nota est,an
gulus

Q 2

let points H and K be joined by line HK. Then, if the two arcs AB and AG are equal, the two lines GL and BM will be equal and adjacent. Similarly line GH will be equal to line BK, because arcs BD and GE are equal, and [will be] parallel [to line BK] by the definition of a surface erected [perpendicularly] upon [another] surface and by [Euclid] XI.4. Hence by [Euclid] I.[not given] the two straight lines BG and KH are equal. Furthermore, GB is known—namely, the chord of given arc GB—and thus line HK will be known. Similarly, the two surfaces LH and MK [have] parallel sides [GH and BK]; therefore, by [Euclid] I.[not given], known [line] GL will be equal to HZ and known [line] BM will be [equal] to KZ. Then by Th. I.[47] above, rectilinear planar $\triangle\ HZK$, having three known sides, will reveal [its] $\angle\ HZK$, whose size determines arc DE. Thus arc DE, which defines the size of $\angle\ BAG$, is concluded known. Therefore spherical $\triangle\ ABG$ has two sides BA and AG known together with $\angle\ BAG$, and hence by Th. [30] above its other angles will not be hidden.

We could proceed differently and more easily, if the two arcs AB and AG are equal, in this way: From point A let perpendicular arc AD descend, which, extended sufficiently, will necessarily intersect side BG of this triangle at point D. BD will be equal to GD by [Th. III.56 above, but more likely I.4 or I.5 of Euclid is intended]. Therefore, since the total arc BG is known, BD will be found. Furthermore, by Th. [19] above, the sine of the complement of BA to the sine of the complement of AD is as the sine of the complement of BD to the whole sine. Thus AD will be known. Hence also $\angle\ ABD$ and its equal AGB [will be known]. Finally, $\angle\ BAG$, etc., will similarly [be found].

However, if one of the two arcs BA and AG is greater than the other yet each is less than a quadrant, then let arc AB, for example, be greater than arc AG; whereupon, in turn, arc GE will be greater than arc BD, and when the former construction is consulted, line GH will be longer than line BK. Therefore let

a portion HN, equal to line BK, be cut by line BN, which, when drawn, must equal line HK as above. Similarly, as GL will be equal to line HZ, so BM [will be equal] to KZ. Furthermore, $\angle\ BNG$ will be a right [angle]; and since the two sides BG and GN of right $\triangle\ GBN$ are known, for chord BG of arc BG is known by hypothesis while GN is the difference of the two known sines BK and GH, then side BN [of $\triangle\ GBN$] will be known by Th. I.[26] above. Thus HK will be declared known. Moreover, we may declare the two sides HZ and ZK [to be] known as before. Therefore the three sides of $\triangle\ HZK$ are known. Hence we may measure $\angle\ HZK$ and the others as previously.

But if each of the arcs AB and AG is greater than a quadrant, let them be extended until they meet at point S,[*] whereby it happens that the two arcs BS and SG become known. For by Th. [III.39 above or perhaps a theorem of I of Euclid] each of the arcs ABS and AGS is a known semicircumference. Therefore $\triangle\ BSG$, [formed] of three sides [that are] known from the already-mentioned proofs, will have a known $\angle\ BSG$, equal to $\angle\ BAG$. Therefore, guided by the previous method, since the two sides AB and AG are known together with $\angle\ BAG$, we will find the other two angles by the directions of Th. [28] above.

Finally, let one of the two arcs AB and AG be greater than a quarter of the circumference while the other is smaller. For example, let AB be greater than a quadrant and AG be less. When the [first] construction is resumed, we will change nothing in it except that we will extend line GH beyond point H until HN becomes equal to the sine BK of arc BD. Similarly, let radius AZ be extended so that BM can rest on it perpendicularly. With these so arranged, we will reason in this way. Chord BG is known because its arc is known. Line G[†] comprises the sine of known arc GE, and line HN is equal to the sine BK of known arc BD. Therefore the total [line] GN is known.

[*]In the figure, for d read s.
[†]For G read GH.

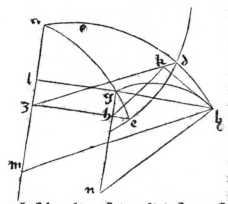

gulus autem b n g rectus est, nã superfi
cies h n b k æquedistantibus contine
tur lineis, & angulus eius apud k rectus
quare per　primi huius latus n b trianguli b n g rectanguli notũ erit, cui
æqualis est linea h k. trianguli itacp pla
ni h 3 k tribus lateribus cognitis angu
lum h 3 k & reliqua queadmodum ante
hac mensurabimus. ¶ Operatioñi au
tem par loci dabimus, nihil em docebi
mus nisi quo pacto reperiantur tria latera triãguli plani, cuius unus angulus
in centro sphæræ quiescẽs respondet an
gulo sphærali quæsito, qualis in figura est triangulus h 3 k. reliqua enim in locis
suis superioribus satis explanasse uidemur. Siue igitur duo arcus a b & a g æqua
les fuerint, siue non, sinus eor accipiemus pro duobus lateribus continentibus an
gulum trianguli plani in centro sphæræ quiescentem, qui respondet angulo sphæ
rali quæsito. pro tertio autem latere trianguli plani cordam arcus tertij constitue-
mus, si æquales occurrant duo arcus angulũ quæsitum continentes, qui si fuerint
inæquales, uterc̈p tñ aut minor quarta, aut maior ea sinus cõplemetor arcuũ, qui
angulum quæsitũ ambiũt, eliciere, & differentiam eor in se multiplicatã ex quadrato cordæ arcus tertij minuas, relictic̈p radicem quadratã pro latere tertio trian
guli plani ponas. Qd si alter eor maior quadrante, reliquus autem minor eo fuerit, sinus complementor huiusmodi collige, & summã eorũ in se ductã, ex quadrato cordæ arcus tertij subtrahas, relicti enim radix quadrata lateri tertio trianguli
plani adnumerabitur. In exemplo. Sit uterc̈p arcuũ a b & a g 20.gr. & arcus b g
36. sinus 20. graduũ est 20521. tantum c̈p numerabo utruncp later k 3 & 3 h. Sinus 36.gr. est 35267. tantũ est latus h k, similiter faciam, si uterc̈p arcuũ a b & a
g occurrat 160.gr. & arcus b g 36. redibunt enim pristini numeri. Sed dabit mihi quispiam latus a b 35.gr. latus a g 36. & latus b g 46. sinus 25.gr. est 25357.
quem dabo lateri k 3. sinus 16. est 16538, lateri 3 h adnumerandus corda b g
46888. Sinus complementi 16. graduũ est 57676. Sinus complementi 25. gra. est
54378. quem minuo ex 57676. manet 3298. hæc in se faciunt 10876804. quæ sublata ex quadrato cordæ b g, scilicet 2198484544. manet 2187607740. quorum
radicem quadratam 46772. constituo pro linea h k. Tandem ponatur latus a b
130 .gra. latus a g 75. & latus b g 65. Sinus 130 .gra. est 45963. pro latere k 3.
Sinus 75.gr. 57956. numerus lateris h 3, & corda 65.gr. est 64476. sinus complementi 130.gr. est 38567. sinus complementi arcus a g est 15529. quem addo ad
38567. colliguntur 54096. hæc in se faciunt 2926377216. quæ subtraho ex quadrato cordæ b g, quod est 4157154576. manet 1230777360. huius radicem quadratã 35082. lateri h k deputabo. Perducti igitur ad hypothesim　primi huius,
quæ nobis angulũ h 3 k cognitũ faciet, unde & angulus d e numerabitur, quod
non erit ambiguũ, si ea quæ in processu　primi huius cõmemorauimus, memo-
riæ mandasti. arcus autem d e quãtitatem anguli b a g determinat, unde & b
a g & reliqui anguli nõ erunt ignoti. Poterimus aũt alio tramite idem attingere.
Nam ppositus triangulus a b g habeat duo latera a b & a g inæqualia, utruc̈p
tamen minus quadrante, & libeat inuenire quantitatem anguli b a g, super pun
cto a facto polo describatur circulus magnus in sphæra d h, cuius circumferen-
　　　　　　　　　　　　　　　　　　　　　　　　　　　　　　　　　　　tiæ oc

Moreover, ∠ *BNG* is a right [angle] for the surface *HNBK* is enclosed by parallel lines, and its angle at *K* is a right [angle]. Therefore, by Th. I.[26] above, side *NB* of right △ *BNG* will be known, [and] line *HK* is equal to [*NB*]. Thus, since three sides of planar △ *HZK* are known, we will measure ∠ *HZK* and the others as before.

Now, we will give little space to the mechanics, for we will teach nothing except how three sides may be found of the planar triangle whose one angle, resting at the center of a sphere, corresponds to the desired spherical angle. *HZK* in the figure is a triangle of this sort. We seem to have explained the other things sufficiently in their [appropriate] places above. Thus, whether the two arcs *AB* and *AG* are equal or not, we will use their sines as the two sides of the planar triangle that contain the angle resting in the center of the sphere. This [angle] corresponds to the desired spherical angle. Furthermore, for the third side of the planar triangle we will take the chord of the third arc if the two arcs containing the desired angle are equal. [But] if [the two arcs] are unequal yet each is either less than a quarter [of a circumference] or greater than one, then find the sines of the complements of the arcs which include the desired angle; subtract their difference–squared from the square of the chord of the third arc; and of the remainder, take the square root for the third side of the planar triangle. But if one of the [two mentioned arcs] is greater than a quadrant while the other is less than one, add the sines of these complements and subtract their sum–squared from the square of the chord of the third arc, for the square root of the remainder will be considered as the third side of the planar triangle.

For example, let each of the arcs *AB* and *AG* be 20° and arc *BG* 36°. The sine of 20° is 20521, and this value will be considered [to be] each of the sides *KZ* and *ZH*. The sine of 36° is 35267. This value is side *HK*. One may do [this] similarly if each of the arcs

AB and *AG* is 160° and arc *BG* 36°, for the former numbers will return.

But someone may give me side *AB* as 35°,* side *AG* as 16°, and side *BG* as 46. The sine of 25° is 25357, which I will assign to side *KZ*. The sine of 16° is 16538, to be considered as side *ZH*, while chord *BG* is 46888. The sine of the complement of 16° is 57676. The sine of the complement of 25° is 54378, which is subtracted from 57676, leaving 3298. This [value] squared makes 10876804, which, subtracted from the square of chord *BG*—namely, 2198484544—leaves 2187607740. The square root of this, 46772, I establish as line *HK*.

Finally, let side *AB* be 130°, side *AG* 75°, and side *BG* 65°. The sine of 130° is 45963, for side *KZ*. The sine of 75° is 57956, the value of side *HZ*, and the chord of 65° is 64476. The sine of the complement of 130° is 38567. The sine of the complement of arc *AG* is 15529, which is added to 38567, giving 54096. This [value] squared makes 2926377216, which is subtracted from the square of chord *BG*, [or] 4157154576, leaving 1230777360. The square root of this, 35082, is assigned to side *HK*.

[In all these examples] therefore [we are] guided to the hypothesis of Th. I.[47] above, which will make ∠ *HZK* known to us, and hence arc† *DE* will be computed. This will not be ambiguous if you committed to memory those things which we related in the procedure of Th. I.[20] above. Moreover, arc *DE* determines the size of ∠ *BAG*, and hence *BAG* and the other angles will not be unknown.

We could arrive at the same thing by a different path. Let the proposed △ *ABG* have two sides *AB* and *AG* that are unequal yet each less than a quadrant, and let it be desired to find the size of ∠ *BAG*. Around point *A* as a pole, let a great circle *DH* be described in the sphere; the circumference of this [circle]

*For 35° read 25°.

†In Latin, for *angulus* read *arcus*.

tiæ occurrat arcus g b prolongatus in puncto h. duo etiam arcus a b & a g ex
tendantur ad puncta e & d, erit uterq; arcuũ a e & a d quadrãs orthogonaliter erectus ad arcũ d h. per autem huius proportio finus arcus g d ad finũ arcus b e eft tanq; finus g h ad finũ arcus b h. eft autem uterq; duog; arcuũ g d &
b e notus, propter fua complementa a g & a b per hypothefim nota. proportio
igitur finus g h ad finũ b h nota habebitur, cũq; differentia duog; arcuũ g h &
b h fit nota, fcilicet arcus g b, erit per huius uterq; eog; arcuũ nota. hinc ex arcu
b h & b e cognitis per huius ꝓpter angulũ e rectum, complementũ arcus e h
& ideo ipfe arcus e h innotefcet. Similiter ex duobus arcubus g d & g h, arcus
d h cognofcetur, quare differentia duog; arcuũ d h & e h fcilicet arcus d e non
ignorabitur, ipfe autem determinat quãtitatem anguli b a g, quare angulus ille
notus erit, hinc & reliqui ut antehac noti fient anguli. Si uero alter quidem duog;
arcuũ a b & a g maior quadrante fuerit, alter aũt minor eo, fit a b maior quadrãte, defcriptoq; ut prius circulo magno fuper a polo, circumferentia eius fecabit arcum g b, quod fiat in puncto, fecabit etiã arcũ b g in puncto, qui fit h, arcus aũt a g continuatus, occurrat ei in puncto d, erit autem uterq; duog; arcuum
e b & g d notus: funt enim cõplementa arcuũ datog;, eft autẽ ꝓportio finus b e
ad finum g d nota, ficut proportio finus b h ad finũ h d per huius. fic proportio finus b h ad finũ h g nota habebitur: cunq; totus arcus b g datus fit per hypothefim, erit per huius uterq; arcuũ b h & h g cognitus. ex duobus aũt arcubus b h & b e cognitis, & propter angulũ e rectum per huius innotefcet arcus
e h. fimiliter duo arcus h g & g d propter angulũ d rectum notificabunt arcũ
d h, fic totus arcus d e determinans quantitatem anguli b a g non latebit, unde angulus b a g cum reliquis trianguli propofiti angulus notus ꝓclamabiť.
Si uero uterq; arcuum a b & a g quadrantem fuperauerit, producantur ipfi donec concurrant, eritq; angulus apud concurfum eorum æqualis ipfi angulo a, fi-
etq; nouus triangulus fuper bafi b g, cuius duo latera minora quadrante nota
erunt, unde reliqua ut fuperius abfoluentur. Operationis autem tenorem præte-
reo, nam ab operationibus · huius atq; pendere dinofcitur.

Quarti libri Triangulorum Finis.

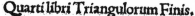

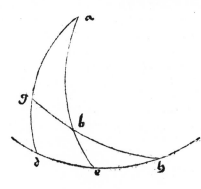

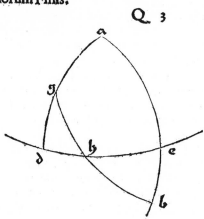

Q 3

meets arc *GB*, extended, at point *H*. Let the two arcs *AB* and *AG* also be extended to points *E* and *D* [respectively]. Each of the arcs *AE* and *AD* will be a quadrant erected orthogonally to arc *DH*. Furthermore by Th. [15] above the ratio of the sine of arc *GD* to the sine of arc *BE* is as that of the sine of *GH* to the sine of arc *BH*. Moreover, each of the two arcs *GD* and *BE* is known because their complements *AG* and *AB* are known through the hypothesis. Therefore the ratio of the sine of *GH* to the sine of *BH* will be known; and since the difference of the two arcs *GH* and *BH* is known— namely, arc *GB* — then by Th. [22] above each of these arcs will be known. Hence, because arc *BH* and [arc] *BE* are known, [and] because ∠ *E* is a right angle, the complement of arc *EH*, and therefore arc *EH* itself, will become known by Th. [25] above. Similarly, from the two arcs *GD* and *GH*, arc *DH* will be found. Therefore the difference of the two arcs *DH* and *EH* — namely, arc *DE* — will not be unknown. Furthermore, [arc *DE*] determines the size of ∠ *BAG*. Therefore that angle will be known, and hence the rest of the angles may be found as before.

However, if one of the two arcs *AB* and *AG* is greater than a quadrant while the other is less, let *AB* be greater than a quadrant. And when, as before, great circle [*DE*] is described around *A* as the pole, its circumference will intersect arc *GB** at point *E*.† It will also intersect arc *BG* at point *H*, while

arc *AG* extended meets it at point *D*. Now, each of the two arcs *EB* and *GD* will be known, for they are the complements of the given arcs. Furthermore, the known ratio of the sine of *BE* to the sine of *GD* is as the ratio of the sine of *BH* to the sine of *HD*‡ by Th. [15] above. Therefore the ratio of the sine of *BH* to the sine of *HG* will be known, and since the total arc *BG* is given by the hypothesis, each of the arcs *BH* and *HG* will be known by Th. [21] above. Moreover, because the two arcs *BH* and *BE* are known and because ∠ *E* is a right [angle], then, by Th. [25] above, arc *EH* will become known. Similarly, the two arcs *HG* and *GD* will identify arc *DH* because ∠ *D* is a right [angle.] Thus the total arc *DE* that determines the size of ∠ *BAG* will not be hidden. Hence ∠ *BAG* will be declared a known angle together with the other [angles] of the proposed triangle.

However, if each of the arcs *AB* and *AG* were to exceed a quadrant, let them be extended until they meet. The angle at their juncture will be equal to ∠ *A*. And a new triangle will be formed upon base *BG* whose two sides will be known [to be] less than a quadrant. From this the other [unknowns] may be found as above.

I am by-passing the course of operation, for it is discerned to follow from the procedures of Ths. [15, 21 or 22, and 25] above.

*For *GB* read *AB*.
†In Latin, for *in puncto* read *in puncto e*.
‡For *HD* read *HG*.

The End of the Fourth Book of Triangles.

LIBER QVINTVS
TRIANGVLORVM.

I.

Si fuerint duo trianguli rectanguli, qui duos angulos acutos & æ-
quales habeant, latera autem rectos subtendentia angulos inæqualia,
erit proportio sinus differentiæ eorum laterum ad sinum differentiæ
duorum laterum rectiis atcɜ acutis æqualibus substratorum, tancɜ pro
portio eius, quod sub sinibus complementorum acutis angulis subten
sorum laterum continetur, ad id, quod sub sinu toto & sinu cõplemen
ti anguli acuti.

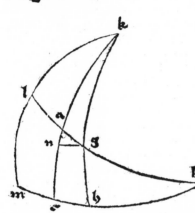

Triangulos huiusmodi ex arcubus circulorum magnoꝶ unius sphæræ, aut duarum
æqualium concludi subauditur, de reliquis
enim in præsentiarum nihil disserimus. Sint
tales duo trianguli a b c & d e f, duos qui
dem angulos a g b & d f e rectos, duos au
tem a b c & d e f acutos æquales habētes:
sitcɜ latus a b trianguli a b c longius late
re d e trianguli d e f, quæ res arguet etiam
latus b c unius trianguli longius esse latere
e f alterius, si tertium huius satis tenes. Dico itacɜ, cɜ sinus rectus differentiæ duorū laterum a b & d e ad sinum differentiæ duorum laterū b c & e f proportionem habet,
quam id, quod sub sinu toto & sinu complementi anguli a b c, siue d e f ad id, quod sub sinubus
complementorū duoꝶ arcuum a c & d f continetur.
Quod ut facilius demonstretur, scindo ex a b arcū g b
æqualem arcui d e, & ex b c arcum b h æqualem arcui e f, continuatiscɜ duobus punctis g & h per arcū
g h, constabit per secundi huius angulū g h b esse
rectum: & ideo per secundi huius duo circuli, quoꝶ sunt arcus a c & g h p polos circuli b c transibunt: duo itacɜ arcus a c & g h continuati paulo superius
concurrant in puncto k, qui necessario polus circuli b c habebitur, & utercɜ arcu
um k c & k h quadrans circumferentiæ magnæ pronunciabitur, intelligatur de
nicɜ circulus magnus transiens per polum k, & polum circuli a b, cuius arcus k
m occurrat duobus a b & b c prolongatis, quoad satis erit uersus sinistrā, huic
quidem in puncto l, isti autem in puncto m: occurret autem orthogonaliter se
cundi huius arguente, eritcɜ per & eiusdem unusquiscɜ trium arcuū k m, b l
& b m quarta circumferentiæ magnæ: unde & arcus l m quantitatem anguli a
b c determinabit, cuius complementū est k l: nemo autem ignorat arcum a g es
se differentiā duoꝶ arcuū a b & d e. item c h excessum arcus c b super f e, demittatur itacɜ ex puncto g perpendicularis arcus g n ad quadrantem k c, proportio igitur sinus a g ad sinum c h componitur ex duabus, proportione scilicet
sinus

FIFTH BOOK OF THE TRIANGLES

Theorem 1

If there were two right triangles, which have two equal acute angles but unequal sides subtending the right angles, then the ratio of the sine of the difference of these sides to the sine of the difference of the two sides underlying the right [angles] and the equal acute [angles] will be as the ratio of the product of the sines of the complements of the sides subtending the acute angles to the product of the whole sine times the sine of the complement of the acute angle.

It is understood that these triangles are constructed of arcs of great circles of one sphere or of two equal [spheres], for we are discussing nothing else in the present [theorems]. Let *ABC* and *DEF* be two such triangles, having two right angles *AGB*** and *DFE* and two equal acute [angles] *ABC* and *DEF*, and let side *AB* of △ *ABC* be longer than side *DE* of △ *DEF*, which fact also makes it clear that side *BC* of the one triangle is longer than side *EF* of the other, if you understand the Third Book above sufficiently. Then the right sine of the difference of the two sides *AB* and *DE* to the sine of the difference of the two sides *BC* and *EF* has the [same] ratio as that of the product of the whole sine times the sine of the complement of ∠ *ABC*, or *DEF*, to the product of the sines of the complements of the two arcs *AC* and *DF*.

In order that this may be shown more easily, from *AB* intercept arc *GB* equal to arc *DE*, and from *BC* [intercept] arc *BH* equal to arc *EF*. When the two points *G* and *H* are joined by arc *GH*, it will be established by Th. [III.36] above that ∠ *GHB* is a right [angle]. Therefore by Th. [III.20 or IV.2] above the two circles, of which *AC* and *GH* are arcs, will pass through the poles of circle *BC*; thus the two arcs *AC* and *GH*, extended a little above, will meet at point *K*, which will necessarily be the pole of circle *BC*. And each of the arcs *KC* and *KH* will be declared a quadrant of the great circumference. Next, let a great circle be understood passing through pole *K* and through the pole of circle *AB*. Arc *KM* of this [circle] meets the two [arcs] *AB* and *BC*, extended sufficiently toward the left— the former [*AB*] at point *L* and the latter [*BC*] at point *M* — and furthermore, it meets [them] orthogonally by the reasoning of Th. [III.17] above. By Ths. [III.11] and [IV.1] above, each of the three arcs *KM*, *BL*, and *BM* will be a quarter of the great circumference. Hence arc *LM* will determine the size of ∠ *ABC*, whose complement is *KL*. Furthermore, no one will be unaware that arc *AG* is the difference of the two arcs *AB* and *DE*. Similarly, *CH* is the excess of arc *CB* over *FE*. Therefore let perpendicular arc *GN* be dropped from point *G* to quadrant *KC*. Then the ratio of the sine of *AG* to the sine of *CH* is composed of two [ratios] — namely, of the ratio

*For *AGB* read *ACB*.

ſinus a g ad ſinum g n,& ex proportione ſinus g n ad ſinū c h,eſt autem per
tertij huius ſinus a g ad ſinum g n ,ſicut ſinus a k ad ſinum k l?ſinus præterea
g n ad ſinum c h ex eodem loco eſt,ut ſinus g k ad ſinū k h. proportio igitur ſi
nus a g ad ſinum c h componitur ex duabus ſcilicet proportione ſinus a k ad ſi
num k l,& proportione ſinus g k ad ſinū totum,ſed ex illis componitur etiã per
ſexti elementoꝝ,quod ſub ſinibus a k & g k continetur ad id,quod ſub ſinu to
to & ſinu k l ſcilicet complementi anguli a b c continetur,uerum igitur enunci
abat theorema præſens.

II.

In omni triangulo ſphærali ex arcubus circulorum magnorum con
ſtante,proportio ſinus uerſi anguli cuiuslibet ad differentiam duorū
ſinuum uerſorum,quorum unus eſt lateris eum angulum ſubtenden=
tis,alius uero differentiæ duorum arcuum ipſi angulo circumiacentiũ
eſt tanꝗ proportio quadrati ſinus recti totius ad id,quod ſub ſinibus
arcuũ dicto angulo circūpoſitorum continetur rectangulum.

Sit huiuſmodi tri=
angulus a b g,duo la
tera habens inæqua=
lia a c maius a b,&
utrunꝗ eorum minus
quadrante,ſuper pun
ctis a & b factis po=
lis deſcribant duo cir
culi magni,quoꝝ cir=
cumferentiæ ſe ſecent
in puncto e:prolonge
turꝗ arcus a b utriꝗ
donec occurret circu=

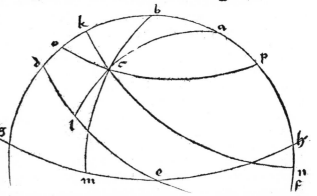

lo ſup a deſcripto in punctis d & f,circulo aut ſup b lineato in punctis g & h,cõſtat
arcū g d æqualē eſſe arcui a b:utrunꝗ aut arcuū e d & e g eſſe quadrantē circūfe
rētiæ magnæ.cõtinuēt deniꝗ duo arcus a c & b c,donec uterꝗ eoꝝ quarta circū
ferentiæ fiat:hic quidem occurrens arcui d e in l puncto,ille aut arcui g e in pū
cto m :factiſꝗ iteꝝ punctis a & b polis,ſuper a quidē ſecundū quantitatem cor
dæ a c deſcribat circulus minor in ſphera k n :ſuper b aut ſecundū diſtantiã b
circulus minor o p,erit itaꝗ arcus b k differentia duoꝝ arcuū a b & a c,& ar=
cus b o æqualis arcui b c,arcus aut d l quantitatē anguli b a c determinabit.
Dico igit,ꝗ pportio ſinus uerſi d l ad differentiã duoꝝ ſinuū uerſoꝝ,quos habēt
arcus b k & b c eſt ut quadrati ſinus recti totius ad id,quod continet ſub duobus
ſinibus rectis arcuū a b & a c.Qꝺ ut apertius demonſtret,altera figuratio aſſu=
menda eſt,in qua ſit circulus g b h,quemadmodũ in prima ſup centro x ,quod &
centrū ſphæræ habebit,ſirꝗ cõis ſectio circuloꝝ g b h & g e h diameter g h,quã
in prima figūra lineare nõ decuit confuſionis uitandæ gratia,ſed & cõis ſectio cir
culoꝝ g b h & d e f ſit linea d f.item duo circuli g b h & k c n ſecent ſe in linea
k n ,duo demū circuli g b h & o c p in linea recta o p cõuincēt,deinde educant
duæ ſemidiametri ſphæræ x a quidē ſecanslinea k n in puncto y, x b aut ſecās
o p in puncto r,conſtat aut per   huius k n eſſe diametrū circuli k c n,& o p
l ineã eſſe diametrū circuli o c p,ꝗ circulus magnus g b h per polos utriuſꝗ eo
rum in

of the sine of *AG* to the sine of *GN* and of the ratio of the sine of *GN* to the sine of *CH*. Moreover, by Th. [IV.15] above, the sine of *AG* to the sine of *GN* is as the sine of *AK* to the sine of *KL*. Besides, the sine of *GN* to the sine of *CH* is as the sine of *GK* to the sine of *KH* from the same [theorem]. Therefore the ratio of the sine of *AG* to the sine of *CH* is composed of two [ratios] — namely, of the ratio of the sine of *AK* to the sine of *KL* and of the ratio of the sine of *GK* to the whole sine. But by VI.[not given] of the *Elements*, because of these [component ratios] the [ratio of the sine of *AG* to the sine of *CH*] is composed also of the product of the sines of *AK* and *GK* to the product of the whole sine times the sine of *KL* — namely, [the sine] of the complement of ∠ *ABC*. Q.E.D.

Theorem 2

In every spherical triangle that is constructed from the arcs of great circles, the ratio of the versed sine of any angle to the difference of two versed sines, of which one is [the versed sine] of the side subtending this angle while the other is [the versed sine] of the difference of the two arcs including this angle, is as the ratio of the square of the whole right sine to the rectangular product of the sines of the arcs placed around the mentioned angle.

Let *ABG** be a triangle of this sort, having two unequal sides, *AC* greater than *AB*, and each of them less than a quadrant. Around points *A* and *B* as poles, let two great circles be described, whose circumferences intersect each other at point *E*; let arc *AB* be extended on both sides until it meets the circle described around *A* at points *D* and *F* and [until it meets] the circle drawn around *B* at points *G* and *H*. It is established that arc *GD* is equal to arc *AB*. Furthermore each of the arcs

ED and *EG* [is established] to be a quadrant of the great circumference. Next let the two arcs *AC* and *BC* be continued until each of them is a quarter of the circumference, the former meeting arc *DE* at point *L* while the latter [meets] arc *GE* at point *M*. Again, since the points *A* and *B* are poles, let a small circle *KN* be described in the sphere around pole *A* with the length of chord *AC* [as polar line], and let a small circle *OP* [be described] around *B* with the length of *B*† [as polar line]. Then arc *BK* will be the difference of the two arcs *AB* and *AC*; arc *BO* will be equal to arc *BC*; and furthermore, arc *DL* will determine the size of ∠ *BAC*. Then it can be said that the ratio of the versed sine of *DL* to the difference of the two versed sines which arcs *BK* and *BC* have is as that of the square of the whole right sine to the product of the two right sines of arcs *AB* and *AC*.

So that this may be shown more clearly, another construction is also to be used. In this, just as in the first construction, let there be a circle, *GBH*, around a center *X*, which is also the center of the sphere. Let the line of intersection of the circles *GBH* and *GEH* be diameter *GH*, which it was not appropriate to draw in the first figure, to avoid confusion. Furthermore, let the intersection of the circles *GBH* and *DEF* be line *DF*. Similarly the two circles *GBH* and *KCN* intersect one another on line *KN*. Finally the two circles *GBH* and *OCP* are shown [to intersect] on straight line *OP*. Next, let two radii of the sphere be drawn — *XA* intersecting line *KN* at point *Y* and *XB* intersecting *OP* at point *R*. Furthermore, it is established by Th. [III.7, with III.14, and III.17] above that *KN* is the diameter of circle *KCN* and line *OP* is the diameter of circle *OCP* because great circle *GBH* passes through the poles of each of them.

*For *ABG* read *ABC*.
†For *B* read *BC*.

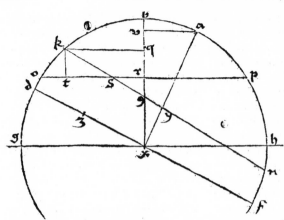

rū incedat, punctusq́ y erit
centrū circuli k c n, & pun
ctus r centrū circuli o c p,
rurſus ex puncto k duæ ꝑ
ueniāt ꝑpendiculares, una
quidē k t ad lineā o p, alia
aũt k 9 ad lineā b x, pun
ctus item a ſit caput ꝑpen
dicularis a u ſupra lineā b
x deſcendentis. erit itaq́ li
nea a u ſinus rectus arcus
a b: b q aũt ſinus uerſus ar
tus b k, & b r ſinus uerſus
arcus b o ſiue b c, horū ſi

nuum uerſoꝛ differentia eſt linea q r æqualis lineæ k t. ſignato iterū puncto s,
in quo cōicant duæ diametri circuloꝛ k c n & o c p ſcilicet lineæ k n & o p, ſci
endū eſt lineā k s eſſe ſinū uerſum arcus k c primæ figuræ, cū em uterq́ circuloꝛ
k c n & o p c orthogonaliter ſit erectus ſupra circulū g b h, per polos eoꝛ ince
dente, ut docet huius, erit eoꝛ ſectio cōis, quæ ꝑtranſit punctū s ex undeci
mi elementoꝛ orthogonaliter erecta ſupra circulū g b h, & ideo per conuerſionē
diffinitionis lineæ ꝑpendiculariter erectæ ſupra ſuperficiē, hæc cōis ſectio lineæ k
n, diametro ſcilicet circuli k c n orthogonaliter incidit, & ideo hāc ſectionē cōem
diameterᴀk n diuidit per medium, eritq́ medietas eius ſinus rectus arcus k c, &
ideo k s ſinus uerſus eiuſdē. cūq́ d f recta ſit diameter circuli d e f. ſumaꝛ ex ea
d 3 linea ſinus ſcilicet uerſus ipſius arcus d l, ad quā ſe habebit k s ſicut ſemidia
meter circuli k c n, q̃ eſt k y ad ſemidiametrū circuli d e f ſcilicet d x: ſunt em
duo arcus k c & d l ſimiles per huius, ꝓpter duos arcus a d & a l à polo duo
rū circuloꝛ k c n & d e f deſcendētes, & dictos arcus includētes. Iā ad portū acce
demus, ubi prius duos triāgulos k t s & a u x æquiāgulos oſtēdemus hoc pacto:
duo anguli t k s & r 9 s, ideoq́ x 9 y ſunt æquales, ꝓpter æquediſtantiā linearū
k t & r 9, quare cū uterq́ anguloꝛ k t 3 & x y 9 ſit rectus, erit reliquus angulus
k s t æqualis reliquo y x 9 ſiue a x u. angulus aũt a u x rectus diſponebatur,
ſunt itaq́ duo trianguli k t s & a u x æquianguli, & ideo per quartā ſexti & ꝑmu
tatim ꝓportio x a ad a u ſicut s k ad k t. proportio aũt d 3 ſcilicet ſinus uerſi
anguli b a c ad lineā k t ſcilicet differentiā duoꝛ ſinuū uerſorū ſupra memorato
rum componiꝛ ex duabus ſcilicet ꝓportione ſinus d 3 ad k s, & ex ꝓportione
k s ad k t: erat aũt d 3 ad k s ſicut d x ſcilicet ſinus totus ad lineā k y, quæ eſt
ſinus rectus arcus a k ſiue a c, & ꝓportio k s ad k t ſicut a x ſinus recti totius
ad lineā a u, quæ eſt ſinus rectus arcus a b, ꝓportio igiꝛ ſinus uerſi arcus d l ſiue
anguli b a c, ad differentiā ſinuū uerſoꝛ ſupradictorū cōponiꝛ ex duabus, ex pro
portione ſcilicet ſinus recti totius ad ſinum rectum arcus a c, & ex ꝓportione ſi
nus recti totius ad ſinū rectum arcus a b, ex eiſdē aũt duabus proportiōibus cō
poniꝛ etiā, ꝓportio quadrati ſinus recti totius ad id, quod cōtinetur ſub ſinibus re
ctis duorū arcuū a b & a c, proportio igiꝛ ſinus uerſi anguli b a c ad differentiā
duoꝛ ſinuū uerſorum, quos habent duo arcus b o & b k eſt, ut ꝓportio quadrati
ſinus recti totius ad id, quod contineꝛ ſub duobus ſinibus rectis duoꝛ arcuū a b &
a c, quod erat demonſtrandū. Hoc tñ non ignorare uideris, q̃ duo ſinus uerſi duo
rum arcuū b k & b o, eam habent differentiā, quā duo ſinus recti cōplementorū
ſuoꝛ

Point Y will be the center of circle KCN and point R the center of circle OCP. Again, from point K let two perpendiculars emerge — one, KT, to line OP and the other, $K9$, to line BX. Similarly, let point A be the top of perpendicular AV descending to line BX. Then line AV will be the right sine of arc AB; moreover, BQ will be the versed sine of arc BK, and BR will be the versed sine of arc BO, or BC. The difference of these versed sines is line QR, equal to line KT. Again, when point S is marked, at which [point] the two diameters of circles KCN and OCP — namely, the lines KN and OP — meet, it is to be understood that line KS is the versed sine of arc KC of the first figure. For since each of the circles KCN and OPC is erected orthogonally upon circle GBH that passes through their poles as Th. [III.17] teaches, then their line of intersection, which passes through point S from XI.[not given] of the *Elements*, will be erected orthogonally upon circle GBH. Therefore by the converse of the definition of a line erected perpendicularly upon a surface, this line of intersection rests perpendicularly on line KN — namely, the diameter of circle KCN. Therefore diameter KN bisects this line of intersection, and half of that [line of intersection] will be the right sine of arc KC. Thus KS is the versed sine of the same [arc KC]. Since straight [line] DF is the diameter of circle DEF, line DZ — namely, the versed sine of arc DL — is taken from this. KS will be to [DZ] as the radius KY of circle KCN to the radius DX of circle DEF. For the two arcs KC and DL are similar by Th. [III.23] above, because the two arcs AD and AL descend from the pole of the two circles KCN and DEF and include the mentioned arcs.

By now we approach the place where we may show first that the two triangles KTS and AVX [are] equal-angled in this way: The two angles TKS and $R9S$, and therefore $X9Y$, are equal because of the parallelism of lines KT and $R9$. Therefore, since each of the angles KTZ and $XY9$ is a right [angle], the remaining $\angle KST$ will be equal to the other $YX9$, or AXV. Moreover, $\angle AVX$ was arranged [to be] a right [angle]. Therefore the two triangles KTS and AVX are equal-angled, and thus, by [Euclid] VI.4 and rearrangement, the ratio of XA to AV is as SK to KT. Furthermore, the ratio of DZ — namely, the versed sine of $\angle BAC$ — to line KT — namely, the difference of the two versed sines mentioned above — is composed of the two [ratios] — namely, of the ratio of sine DZ to KS and of the ratio of KS to KT. Moreover, DZ to KS was as DX — namely, the whole sine — to line KY, which is the right sine of arc AK, or AC. And the ratio of KS to KT was as that of AX — the whole right sine — to line AV, which is the right sine of arc AB. Therefore the ratio of the versed sine of arc DL, or $\angle BAC$, to the difference of the versed sines mentioned above is composed of two ratios — namely, of the ratio of the whole right sine to the right sine of arc AC and of the ratio of the whole right sine to the right sine of arc AB. Now from these two ratios is also composed the ratio of the square of the whole right sine to the product of the right sines of the two arcs AB and AC. Therefore the ratio of the versed sine of $\angle BAC$ to the difference of the two versed sines [of] the two arcs BO and BK is as the ratio of the square of the whole right sine to the product of the two right sines of the two arcs AB and AC. Q.E.D.

Yet you should not neglect [the fact] that the two versed sines of the two arcs BK and BO have the [same] difference which the two right sines of their complements [have].

ſuorum:unde ſi in quopiã negocio tuo opus fuerit hoc theoremate, poteris ſupra=
dictã differentiam inuenire, ſubtrahendo ſinum rectũ cõplemẽti altẽrius duorum
arcuũ b. k & b o ex ſinu recto cõplementi reliqui eoꝝ. Aſſumpſimus autẽ duos ar
cus angulũ, de quo ſermo habitus eſt, minores quadrante, quo demonſtratio no=
ſtra planior putareṫ:nam ſi fuerit uterꝗ eoꝝ maior quadrãte, intelligant prolon=
gari, donec concurrant ãngulum alium æqualem priori comprehendendo : fiet
itaꝗ nouus triangulus ſupra arcum tertium, cuius duo latera quadrante minora
habebuntur. Quòd ſi alter eorum minor quadrante, alter uero maior eo præbea
tur, tametſi figurationem parumper mutari oporteat, una tamen eadem ſyllogiſ
mi forma ueritatem theorematis concludet:hoc uno attento, ꝗ arcus quiſlibeṫ cũ
eo, qui ſibi ex ſemicirculo deficit, eundem ſinum rectum accipit. Satis ergo certitu
dinem propoſitionis noſtræ oſtendiſſe uidemur.

III.

Datis tribus lateribus trianguli ſphæralis ex arcubus circulorum
magnorum conſtantis, omnes angulos eius dimetiri.

Etſi propoſitum illud exꞔqui liceat per huius, tamen quo iucundior eſſet ue
ritatis contemplatio, dum per plures ac diuerſas uias ad eandem metam perueni=
tur, libuit præcedens theorema propoſito noſtro ſuppeditare. Talis ergo ſit triangulus a b g , ex arcubus circulorum magnorum conſtans, propoſitum eſt inueni
re angulum eius b a c , aut alium quemlibet, ſubijciamus tria latera eius inæqua=
lia, nam ſi duo eius quæcunꝗ latera fuerint æqualia, procedendum erit iuxta mo
nita huius. Cum itaꝗ ex præcedenti ſit proportio quadrati ſinus recti totius
ad id, quod ſub ſinibus rectis duorum arcuum a b & a c continetur, tãꝗ proportio ſinus uerſi anguli b a c quæſiti differentiam duorum ſinuum uerſoꝝ, quo
rum unus eſt ipſius arcus b c , alter uero differentiæ duorum arcuum a b & a c ,
& tres harum quantitatum ſunt notæ propter hypotheſim, erit & quarta cognita
ſcilicet ſinus uerſus anguli b a c , hinc arcus ſuus, qui determinat quantitatem an
guli b a c , & ideo angulus ipſe menſuratus offeretur, pro reliquis autem duobus
angulis cognoſcendis nihil noui præcipimus, quoniam ex angulo b a c iam co=
gnito cum latere b c eum reſpiciente, reliquiſꝗ lateribus notis argumento
huius, quod reliquum eſt enitemur.

IIII.

Quod præcedens tradidit alio ſyllogiſmo concludere.

Habeat em̃ propoſitus trian=
gulus a b c, duo latera a b & a c
inæqualia , quadrante diuiſim minora, libeatꝗ inuenire quantita=
tem anguli b a c, ſuper puncto a
facto polo, deſcribatur circulus ma
gnus in ſphæra d h, cuius circum=
ferentiæ occurrat arcus b c prolõ
gatus in puncto h : duo inſuper ar
cus a b & a c extendãtur uſꝗ ad
arcum d h, cui incidant in pũctis
e & d , erit itaꝗ uterꝗ arcuum a
e & a d quadrans orthogonali=
ter erectus ſupra arcũ d h, duo au

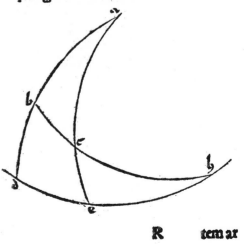

Hence, if there is [ever] a need for this theorem in any of your work, you could find the above-mentioned difference by subtracting the right sine of the complement of one of the two arcs *BK* and *BO* from the right sine of the complement of the other of them.

Moreover, we took the two arcs [including] the angle, about which this discourse is concerned, [to be] less than a quadrant, so that our proof would be considered clearer. Now if each of them is greater than a quadrant, they are understood to be extended until they meet to form another angle equal to the previous [one]. Then a new triangle will be made upon the third arc, [and] the two sides of this [triangle] will be less than a quadrant. But if one of them is given less than a quadrant while the other is given greater than one, then, although the construction would have to be changed a little, nevertheless one [and] the same form of deduction will conclude the truth of the theorem. When one is attentive to this, then any arc has the same right sine as that [supplemental arc] which is left from a semicircle. Therefore we seem to have proved the certainty of our proposition sufficiently.

Theorem 3

When three given sides of a spherical triangle are constructed from the arcs of great circles, all the angles of this [triangle] may be measured.

Although one may execute this proposition by Th. [IV.34] above, nevertheless since contemplation of the truth is more pleasant when one arrives at the same goal by several and different ways, the preceding theorem may be applied to our theorem. Therefore let *ABG** be such a triangle that is constructed from the arcs of great circles. The purpose is to find its ∠ *BAC*, or any other [angle].

We have taken its three sides [to be] unequal, for if any two sides of [the triangle]

were equal, the procedure would be like the instruction of Th. IV.34 above [the second method]. Therefore, since from the preceding the ratio of the square of the whole right sine to the product of the right sines of the two arcs *AB* and *AC* is as the ratio of the versed sine of desired ∠ *BAC* to the difference† of the two versed sines, one of which is that of arc *BC* while the other is that of the difference of the two arcs *AB* and *AC*, and since three of these quantities are known from the hypothesis, then the fourth may be found — namely, the versed sine of ∠ *BAC*. Hence its arc, which determines the size of ∠ *BAC*, and therefore the angle itself, will be measured. Furthermore, we will teach nothing new for the identification of the other two angles, because, [with] ∠ *BAC* already known together with side *BC* opposite it and with the other sides known, we may find what is left by the reasoning of Th. [IV.30, combined with Th. IV.8] above.

Theorem 4

What the preceding taught [is] concluded by another syllogism.

Let the proposed △ *ABC* have two unequal sides *AB* and *AC*, less than a quadrant division, and let it be desired to find the size of ∠ *BAC*.

Around point *A* as pole, let a great circle *DH* be described in the sphere. Arc *BC* extended meets the circumference of this [circle] at point *H*. Let the two arcs *AB* and *AC* be extended above continuously to arc *DH*, which they intersect at points *E* and *D*. Then each of the arcs *AE* and *AD* will be quadrants erected orthogonally upon arc *DH*.

Furthermore, the two

*For *ABG* read *ABC*.

†In the Latin text, for *quaesiti differentiam* read *quaesiti ad differentiam*.

tem arcus b d & c e noti erunt,sunt etem complementa duorum arcuum a b
& a c per hypothesim notorū,sed proportio sinus recti b d ad sinum rectum ce
est per 　huius,ut sinus recti arcus b h ad sinum rectum h c. proportio igitur
sinus recti b h ad sinum rectum h c nota est,cunq̃ differentia duorum arcuum
b h & h c nota sit,scilicet arcus b c,erit per 　huius uterq̃ eorum cognitus,de
inde ex duobus arcubus h c & c e notis,& angulo e recto per 　huius cognoscetur arcus h e,similiter duo arcus h b & b d noti cum angulo d recto notificabunt arcum h d ;arcus igitur h e demptus ex arcu h d , relinquet arcum d e
cognitum,qui determinat angulum a b c ,unde & ille notus habebitur :reliquos
autem duos angulos latera sibi opposita per 　huius notos elicient. Si uero al-

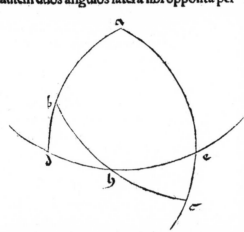

ter duorum arcuum a b & c maior quadrante fuerit, reliquus uero
minor eo,sit a c maior, descripto-
q̃ ut prius circulo magno super a
polo, circumferentia eius secabit
duos arcus a c & b c ; hunc ergo
secet in e ,illum uero in puncto h,
prolongeturq̃ arcus a b , donec
occurret dictæ circumferentiæ in
puncto d ,erit iterum uterq̃ arcuum b d & c e notus,quoniã sunt
complementa duorum arcuum a
b & a c datorum , est autem per
huius proportio sinus recti b
d ad sinum rectum c e ,sicut sinus

recti b h ad sinum rectum h c propter duos angulos d & e rectos,sic ergo proportio sinus recti b h ad sinum rectum h c nota habebitur.cunq̃ totus arcus b c
sit notus per hypothesim,erit per 　huius uterq̃ arcuum b h & h c cognitus :
ex duobus autem autem arcubus h b & b d cognitis cum angulo d recto cognoscetur arcus d h per 　huius:similiter duo arcus h c & c e cum angulo e
recto,arcum h e notum suscitabunt:duo tandem arcus d h & h e collecti, totũ
arcum d e notificabunt,qui determinat quantitatem anguli b a c ,unde & ipse
notus concludetur,cætera ut antehac perficientur.Quòd si uterq̃ arcuum a b &
a c quadrantem superauerit,intelligantur prolongati donec concurrent, facien-
tes angulum nouum æqualem ipsi angulo a quæsito,fiet itaq̃ alius triangulus su
pra arcum b c ,cuius duo latera minora quadrante nota erunt. per modum ergo
prædictum angulus duobus illis lateribus contentus innotescet , qui est æqualis
angulo a ,unde & ipse angulus a notus enunciabitur. Est præterea alius modus
inueniendi angulum trianguli sphæralis quemcunq̃ uoles ex tribus lateribus datis,intelligantur enim duci tres cordæ ipsorum arcuum datorũ,tres quoq̃ semidi
ametri sphæræ egrediantur ad tria puncta angularia ipsius trianguli:habebis igi
tur pyramidem supra basim trilateram,cuius sex lineæ notæ sunt, poteris igitur
aliunde discere inclinationem unius superficiei lateralis supra aliam , superficiem
inquam quæ clauditur duabus semidiametris sphæræ,& una trium cordarum di-
ctarum:quantitas enim huius inclinationis angulum duorum arcuum, quorum
cordæ assumptæ sunt manifestabit.in hac autem inquisitione sinus recti arcuum
datorum,ac sinus recti complementorum suorum maxime utiles erunt,ne tamen
prolixus nimium uidear,post tres uias bonas iam absolutas,hanc quartam præ-
tereundam

arcs *BD* and *CE* will be known, for indeed they are the complements of the two arcs *AB* and *AC*, [which are] known by the hypothesis. But the ratio of the right sine of *BD* to the right sine of *CE*, by Th. [IV.15] above, is as that of the right sine of arc *BH* to the right sine of *HC*. Therefore the ratio of the right sine of *BH* to the right sine of *HC* is known. And since the difference of the two arcs *BH* and *HC* is known — namely, arc *BC* — then by Th. [IV.22] above each of them may be found. Next, since the two arcs *HC* and *CE* are known and since ∠ *E* is a right [angle] by Th. [IV.1] above, then arc *HE* will be known by [Th. IV.25]. Similarly, the two known arcs *HB* and *BD*, together with right ∠ *D*, will identify arc *HD*. Therefore arc *HE*, subtracted from arc *HD*, leaves arc *DE* known, which determines ∠ *ABC*.* Hence [∠ *BAC*] will be known. Furthermore, the sides opposite [the other angles] will make [those] other two angles known by Th. [IV. 28] above.

However, if one of the two arcs *AB* and *C†* is greater than a quadrant while the other is less than one, let *AC* be the larger. As before, when a great circle is described around *A* as the pole, its circumference will intersect the two arcs *AC* and *BC*. Therefore let it intersect the former at *E* and the latter at point *H*. Let arc *AB* be extended until it meets the mentioned circumference at point *D*. Again, each of the arcs *BD* and *CE* will be known, because they are the complements of the two given arcs *AB* and *AC*. Furthermore, by Th. [IV.15] above, the ratio of the right sine of *BD* to the right sine of *CE* is as that of the right sine of *BH* to the right sine of *HC* because the two angles *D* and *E* are right [angles]. Therefore in this manner the ratio of the right sine of *BH* to the right sine of *HC* will be known. And since the total arc *BC* is known by the hypothesis, then, by Th. [IV.21] above, each of the arcs *BH* and *HC*

will be known. Moreover, since the two arcs *HB* and *BD* are known together with right ∠ *D*, arc *DH* may be found by Th. [IV.25] above. Similarly, the two arcs *HC* and *CE*, together with right ∠ *E*, make arc *HE* known. Finally, the two arcs *DH* and *HE* combined will identify arc *DE*, which determines the size of ∠ *BAC*. Hence this [angle] is concluded known; the others may be determined as before.

But if each of the arcs *AB* and *AC* were to exceed a quadrant, they may be understood [to be] extended until they meet, forming a new angle equal to the desired ∠ *A*. Then another triangle will exist upon arc *BC*. The two sides of this [new triangle], [which are] less than a quadrant, will be known. Therefore, by the method previously mentioned, the angle included by these two sides will be known, which [angle] is equal to ∠ *A*. Hence ∠ *A* itself will be declared known.

Besides, there is another way of finding any angle you wish of a spherical triangle when three sides are given. For three chords of those given arcs may be understood to be drawn. Also let three radii of the sphere emerge to the three vertices of this triangle. Then you will have a pyramid upon a three-sided base. The six lines of this [triangular pyramid] are known. Thus you can find the inclination of one lateral surface upon another from some other direction — that is, upon the surface which is enclosed by two radii of the sphere and one of the three mentioned chords. Now, the amount of this inclination will reveal the angle of the two arcs whose chords were taken [as sides of the pyramid]. Moreover, in this investigation the right sines of the given arcs and the right sines of their complements will be most useful. Lest [this] still seem very much too broad, after three good ways have already been related, I have decided this fourth

*For *ABC* read *BAC*.
†For *C* read *AC*.

tereundam arbitratus sum , præsertim cum ex alijs scriptis meis plané colligi possit.

V.

Datis duobus angulis trianguli sphæralis cum aggregato duorum laterum eis oppositorum, utrunc̨ʒ eorum secernere.

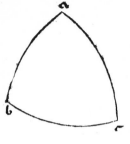

Triangulus a b c , duos angulos a b c & a c b notos habeat, congeriemc̨ʒ duorum laterum a b & a c cognitam. Quærimus utrunc̨ʒ eorum seorsum, quoni am per   huius proportio sinus recti arcus a b ad si num rectum a c est,ut sinus recti anguli a c b ad si num rectum anguli a b c : illa autem nota est propter angulos datos,sinus ergo rectus a b ad sinum rectum a c proportionem habebit datam, cunc̨ʒ aggregatum ex istis arcubus sit datum,erit per   huius uterc̨ʒ eorū separatim cognitus,quod erat inueniendum,pro reliquo autem latere, reliquoc̨ʒ angulo cognoscendis   huius repetendam censeo.

VI.

Datis duobus angulis trianguli sphæralis ex arcubus circulorum magnorum constantis , cum differentia laterum eis oppositorum, utrunc̨ʒ eorum secernere.

Hæc ex   quemadmodum præcedens ex   huius pendere dinoscitur : erit enim proportio sinus recti unius quæsitorum arcuum ad sinum rectum alterius cognita,propter angulos datos ratiocinante   huius;cunc̨ʒ differentiam eorū præbuerit notam hypothesis,uterc̨ʒ eorum proculdubio cognitus emerget.

VII.

Si ab angulo quolibet trianguli sphæralis ad latus sibi oppositum descendat,arcus circuli magni angulum à quo ducitur diuidēs per me dium,sinus recti duorum arcuum angulo diuiso circumpositorum , & sinus recti portionum lateris diuisi eandem proportionem acce ptabunt.

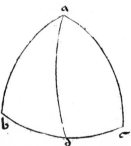

In triangulo tali a b c ducatur arcus a d ex pun cto a diuidens angulum quidem b a c per æqualia, arcum autem b c in duas portiones b d & d c. Dico. q̨ʒ proportio sinus recti a b ad sinum rectum a c est, ut sinus recti b d ad sinum rectum d c. Erit enim per huius sinus rectus a b ad sinum rectum b d sicut sinus rectus anguli a d b ad sinum rectum anguli b a d ,item proportio sinus recti a c ad sinum rectum c d sicut sinus recti anguli a d c ad sinum rectum anguli c a d:sinus autem anguli b a d æqualis est sinui recto anguli c a d ,sinus denic̨ʒ anguli a d b æqualis,imò idem est sinui recto anguli a d c :hi enim duo anguli duobus rectis æquantur,quemadmodum enim duo arcus semicircumferentiæ cō iūctim æquales unum & eundem suscipiunt sinum rectum,ita & duo anguli duo

R 2   bus re

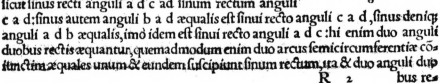

should be forgone, especially since it can clearly be gathered from my other writings.

Theorem 5

When two angles of a spherical triangle are given together with the sum of the two sides opposite them, each of these [sides] may be discerned.

Let △ *ABC* have two known angles *ABC* and *ACB* and [let it have] a known sum for the two sides *AB* and *AC*. We seek each of these [sides] individually.

By Th. [IV.17] above the ratio of the right sine of arc *AB* to the right sine of *AC* is as that of the right sine of ∠ *ACB* to the right sine of ∠ *ABC*. Furthermore, the latter [ratio] is known because the angles were given. Thus the right sine of *AB* to the right sine of *AC* will have a known ratio, and since the sum of these arcs is given, then, by Th. [IV.21] above, each of them will be known separately. Q.E.D. Furthermore, for determination of the other side and the other angle, I advise that Th. [IV.32] be recalled.

Theorem 6

When two angles of a spherical triangle that is constructed from the arcs of great circles are given together with the difference of the sides opposite them, each of these [sides] may be distinguished.

This is recognized to follow from Th. [IV. 22] above just as the preceding [did] from Th. [IV.21] above. For the ratio of the right sine

of one of the desired arcs to the right sine of the other will be known, by the reasoning of Th. [IV.17] above, because the angles were given. And since the hypothesis gave the difference of these [arcs], each of them will be found [by Th. IV.22 above], without doubt.

Theorem 7

If, from any angle of a spherical triangle to the side opposite that [angle], an arc of a great circle descends, bisecting the angle from which it is drawn, then the right sines of the two arcs placed [on either side of] the divided angle and the right sines of the portions of the divided side will have the same ratio.

In such a △ *ABC* let arc *AD* be drawn from point *A*, bisecting ∠ *BAC* and [dividing] arc *BC* into two portions *BD* and *DC*. Then the ratio of the right sine of *AB* to the right sine of *AC* is as that of the right sine of *BD* to the right sine of *DC*.

By Th. [IV.17] above the right sine of *AB* to the right sine of *BD* is as the right sine of ∠ *ADB* to the right sine of ∠ *BAD*. Similarly, the ratio of the right sine of *AC* to the right sine of *CD* is as that of the right sine of ∠ *ADC* to the right sine of ∠ *CAD*. Furthermore, the sine of ∠ *BAD* is equal to the right sine of ∠ *CAD*. Then the sine of ∠ *ADB* is equal to and, in fact, [is] the same as, the right sine of ∠ *ADC*, since these two angles are equal to two right [angles]. For just as the two arcs that, by addition, are equal to a semicircumference have one and the same right sine, so also the two angles

bus rectis coniunctim æquales in finu recto communicare oportet, unam igitur
habent proportionem finus rectus a b ad finum rectum b d,& finus rectus a c
ad finum rectum c d; permutatis itaq; terminis uerum enunciaffe propofitio=
nem confiteberis.

VIII.

Si fuerint duo trianguli rectanguli,quorum angulus acutus unius
datus fuerit æqualis angulo acuto alterius, differentia quoq; laterum
rectos fubtendentium æqualis differentiæ duorum arcuum, qui re=
ctis & acutis datis fubfternuntur,fueritq; latus unum quodcunq; alte=
rius duorum triangulorum cognitum,reliqua omnia cum ipfa diffe=
rentia prædicta innotefcent.

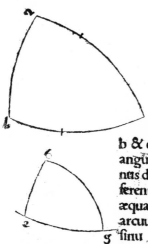

Sint duo trianguli tales a b c & d e g, duos
rectos habentes a b c & d e g,duosq; acutos da
tos a c b & d g e,fitq; latus a c unius longius
latere d g alterius,differentia autem duorum ar
cuum a c & d g æqualis differentiæ duorum ar
cuum b c & e g,q̃uis non fit nota: fit demum
unus fex arcuum ex duobus triangulis datus.Di
co q̃ omnes reliqui arcus innotefcent. Sit arcus
b c uerbi gratia datus,ex quo & duobus angulis
b & c notis   arguente,reliqui duo arcus eiufdem trianguli notificabuntur.eft autem per huius proportio fi
nus differentiæ duorum arcuum a c & d g ad finum differentiæ duorum arcuum b c & e g, quæ eft proportio
æqualitatis,ficut eius,quod fub finibus complementorum
arcuum a b & d e continetur ad id,quod fub finu toto &
finu complementi anguli a c b fiue d g e continetur:
hæc igitur duo fub prædictis finibus contenta funt æqua=
lia:quod autem fub finu toto & finu complementi anguli a c b continetur,eft co
gnitum,propter finum totum & finum complementi anguli a c b notos:quam=
obrem quod fub finibus complementorum arcuum a b & d e continetur notũ
erit,eft autem complementum arcus a b notum propter ipfum arcum a b pri=
us menfuratum:hinc finus huius complementi,& ideo per   huius finus cõple=
menti arcus d e innotefcent,quo demũ cognofcemus cõplementũ arcus d e, &
inde ipfum arcum d e,qui tandem cum duobus angulis d e g & d g e intercedẽ
te   huius reliqua latera trianguli fui manifeftabit,hinc etiã differentia duorũ
arcuum a c & d g,quæ ponebatur æqualis differentiæ duorum arcuum b c &
e g nota pronunciabitur,quæ fuerunt demonftranda.

IX.

Data differentia duorum arcuum,fi quod fub duobus eorum fini=
bus continetur,rectangulum fuerit datum,utriufq; arcus attingere
noticiam.

Duorum arcuum a b & b c differentia a c fit data,quodq; fub finu arcus a
b qui fit a e,& finu arcus b c,quem uides c z continetur,fit datum. Quærimus
utrunq;

that, by addition, are equal to two right [angles] must share in the right sine. Therefore the right sine of *AB* to the right sine of *BD* and the right sine of *AC* to the right sine of *CD* have one [and the same] ratio. Then when the terms are interchanged, you will agree that the proposition stated the truth.

Theorem 8

If there are two right triangles, of which the acute angle of one is given equal to the acute angle of the other and the difference of the sides subtending the right [angles] is equal to the difference of the two arcs which lie beneath the right [angles] and the given acute [angles], and if any one side of one of the two triangles is known, then all the other [unknowns] as well as the aforementioned difference may be found.

Let there be two such triangles *ABC* and *DEG* having two right [angles] *ABC* and *DEG* and two given acute angles *ACB* and *DGE*, and let side *AC* of one [triangle] be longer than side *DG* of the other. Furthermore, let the difference of the two arcs *AC* and *DG* be equal to the difference of the two arcs *BC* and *EG*, although [the value of the difference] is not known. Finally let one of these six arcs of the two triangles be given. Then all the other arcs may be found.

Let arc *BC*, for example, be given. Since this and the two angles *B* and *C* are known, then, by the reasoning of Th. [IV.27] above, the other two arcs of this same triangle may be identified. Moreover, by Th. [1] above, the ratio of the sine of the difference of the two arcs *AC* and *DG* to the sine of the difference

of the two arcs *BC* and *EG*, which [in this case] is the ratio of equality, is as that of the product of the sines of the complements of arcs *AB* and *DE* to the product of the whole sine times the sine of the complement of ∠ *ACB*, or *DGE*. Therefore the latter two products of the aforementioned sines are equal. Furthermore, the product of the whole sine times the sine of the complement of ∠ *ACB* is known because the whole sine and the sine of the complement of [given] ∠ *ACB* are known. Whereupon the product of the sines of the complements of arcs *AB* and *DE* will be known. Furthermore, the complement of arc *AB* is known because arc *AB* itself was measured previously. Hence the sine of this complement, and therefore by Th. [I.17] above the sine of the complement of arc *DE*, will be known. Finally from this we know the complement of arc *DE* and thence arc *DE* itself. [*DE*], together with the two angles *DEG* and *DGE*, will at last reveal the other sides of its triangle [*DEG*] with the help of Th. [IV.27] above. Hence the difference of the two arcs *AC* and *DG*, which was given equal to the difference of the two arcs *BC* and *EG*, will be declared known. Q.E.D.

Theorem 9

If the difference of two arcs is given, [and] if the rectangular product of their two sines is given, [one may] arrive at the identity of each arc.

Let the difference *AC* of the two arcs *AB* and *BC* be given, and let the product of the sine, *AE*, of arc *AB* times the sine, *CZ*, of arc *BC* be given. We seek

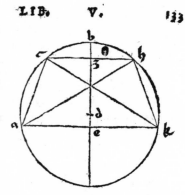

utruncp arcuum a b & b c . Intelligo autem rectangulum prædictum esse datum respectu quadrati semidiametri circuli:continuatis ita cp duobus sinibus a e & c 3 ,donec in duobus punctis h & k circumferentiæ desinant , ducantur cordæ a c , a h , k c & k h . cum igitur quod sub a e & c 3 continetur datum sit, erit quod sub duplis earum continetur datum, hoc etenim ad illud quadruplum conuincitur ex sexti elementorū,cui si addiderimus quadratum cordæ a c, id est,quod sub a c & h k æqualibus quidem propter æquedistantiam cordarum a k & c h : notis autem propter arcum a c ex hypothesi notum, colligetur quadratum cordæ a h cognitum : est nancp a h diameter quadranguli a c h k circulo inscripti æqualis c k diametro eiusdem:quod autem sub duabus diametris huiusmodi quadranguli continetur,æquum est ei,quod sub binis lateribus eius oppositis concluditur:hinc corda a h & arcus eius a h innotescent, ex quo si dempseris arcum a c notum,relinquetur arcus c h cognitus cum eius dimidio c b, cui si addideris arcum a c notum,resultabit totus arcus a b cognitus,cuius obtentu hactenus cursum est.

X.

Si fuerint duo trianguli rectanguli,quorum angulus acutus unius æqualis angulo acuto alterius,duo autem latera rectos angulos subtendentia habuerint differentiam notam , itemcp duo latera rectis & acutis datis subiacentia differentiam cognitam habuerint, omnia eorum latera innotescent .

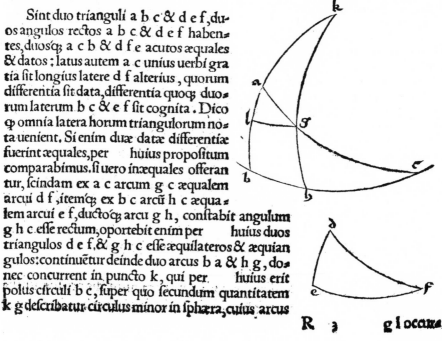

Sint duo trianguli a b c & d e f,duos angulos rectos a b c & d e f habentes,duoscp a c b & d f e acutos æquales & datos : latus autem a c unius uerbi gratia sit longius latere d f alterius , quorum differentia sit data,differentia quocp duorum laterum b c & e f sit cognita . Dico cp omnia latera horum triangulorum nota uenient. Si enim duæ datæ differentiæ fuerint æquales,per huius propositum comparabimus.si uero inæquales offerantur, scindam ex a c arcum g c æqualem arcui d f,itemcp ex b c arcū h c æqualem arcui e f,ductocp arcu g h, constabit angulum g h c esse rectum,oportebit enim per huius duos triangulos d e f,& g h c esse æquilateros & æquiangulos:continuētur deinde duo arcus b a & h g, donec concurrent in puncto k, qui per huius erit polus circuli b c, super quo secundum quantitatem k g describatur circulus minor in sphæra,cuius arcus

R 3 g l occur

· 282 ·

each of the arcs *AB* and *BC*.

Now I understand the aforementioned rectangular [product] to be given in terms of the square of the radius of the circle, so that, when the two sines *AE* and *CZ* are extended until they stop at the two points *H* and *K* on the circumference, the chords *AC*, *AH*, *KC*, and *KH* may be drawn. Then since the product of *AE* times *CZ* is given, the product of two times [each of] them will be given, for indeed this latter [product] is proven from VI.[18] of the *Elements* [to be] fourfold the former [product]. If to this [fourfold product] we were to add the square of chord *AC* — that is, the product of *AC* times *HK*, [these two chords being] equal because chords *AK* and *CH* are parallel and being known because arc *AC* is known from the hypothesis — then the square of chord *AH* may be considered known, for *AH* is the diagonal of quadrangle *ACHK* inscribed in the circle [and is] equal to diagonal *CK* of the same [quadrangle]. Furthermore, the product of the two diagonals of this quadrangle is equal to the product of the two pairs of sides, [each pair] opposite its [respective diagonal]. Hence chord *AH* and its arc *AH* will become known. If you subtract known arc *AC* from this [arc *AH*], arc *CH* is left known together with its half *CB*. If you add known arc *AC* to this [arc *CB*], the total arc *AB* will result. Q.E.D.

Theorem 10

If there were two right triangles, of which the acute angle of one is equal to the acute angle of the other and [of which] the two sides subtending the right angles have a known difference, and similarly if the two sides underlying the right [angles] and the given acute [angles] have a known difference, all the sides of these [two triangles] may be found.

If *ABC* and *DEF* are two triangles having two right angles *ABC* and *DEF* and two equal and given acute angles *ACB* and *DFE*, and if side *AC*, for example, of one is longer than side *DF* of the other [and] their difference is given, and if the difference of the two sides *BC* and *EF* is also known, then all the sides of these triangles will become known.

If the two given differences are equal, we will settle the proposition by Th. [8] above. However, if they are given unequal, then cut from *AC* an arc *GC* equal to arc *DF*; similarly, from *BC* cut arc *HC* equal to arc *EF*. When arc *GH* is drawn, ∠ *GHC* will be established to be a right [angle], for by Th. [III.36] above the two triangles *DEF* and *GHC* must be equal-sided and equal-angled. Then let the two arcs *BA* and *HG* be extended until they meet at point *K*, which by Th. [IV.2] above will be the pole of circle *BC*. Around this [pole] with the length of *KG* [as polar line], let a small circle be described in the sphere. Arc *GL* of this [circle]

g l occurrat arcui b k in l.Quia autem per huius proportio finus a g ad fi
num b h eſt, ut ius quod ſub ſinibus arcuum a k & g k ad id, quod ſub ſinu toto
& ſinu complementi anguli a c b continetur, tres autem harum quantitatum no
tæ ſunt, duos enim arcus a g & b h dedit hypotheſis, quod autem ſub ſinu toto
& ſinu complementi anguli a c b continetur notum eſt, propter angulum a c b
datum: quare quod ſub ſinibus duorum arcuum a k & g k continetur, notũ ha
bebitur. eſt autem per huius quadratum ſinus totius ad id, quod ſub ſinibus ar
cuum a k & g k continetur, tanꝗ ſinus uerſi anguli a k g, ſiue arcus b h eum
determinantis ad differentiam duorum ſinuum uerſorum, quos habent duo arcus
a g & a l: cunꝗ tres harum quantitatum ſint notæ, ut patuit, erit & differentia
dictorum ſinuum uerſorum cognita. cunꝗ arcus a g per huius ſit maior ar
cu a l, & ipſe notus eſt p hypotheſim, erit eius ſinus uerſus cognitus, à quo ſi dem
pſeris prædictam differentiam duorum ſinuum uerſorum, manebit ſinus uerſus ar
cus a l inuentus: hinc arcus a l non poterit latere, qui eſt differentia duorum ar
cuum a k & g k: quod autem ſub ſinibus a k & g k continetur, notum pridẽ
concludebatur, & iam differentiam eorum arcuum notam reddidimus, quare ex
præmiſſa uterꝗ eorum cognitus offeretur, hinc ſua complementa, arcus uidelicet
a b & g h innoteſcent: ex arcu deniꝗ g h duobuſꝗ angulis g h c & g c h da
tis huius ratiocinante, uterꝗ arcuum g c & h c, qui ſunt æquales duobus d
e & d f cognoſcetur, quibus ſi adiecerimus duos arcus a g & b h, ex hypotheſi
notos reſultabunt duo arcus a b & a c cogniti: trina igitur latera poſitorum
trianguloruꝗ nota fecimus, quod erat proclamandum.

XI.

Sinu uerſo alicuius arcus ad ſinum rectum eiuſdem proportionem
habente notam, arcum ipſum innoteſcere.

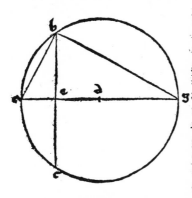

Sit in circulo a b g c diametrum a g ha
bente, corda b c, quam diameter per medium
ſecet in e, cõſtabit itaꝗ a e eſſe ſinum uerſum
arcus a b & b e ſinum rectum eiuſdem: det er
go quiſpiam nobis proportionem a e ad b e,
Dico ꝗ arcus a b cognitus reddet. Ductis e
nim duabus cordis a b & b g, erunt per 30.
tertij & 8.ſexti elementorum duo trianguli par
tiales a b e & e b g ſibi inuicem & toti trian
gulo a b g ſimiles, & per corollarium eiuſdem
octauæ linea b e medio loco proportionalis in
ter g e & e a, cunꝗ proportio b e ad e a ſit
cognita, erat enim a e ad e b data, erit & g e
ad e a data proportio, & coniunctim totius di
ametri g a ad ſinum uerſum ꝗ e proportio fiet, diametrum autem circuli ꝓpter
arcus circumferentiæ metiendos notam ſupponimus, quare & ſinus uerſus a e no
tus habebitur, qui tandem arcum ſuum a b non ſinet ignotum.

XII.

Si fuerint duo trianguli rectanguli, quorum angulus acutus unius
fuerit æqualis angulo acuto alterius dato, differentia etiam laterum re
ctis an

meets arc *BK* at *L*. Furthermore, because by Th. [1] above the ratio of the sine of *AG* to the sine of *BH* is as that of the product of the sines of the arcs *AK* and *GK* to the product of the whole sine times the sine of the complement of ∠ *ACB*, and moreover, because three of these quantities are known, for the hypothesis gave the two arcs *AG* and *BH* and the product of the whole sine times the sine of the complement of ∠ *ACB* is known because ∠ *ACB* was given, therefore the product of the sines of the two arcs *AK* and *GK* will be known. Furthermore, by Th. [2] above, the square of the whole sine to the product of the sines of arcs *AK* and *GK* is as [the ratio] of the versed sine of ∠ *AKG*, or arc *BH* determining that [angle], to the difference of the two versed sines [of] the two arcs *AG* and *AL*. And since three of these quantities are known, as is evident, then the difference of the mentioned versed sines will also be known. Since arc *AG*, by Th. [III.42] above, is greater than arc *AL*, and [since *AG*] itself is known by the hypothesis, its versed sine will be known. If from this [versed sine of *AG*] you subtract the aforementioned difference of the two versed sines, the versed sine of arc *AL* will be left uncovered. Hence arc *AL* cannot be hidden, [and] it is the difference of the two arcs *AK* and *GK*. Furthermore, the product of the sines of *AK* and *GK* was already concluded [to be] known, and we have already found the difference of these arcs. Therefore from the previous [theorem] each of these [arcs *AK* and *GK*] is known. Hence their complements — namely, arcs *AB* and *GH* — may become known. Finally, since arc *GH* and the two angles *GHC* and *GCH* are given, then, by the reasoning of Th. [IV.27] above, each of the arcs *GC* and *HC*,

which are equal to the two [arcs] *DE* and *DF*, will be found. If we were to add to these [arcs] the two arcs *AG* and *BH*, known from the hypothesis, the two arcs *AB* and *AC* will result. Therefore we have found three sides apiece of the proposed triangles. Q.E.D.

Theorem 11

When the versed sine of some arc has a known ratio to the right sine of the same [arc], the arc itself becomes known.

In circle *ABGC* having diameter *AG*, let *BC* be a chord, which the diameter bisects at *E*. Thus it will be established that *AE* is the versed sine of arc *AB*, and *BE* is the right sine of the same [arc]. Therefore, if someone gives us the ratio of *AE* to *BE*, then arc *AB* may be found.

When the two chords *AB* and *BG* are drawn, then, by III.30 and VI.8 of the *Elements*, the two triangular portions *ABE* and *EBG* will be similar to each other and to △ *ABG*. And by the corollary of the same [VI].8, line *BE* will be the mean proportional between *GE* and *EA*. Since the ratio of *BE* to *EA* is known, for *AE* to *EB* was given, then the ratio of *GE* to *EA* will also be given, and by addition the ratio of the total diameter *GA* to the versed sine *AE* is [known]. Moreover, we consider the diameter of the circle [to be] known because the arcs of the circumference are measured. Therefore the versed sine *AE* may be found, which finally permits its arc *AB* to be known.

Theorem 12

If there are two right triangles, of which an acute angle of one is equal to a given acute angle of the other, and if the difference of the sides opposite the right

ctis angulis oppositorum fuerit data, cum differentia laterum acutis angulis subtensorum, omnia latera triangulorum cognita reddere.

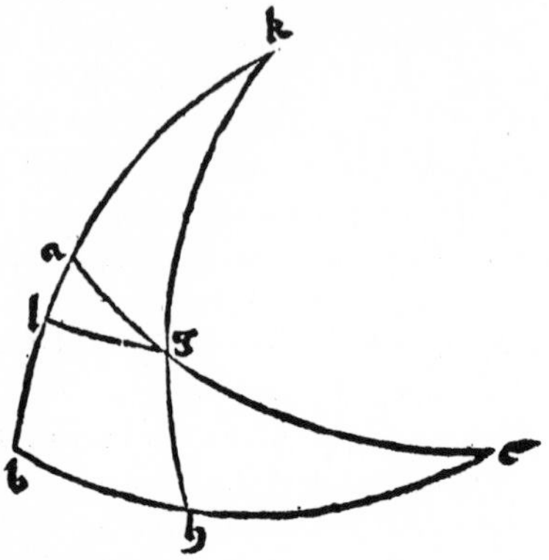

Resumpta figuratione huius datos, supponamus duos arcus a g & a l differentias uidelicet arcuum angulos datos subtendentium. Quærimus omnia latera duorum triangulorum hoc pacto: Proportio quadrati sinus totius ad id, quod sub sinu toto & sinu complementi anguli a c b continetur, per primam sexti est ut proportio sinus totius ad sinum complementi anguli a c b, sumpto sinu toto tanquam altitudine communi ambobus rectangulis: hæc autem componitur ex duabus proportionibus, scilicet proportiõe quadrati sinus totius ad id, quod sub sinibus a k & g k continetur, & proportione eius, quod sub sinibus a k & g k ad id, quod sub sinu toto & sinu complementi anguli a c b continetur: prima harum componentium per huius est, ut sinus uersi arcus b h ad differentiam duorum sinuum uersorum, quorum unus est ipsius arcus a g, alter uero arcus a l: secunda uero cõponens est, ut sinus recti a g ad sinum rectum b h: quare

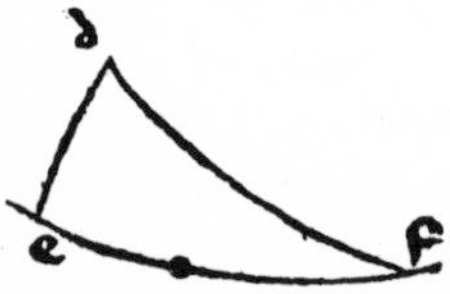

proportio sinus totius ad sinum complementi anguli a c b componitur ex duabus, proportione scilicet sinus uersi b h ad differentiam duorum sinuum uersorũ, quos diximus, & ex proportione sinus recti a g ad sinum rectum b h, & ideo etiam proportio sinus totius ad sinum complementi anguli a c b componitur ex proportionibus duabus, scilicet proportione sinus recti a g ad differentiam duorum sinuum uersorum prædictorũ, & proportione sinus uersi b h ad sinum rectũ eiusdem arcus b h. hæc autem proportio composita est cognita propter sinum totum, & sinum complementi anguli a c b dati, cognitos, prima denique componẽs est nota: est enim arcus a g datus, & ideo sinus eius rectus cognitus: itẽ arcus a l est datus, ipse enim est differentia duorum arcuum a k & g k, siue duorũ a b &

R 4 g h, qua-

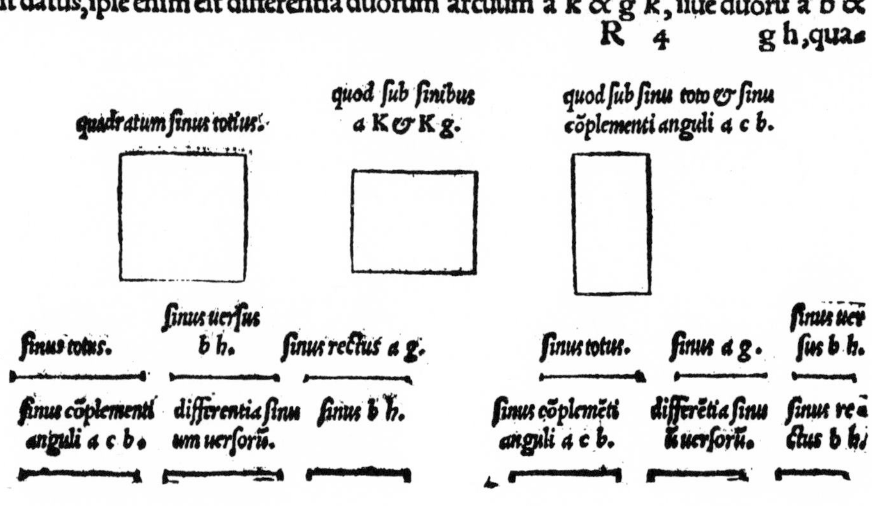

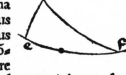

angles is given together with the difference of the sides subtending the acute angles, then all the sides of the triangles may be found.

When the construction of Th. [10] above is repeated, let us consider the two arcs *AG* and *AL* as given — namely, the differences of the arcs subtending the given angles. We seek all the sides of the two triangles in the following way.

The ratio of the square of the whole sine to the product of the whole sine times the sine of the complement of ∠ *ACB*, by [Euclid] VI.1, is as the ratio of the whole sine to the sine of the complement of ∠ *ACB*, when the whole sine is taken as the altitude common to both rectangular [products]. Moreover, this is composed of two ratios — namely, of the ratio of the square of the whole sine to the product of the sines of *AK* and *GK* and of the ratio of the product of sines *AK* and *GK* to the product of the whole sine times the sine of the complement of ∠ *ACB*. The first of these component [ratios] is, by Th. [2] above, as [the ratio] of the versed sine of arc *BH* to the difference of the two versed sines, one of which is that [versed sine] of arc *AG*, while the other is that of arc *AL*. Indeed, the second component [ratio] is as [the ratio] of the right sine of *AG* to the right sine of *BH*. Therefore the ratio of the whole sine to the sine of the complement of ∠ *ACB* is composed of two [ratios] — namely, of the ratio of the versed sine of *BH* to the difference of the two versed sines which we have mentioned, and of the ratio of the right sine of *AG* to the right sine of *BH*. And thus the ratio of the whole sine to the sine of complement of ∠ *ACB* is composed of two ratios — namely, of the ratio of the right sine of *AG* to the difference of the two aforementioned versed sines and of the ratio of the versed sine of *BH* to the right sine of the same arc *BH*. Furthermore, this composite ratio is known because the whole sine and the sine of the complement of given ∠ *ACB* are known. Then the first component is known, because arc *AG* is given and thus its right sine is known, and also because arc *AL* is given since it is the difference of the two arcs *AK* and *GK*, or of the two *AB* and *GH*.

g h, quare uterq; arcuum a g & a l finum uerfum accipiet notum, quorū finuū uerforum differentia non latebit:fic igitur prima proportio componens notos habet terminos,ea autē proportiōe fubtracta ex ipfa proportione, relinquet fecunda componens proportio cognita,quæ erat finus uerfi b h ad finum rectum eiufdem,quare per   huius arcus ipfe b h non ignorabitur.ex duobus autem arcubus a g & b h cognitis,reliqua quæ proponebantur quærenda, argumēto huius abfoluentur,quorum gratia contemplati fumus.

(margin: fubtracta)

XIII.

Si duorum arcuum differentia data fuerit,cum differentia finuum eorum uterq; notus refultabit.

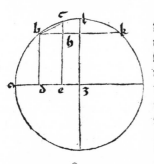

Sint duo arcus a b & a c,quorum differentia b c fit data,duo autem finus eorum b d & c e differentiam habeāt cognitam.Dico ꝗ uterq; eorum innotefcet.Duco enim cordam b k orthogonaliter incidentem finui recto c e in puncto h, eritq; c h differentia duorum finuum,fubtendatur etiam arcui b c corda fua b c,quā oportet effe nota propter arcū ipfum datum: fic ex duabus lineis b c & c h notis angulo apud h recto exiftente,per   primi huius angulum c b h metiemur,ac fi in centro alicuius circuli quiefceret:nunc autem,quoniam in circumferentia fupra arcū cōfiftit,erit & arcus eū fubtēdens ,fcilicet arcus c k notus,cui fi adiunxerimus arcum b c ex hypothefi notum,totus arcus b k notus redundabit,& ideo dimidius arcus b l,inde quoq; complementum fuum a b innotefcent:ex arcu autem a b & b c notis conflabitur totus arcus a c notus,fic utrunq; memoratorum arcuum notum effecimus,quod erat abfoluendum.

XIIII.

In omni triangulo rectangulo duos habente acutos angulos, ficut finus complemēti anguli cuiufuis acuti ad finum totum,fic, quod fub finu complementi lateris fibi oppofiti,& fub finu reliqui acuti continetur ad quadratum finus totius.

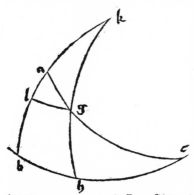

Triangulus a b c habeat angulùm a rectum,& reliquos duos acutos.Dico finum complementi anguli c acuti effe ad finum totum ficut quod fub finu complementi a b & finu anguli continetur ad quadratum finus totius,quod fic demonftratur. Producatur uterq; arcuum c a & c b, ut fiant duo quadrātes c e & c d,fimiliter b g & b h fiāt quadrantes,continuenturq; puncta d & e per arcum d e, qui extenfus concurrat cum h g educto in f,erit ergo uterq; arcuum d f.& h f quarta circuli propter angulos d & h rectos.deniq; duo arcus e d & a b inferius educti concurrant in k.Iam k d ad d f componitur ex duabus k b ad b g, & g h ad h f,de finibus loquor,Sed k d eft complementum anguli a c b, quoniam k eft po-

Therefore each of the arcs *AG* and *AL* has a known versed sine, [and] the difference of these versed sines will not be hidden. In this way, therefore, the first component ratio has known terms. Moreover, when this [component] ratio is subtracted from the ratio itself, the second component ratio is left known; this was [the ratio] of the versed sine of *BH* to the right sine of the same [arc]. Therefore, by Th. [11] above, arc *BH* itself will not be unknown. Since, moreover, the two arcs *AG* and *BH* are known, the others which were to be sought may be found by the reasoning of Th. [10] above, whose example we have considered carefully.

Theorem 13

If the difference of two arcs is given together with the difference of their sines, then each of them may be found.

Let *AB* and *AC* be two arcs whose difference, *BC*, is given, and let their two sines *BD* and *CE* have a known difference. Then each of them will become known.

Draw chord *BK*, resting orthogonally on the right sine *CE* at point *H*. *CH* will be the difference of the two sines. Let chord *BC* be subtended by arc *BC*. [Chord *BC*] must be known because its arc is given. Thus, since the two lines *BC* and *CH* are known [and] since the angle at *H* is a right [angle], then, by Th. I.[27] above, we will measure ∠ *CBH*, as if it were resting in the center of some circle; but now because it is placed on the circumference above the arc, then the arc subtending it — namely, arc *CK* — will also be known. If we were to add arc *BC*, known from the hypothesis, to this [arc *CK*], the total arc *BK* will be found, and thus half the arc, *BL*. Consequently its complement *AB*

also becomes known. Moreover, since arc *AB* and [arc] *BC* are known, the total arc *AC* will be known. Thus we have found each of the mentioned arcs, which was to be accomplished.

Theorem 14

In every right triangle that has two acute angles, as the sine of the complement of either acute angle is to the whole sine so the product of the sine of the complement of the side opposite that [angle] times the sine of the other acute angle is to the square of the whole sine.

If △ *ABC* has ∠ *A* a right [angle] and the two others acute [angles], then the sine of the complement of acute ∠ *C* to the whole sine is as the product of the sine of the complement of *AB* times the sine of angle *B**
to the square of the whole sine, which is proven thus.†

Let each of the arcs *CA* and *CB* be extended so that they become two quadrants *CE* and *CD*. In the same way, let *BG* and *BH* become quadrants. And let points *D* and *E* be joined by arc *DE*, which, extended, meets with *HG*, extended, at *F*. Therefore each of the arcs *DF* and *HF* will be a quarter of a circle because angles *D* and *H* are right [angles]. Next, the two arcs *ED* and *AB* extended above‡ meet at *K*. Now *KD* to *DF* is composed of two [ratios], *KB* to *BG* and *GH* to *HF* — I am [actually] speaking of the sines [of these arcs]. But *KD* is the complement of ∠ *ACB* since

---

*In Latin, for *anguli* read *anguli b*.

† The figures given for this theorem are incorrect and should be ignored. The first figure given for Th. 15 on page 290 is the true figure for this theorem.

‡In Latin, for *inferius* read *inserius*.

k eſt polus circuli c e ,ppter angulos a & e rectos.
k b autem eſt complementum lateris a b oppoſiti
angulo a c b , & g h eſt quantitas anguli a b c ,
unuſquiſcᷜ autem arcuum d f , b g & h f eſt qua⸗
drans circuli.Cum itacᷜ proportio eius,quod ſub an⸗
tecedentibus componentium continetur rectangulum ad id,quod ſub conſequen
tibus eorum continetur,id eſt ad quadratum ſinus totius ex eiſdem proportioni⸗
bus componentibus componitur,patet propoſitio.

X V.
Dato triangulo ſphærali circulum circumſcribere.

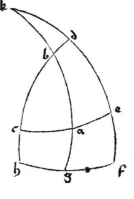

Modus circumſcribendi & inſcribendi circu⸗
los eſt,ut in rectilineis triangulis, diuidendo ſcili⸗
cet latera per æqualia,aut angulos etc.uerum dia⸗
metrum circuli circumſcripti aut inſcripti inueſti
gare,alia requirit media.Dupliciter nancᷜ poteſt
inueniri diameter circuli circumſcripti , aut ſcili⸗
cet per tres cordas notas,aut per arcus ipſos & ſci
entiam triangulorum ſphæralium,de inſcriptione
non ſic,nam quicuncᷜ circulus circumſcribit ſphæ
ricum triangulum, is etiam rectilineum circum⸗
ſcribit,ex tribus cordis trium arcuum conſtantē,
quod inſcripto circulo non accidit. Triangulo a
b c ſphæralo circumſcriptus eſto a b c circulus,
cuius ſemidiametrum quærimus per ratio⸗
nem arcuum. Sit a d arcus perpendicula⸗
ris ad b c, b c per medium diuidatur in f
puncto,unde exeat perpendicularis f e ,in
quo neceſſe eſt eſſe polum circuli circumſcri
pti.diuidatur item a c per medium in g,eᷓ
ductuſcᷜ perpendicularis occurrat arcui f e
in h ,qui erit polus circuli circumſcripti. ex
tribus autem datis lateribus angulum c ha
bebis,& deinde propter f c notum perpen⸗
dicularis quocᷜ f e cum angulo f e c inno⸗
teſcet,& cum arcu e c , hinc e g notificabi
tur.& deinde propter angulum e notum ar
cus g h patefiet.cuncᷜ & g c notus ſit, erit
etiam h c notus,qui inter polum & cuſpi⸗
dem anguli c , id eſt circumferentiam circuli circumſcripti comprehenditur.

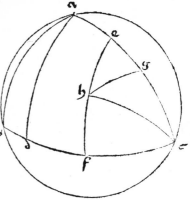

Quinti & ultimi libri Triangulorᷓ finis.

NORIMBERGAE APVD IO. PETREIVM,
ANNO CHRISTI M. D. XXXIII.

K is the pole of circle *CE* because angles *A* and *E* are right [angles]. Moreover, *KB* is the complement of side *AB* opposite ∠ *ACB*, and *GH* is the value [in degrees] of ∠ *ABC*. Furthermore, each of the arcs *DF*, *BG*, and *HF* is a quadrant of the circle. Therefore since the ratio of the rectangular product of the antecedent [terms, *KB* times *GH*] of the component [ratios] to the product of the consequent [terms] of these [component ratios] — that is, to the square of the whole sine — is composed of the same component ratios [as those to be found], the proposition is revealed.

Theorem 15

To circumscribe a circle around a given spherical triangle.

The method of circumscribing and inscribing circles is, as in [the case of] rectilinear triangles, to bisect the sides [for circumscribing] or the angles [for inscribing], and so on. But to find the diameter of a circumscribed or an incribed circle demands other means.

The diameter of a circumscribed circle can be found in two ways — namely, either through three known chords or through the arcs alone and a knowledge of spherical tri-angles. [This is] not so concerning an inscribed [circle]; whatever circle circumscribes a spherical triangle will also circumscribe the rectilinear [triangle] that is constructed from the three chords of the three arcs [of the spherical triangle], [but] this does not happen with an inscribed circle.

Around spherical △ *ABC*, let circle *ABC* be circumscribed, whose radius we seek through consideration of the arcs. Let arc *AD* be perpendicular to *BC*, and let *BC* be bisected at point *F*. From this, let a perpendicular, *FE*, emerge; the pole of the circumscribed circle must be on this [perpendicular *FE*]. Similarly, let *AC* be bisected at *G*, and let a perpendicular, drawn [from point *G*] meet arc *FE* at *H*, which will be the pole of the circumscribed circle. Since the three sides are given, you may obtain ∠ *C*. And, next, because *FC* is known, perpendicular *FE*, as well as ∠ *FEC* and arc *EC*, will also become known. Hence *EG* may be identified. And then because ∠ *E* is known, arc *GH* will be found. Since *GC* is known, *HC* will also be known, [and] this is included between the pole and the vertex of ∠ *C* — that is, [between the pole and] the circumference of the circumscribed circle.

The Fifth and last Book of the Triangles
is ended.

Printed at Nürnberg by John Petreus
in the Year of Christ 1533.

REFERENCE MATTER

BIBLIOGRAPHY

Listed here are those works used for preparation of the "Introduction" and for occasional assistance in the translation.

WORKS PUBLISHED BEFORE 1800

Adam, Melchior (ed.). *Vitae Germanorum philosophorum.* 3rd ed. Frankfurt am Main: John Maximilian à Sande, 1706.

Alfonso X (of Castile). *Tabule astronomice Alfonsi regis.* Ed. by Johann Lucilius Santritter. Venice: Hamman. 1492.

Apian, Peter. *Cosmographia.* Ed. by Gemma Frisius. Antwerp: Christopher Plantinus, 1574.

Brahe, Tycho. *De nova stella.* Hafniae: Laurentius Benedicti, 1573. Reproduction, Copenhagen: Regia Societas Scientiarum Danica, 1901.

———. *Opera omnia.* Frankfurt, 1648.

Finé, Oronce. *Opere divise.* Venice: Senese, 1587.

Gassendus, Peter. *Tychonis Brahei vita.* 2nd ed. The Hague: Adrian Vlacq, 1655.

Hutton, Charles. *A Mathematical and Philosophical Dictionary.* 2 vols. London: J. Johnson and G. G. and J. Robinson, 1795–96.

Kepler, John. *Tabulae Rudolphinae.* Ulm, 1627.

Martin, Benjamin. *Biographia philosophica.* London: W. Owen, 1764.

Nunes, Peter. *Tratado da Sphera,* 1537, reprinted in *Obras,* vol. I. Lisbon: Imprensa Nacional de Lisboa, 1940.

Peurbach, George. *Theoricae novae planctarum.* Ed. by Erasmus Reinhold. Wittenberg, 1542.

Sacrobosco, John of. *De sphaera.* Ed. by Philip Melanchton. Wittenberg: I. Crato, 1563.

Sherburne, Edward. "Catalogue of Astronomers," appendix to *The Sphere of Marcus Manilius.* London: Nathanael Brooke, 1675.

Weidler, Johann Friedrich. *Historia astronomiae.* Wittenberg, 1741.

WORKS PUBLISHED AFTER 1800

Ball, W. W. Rouse. *A Short Account of the History of Mathematics.* 4th ed. (1908). New York: Dover, 1960.

Berry, Arthur. *A Short History of Astronomy.* New York: Scribner's, 1910.

Boas, Marie. *The Scientific Renaissance, 1450–1630.* New York: Harper, 1962.

Braunmühl, Anton von. *Vorlesungen über Geschichte der Trigonometrie,* 2 vols. Leipzig: Teubner, 1900–1903.

Cajori, Florian. *A History of Mathematics.* 2nd ed. New York: Macmillan, 1919.

———. *A History of Mathematical Notations.* 2 vols. Chicago: Open Court, 1928–30.

Cantor, Moritz. *Vorlesungen über Geschichte der Mathematik.* 4 vols. Leipzig: Teubner, 1894–1908.

Delambre, J. B. J. *Histoire de l'Astronomie du Moyen Age.* Paris: Courcier, 1819.

Dreyer, J. L. E. *A History of Astronomy from Thales to Kepler.* 2nd ed. Ed. by W. H. Stahl. New York: Dover, 1953.

Bibliography

————. *Tycho Brahe*. Edinburgh: A. & C. Black, 1890.

Fink, Karl. *A Brief History of Mathematics*. Chicago: Open Court, 1910.

Heath, Thomas. *A History of Greek Mathematics*. 2 vols. Oxford: Clarendon Press, 1921.

Hofmann, Josef. *The History of Mathematics*. New York: Philosophical Library, 1957.

Hooper, Alfred. *Makers of Mathematics*. New York: Random House, 1948.

Karpinski, Louis Charles. *Robert of Chester's Latin Translation of the Algebra of al-Khowarizmi*. New York: Macmillan, 1915.

King, Henry C. *The Background of Astronomy*. New York: Watts, 1957.

Koestler, Arthur. *The Sleepwalkers*. New York: Macmillan, 1959.

Morison, Samuel Eliot. *Admiral of the Ocean Sea*. 2 vols. New York: Time Inc., 1962.

Neugebauer, Otto. *Vorlesungen über Geschichte der Antiken Mathematischen Wissenschaften*. Berlin: J. Springer, 1934.

Neugebauer, Otto, and A. Sachs (eds.). *Mathematical Cuneiform Texts. American Oriental Studies*, vol. XXIX. New Haven: American Oriental Society, 1945.

Olschki, Leonardo. *The Genius of Italy*. New York: Oxford University Press, 1949.

Pannekoek, A. *A History of Astronomy*. New York: Interscience, 1961.

Sarton, George. *Introduction to the History of Science*. 3 vols. Baltimore: Williams & Wilkins, 1927-48.

————. *Six Wings: Men of Science in the Renaissance*. Bloomington: Indiana University Press, 1957.

Scott, J. F. *A History of Mathematics*. London: Taylor & Francis, 1958.

Singer, Charles. *A Short History of Scientific Ideas to 1900*. Oxford: Clarendon Press, 1959.

Smith, David Eugene. *History of Mathematics*. 2 vols. New York: Dover, 1958.

————. *Source Book in Mathematics*. New York: McGraw-Hill, 1929.

Thorndike, Lynn. *A History of Magic and Experimental Science*. 6 vols. New York: Macmillan, 1923-34.

Wolf, A. *A History of Science, Technology, and Philosophy in the 16th and 17th Centuries*. London: Allen & Unwin, 1950.

Zeller, Mary Claudia. *The Development of Trigonometry from Regiomontanus to Pitiscus*. Ann Arbor: Edwards Bros., 1944.

Zinner, Ernst. *Entstehung und Ausbreitung der Copernicanischen Lehre*. Erlangen: Mencke, 1943.

————. *Leben und Wirken des Johannes Müller von Königsberg genannt Regiomontanus*. Munich: C. H. Beck, 1939.

JOURNAL ARTICLES

Bond, John David. "The Development of Trigonometric Methods down to the Close of the XVth Century," *Isis*, IV (October 1921), 295-323.

Curtze, Maximilian. "Urkunde zur Geschichte der Trigonometrie im Christlicher Mittelalter," *Bibliotheca Mathematica* (Ser. 3), I (1900), 321-416.

Karpinski, Louis C. "Bibliographical Check List of All Works on Trigonometry Published up to 1700 A.D.," *Scripta Mathematica*, XII (1946), 267-83.

————. "The Unity of Hindu Contributions to Mathematical Sciences," *Scientia*, XLIII (June 1928), 381-88.

INDEX

Angle, planar: complement of, 31; in finding areas, 35; finding acute, 61, 63; finding third, of triangle, 63–65; size of, from sides, 97; and ratio of sides, 97–99, 105–7. *See also* Triangle, planar

—spherical: exterior, 191–93; sine of, 243–45; complement of, 249; mentioned, 145, 149, 179–89 *passim*. *See also* Triangle, spherical

Approximation, 35

Arc: complement of, 31, 249; measure of, defined, 61; of circle, 167–73, 189–91, 203–13; orthogonal, 189–91; sine of, 237–43, 249, 281–83, 289. *See also* Circles; Triangle, spherical

Area: of triangle, 8, 133; of square, 33; approximation of, 35; of rectangle, 51

Axioms, 33

Bessarion, Cardinal, 9n36, 13, 27

Brahe, Tycho, 3

Circles: arcs of, 31; chords of, 31; construction of, 31, 157–59, 291; circumscribed, 129, 135, 291; within sphere, 153, 163–65, 167–71; great, within sphere, 153–55, 157–63, 165, 171–73, 177–79; and polar lines, 153, 173–75; great, and quadrants, 155, 203; orthogonal intersection of, 161–63; parallel, within sphere, 163–67; small, within sphere, 175–79; intersection of, within sphere, 219–23. *See also* Arc; Sphere

Cosines, Law of, 8, 271–75

Degree: circumferential, 61; measure defined, 61, 71n

Denominator, 71n

Designation, 31 and n

De triangulis: early editions of, 9; previous translation of, 9; comments on present translation of, 9–10; writing of, 14, 27; in astronomy, 27, 69, 245, 261

Difference of unequal quantities, 37

Distance: between triangle vertices, 139–41; between arcs within sphere, 167–71, 175–77

Division, 31, 53

Equations: algebraic, 8, 119 and n, 129; simultaneous solution of, 37n

Fractions: common, 71; sexagesimal, 71n

Gassendus, Peter, 10–11

Heliocentric theory, 3 and n6

Mean proportional, 35

Measure: of a quantity, 31; common, 37

Minutes, 71n

Müller, Johannes (or John, or Johann). *See* Regiomontanus

Multiplication: defined, 31; product from, as area, 51, 93n

Numbers: known, 31; compared with quantities, 37n; denominate, 43–45; common, 45–47

Parallelogram, rectangular, 51

Perpendicular. *See* Sphere; Triangle, planar; Triangle, spherical

Peurbach, George, 11–12, 27

Pirkheimer, Willibald, 8, 23

Planes: intersecting sphere, 143–45; intersection of two, 157

Product, 51 and n, 271, 283

Proportion: mean proportion in, 35; terms of ratios in, 39–43; fourth proportional in, 57–59. *See also* Ratios

Pythagorean theorem, 65

Quantity: defined, 31; compared with number, 37n; known, 39

Ratios: given, 31; equal, 31, 37–39; unequal, 31; antecedent term in, 39; consequent term in, 39; solutions of, 39, 47–51. *See also* Proportion

Rectangle: as product, 51, 271, 283; finding side of, 51–57; antecedent and consequent sides of, 57n

Regiomontanus: posthumous reputation of, 3–4;

· 297 ·

campanatus 286

I.16 (p. 36) cited by Guido Ubaldo p 245 in Euoli

36 every ratio can be found in numbers
5 Menelaus